COLOR PROTOCOLS

TECHNOLOGIES OF RACIAL ENCODING IN CHROMATIC MEDIA

EDITED BY CAROLYN L. KANE AND LIDA ZEITLIN-WU

The MIT Press
Cambridge, Massachusetts
London, England

The MIT Press
Massachusetts Institute of Technology
77 Massachusetts Avenue, Cambridge, MA 02139
mitpress.mit.edu

The MIT Press would like to thank the anonymous peer reviewers who provided comments on drafts of this book. The generous work of academic experts is essential for establishing the authority and quality of our publications. We acknowledge with gratitude the contributions of these otherwise uncredited readers.

This book was set in Stone Serif and Stone Sans by Westchester Publishing Services. Printed and bound in the United States of America.

Library of Congress Cataloging-in-Publication Data

Names: Kane, Carolyn L. editor | Zeitlin-Wu, Lida editor
Title: Color protocols : technologies of racial encoding in chromatic media / [edited by] Carolyn L. Kane and Lida Zeitlin-Wu.
Description: Cambridge, Massachusetts : The MIT Press, [2025] | Includes bibliographical references and index.
Identifiers: LCCN 2024056762 (print) | LCCN 2024056763 (ebook) | ISBN 9780262553506 paperback | ISBN 9780262384889 pdf | ISBN 9780262384896 epub
Subjects: LCSH: Race | Human skin color—Social aspects | Color—Social aspects | Racism | Racism in mass media
Classification: LCC HT1521 .C537 2025 (print) | LCC HT1521 (ebook) | DDC 305.8—dc23/eng/20250515
LC record available at https://lccn.loc.gov/2024056762
LC ebook record available at https://lccn.loc.gov/2024056763

10 9 8 7 6 5 4 3 2 1

EU Authorised Representative: Easy Access System Europe, Mustamäe tee 50, 10621 Tallinn, Estonia | Email: gpsr.requests@easproject.com

COLOR PROTOCOLS

CONTENTS

PART II: IMAGES

PART III: ALGORITHMS

INTRODUCTION: COLOR CODES

What comes to mind when you hear the word "color"? Some might picture paint swatches at the hardware store, digital filters, or the whimsical names of Crayola crayons. For others, the term might evoke a more damning and less aesthetically pleasing history: one in which "people of color" have been systemically othered through *colorism,* or the socially engrained privileging of lighter skin tones over darker ones. Both associations are legitimate, and yet each represents only one thread of color's complex sensory, technological, and political narratives. Crayola crayons illustrate this dual understanding of color as both race and hue. On the one hand, Burnt Sienna, Gold Ochre, and Olive Green; on the other, Flesh, a peachy beige designed to stand in for "normal" or "standard" (i.e., white) skin tones.[1] This opening question—"What comes to mind when you hear the word 'color'?"—is thus more loaded than it appears. To do justice to the myriad meanings of "color" and its undeniable entwinement with legacies of systemic racism, we need to enter precarious and sometimes uncomfortable terrain.

As media studies scholars describing our research on color to others, we often receive the following response: "Color, as in . . . *abstract* color? Or, color, as in race?" The simple yet common riposte discloses two fundamentally distinct understandings of the term, both embodied by the same word. In the first instance, "abstract" color is taken as a measurable set of wavelengths in the visible light spectrum. In the second instance, "color" denotes skin pigmentation as a culturally encoded marker of race. This discrepancy leads to a host of problems. For example, a quick library catalog search for "color" generates numerous titles that appear to be about color media (in the first sense) but on second glance, are primarily about race, or vice versa. Color's double-faceted nature—as both a racializing tactic and a property of light or pigment—often means that this complex ambiguity goes unacknowledged, dismissed as a mere incident of linguistic overlap. As this book reveals, however, this overlap is far from coincidental.

Another set of problems emerge when acknowledging that the two disparate, yet interconnected definitions are rarely discussed in tandem. Where most people would immediately understand one definition as socially and politically charged, color's other, allegedly more scientific definition appears, by contrast, innocuous and ideologically neutral. But this assumption could not be more incorrect. Treating the two uses of the term as wholly distinct glosses over the ways in which color's technical and sensory properties are intimately linked to race, and how race is, in turn, deeply linked to histories of color indexing and ordering. How do we begin to parse this fraught relationship between color-as-race and color-as-hue through the lens of media studies and the history of technology?

Color Protocols: Technologies of Racial Encoding in Chromatic Media offers a series of responses to these thorny issues. It begins with the premise that color is central to the history of systemic racism and, in turn, that the encoding of race vis-à-vis color is an intrinsic aspect of chromatic technologies (broadly understood). One of the book's central aims, then, is to catalyze a long-overdue conversation about color as a racial technology by reframing pressing issues in critical race studies as they intersect with the history of modern color media, from the linguistic to the optic and informatic. As such, this book offers the first systematic introduction to color's racialized double valence within a media studies context. Previous edited volumes on color have focused on only one side of its aforementioned double valence, addressing it either as abstract and material qualia of art, design, media, or architecture histories[2] or as a facet of racialized divides.[3] In contrast, *Color Protocols* builds on both of these legacies to bring aesthetic and technoscientific histories of color in the arts, social sciences, and humanities into conversation with the history and critical theory of race, ethnicity, and indigeneity.

Further, while terms like "white privilege," "power," and "systemic racism"—previously the domain of the humanities and social sciences—have clearly exploded in mainstream news outlets and pop culture, these concepts have not yet been adequately addressed within science, technology, engineering, and mathematics (STEM) fields or the creative industries, including graphic design, architecture, and related art and design disciplines. This book responds to this discrepancy by featuring scholars and practitioners across the arts, sciences, and humanities who bridge this gap by weaving a twofold analysis of color in the histories of media, the human sciences, philosophy, cultural studies, and science and technology studies. In so doing, *Color Protocols* speaks to readers who actively embrace color—whether in computer graphics interfaces or in the chemistry lab—and wish to gain deeper insight into the ways that color is intrinsically bound to broader histories and cultures of racism.

As we previously described, working on color—arguably one of the most interdisciplinary and capacious topics—can bring with it a slew of frustrations and challenges. At times, it feels as though the reluctance to address chromatic media's racialized double bind stems from a desire to avoid the precarious relationship between color and race, and yet to leave race entirely out of the scholarship on color media is also a form of historical and political erasure. The eighteen curated essays in this book are written by scholars who have come up against similar trepidations in their respective fields, from the aftermath of the transatlantic slave trade to the informatically coded digital present. This chronological trajectory is reflected in the book's tripartite organization: part I, "Grammars," part II, "Images," and part III, "Algorithms," each of which is preceded by an introductory essay by the editors that provides a theoretical and historical backdrop for each chromatic paradigm. Each section contains both original essays and republished excerpts that are fundamental to understanding how the history of color technology and its racialized double valences have played out across multiple fields and media platforms over the last century and a half. In this sense, we envision this book as a hybrid reader and anthology—one that sheds new light on older writings through sometimes surprising transhistorical juxtapositions. For instance, some of the republished excerpts, like Frederick Douglass's "The Color Line" in part I, might at first seem out of place in a book about chromatic media. Put in dialogue with the introductory essays and the other pieces in the volume, however, Douglass's still all-too-relevant musings on color as both metaphor and lived experience give readers fresh eyes through which to approach this piece.

A caveat for the reader: given the historical scope of this book (from roughly the late nineteenth century to the present), we must underscore that we are in no way arguing that racism did not exist prior to modernity.

However, we do contend that the intimate relationship between color and race is inseparable from two simultaneous historical shifts: (1) the rise of "race science" in the eighteenth and nineteenth centuries (alongside the global dye and textile industries, from the East India Company to I. G. Farben), and (2) the growth of psychophysics as a branch of behavioral science and its adoption in numerous engineering protocols and mathematically based color systems.

This explains why, in part, this book's chronological structure mirrors the rightly contested configuring of media history into temporal periods arranged from the past to the present, often by virtue of progress myths and imperial legacies, many of which our authors seek to deconstruct.[4] Still, we maintain that this organizational strategy more readily enables a transhistorical reading of the ways in which colorism and systemic racism have been (often imperceptibly) woven into the emergence of media and culture today. For all the bright and celebratory colors in our media-saturated landscapes, the irony continues to escape us that we fail to see how our cherished color technologies turn on deeply entrenched legacies of chromophobia and the long-standing subordination of bodies of color.

In sum, *Color Protocols* does not offer a comprehensive history of systemic racism or color science. Rather, each essay in parts I–III identifies a key moment over the last century and a half in which racial capitalism and the history of modern color technology have come to mutually inform one another, making real-life, embodied color that much harder to see. Throughout each section, the perennial question looms: What is at stake when ephemeral categories like color, race, and ethnicity are transformed into a common language, lexicon, or industry standard? And, more critically, how can we avoid the epistemic and ontological violence that seems inevitable when organizing color into a series of grammars, syntaxes, indexes, and protocols?

A NOTE ON PROTOCOLS

The precise nature of racial encoding in each color technology is particular to culture and context. Once a new technology is made available for mass consumption, it is embedded within industry standards and etiquettes for common use. This process encompasses one meaning of the term "protocol," which we refer to in the title of this volume. Internet computer

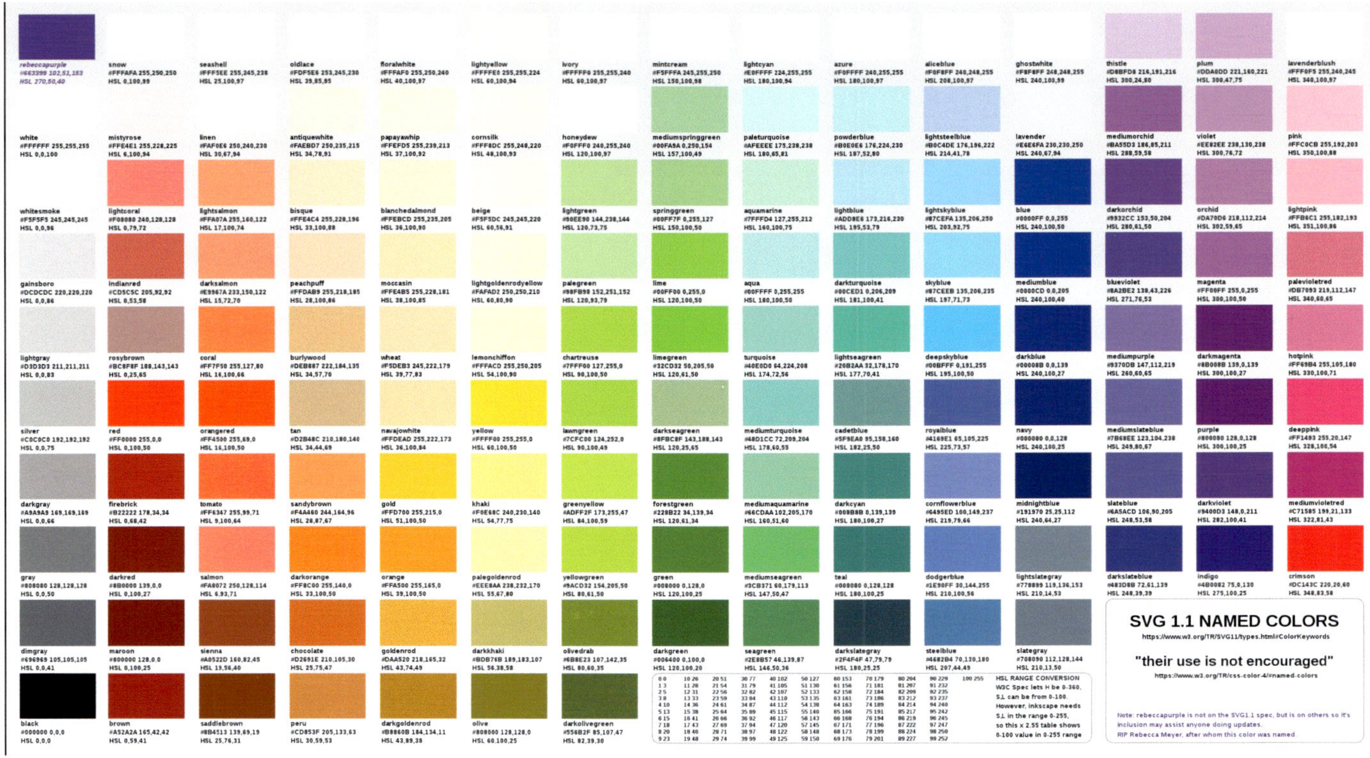

FIGURE I.1

XHTML "extended color" chart (SVG1.1) with HEX and HSL codes for Internet-based text, background color, frame borders, tables, and individual table cells. These color codes were specified by the World Wide Web Consortium (W3C) in 2011. To use color online effectively, one must pick from these colors, making it the HEX system's color protocol. Image courtesy of Tom Cowap and the Creative Commons.

FIGURE I.2
Samples of the Pantone Matching System. This proprietary color space, owned by X-Rite, is an industry standard in numerous creative industries (graphic design, fashion, and printing) and is applied to mainly physical platforms like paint, textiles, nylon, and plastic.

protocols, for instance, typically describe a fixed and mechanized arrangement that is necessary for normative network functioning and governs how specific technologies are agreed on, adopted, and implemented around the world. Examples include the World Wide Web's HyperText Transfer Protocol (HTTP) and HyperText Markup Language (HTML), developed in 1990 to set the conditions of possibility for most uses of the Internet.[5]

A second, more prosaic definition of protocols involves a series of culturally encoded habits and practices. Circa 1994, information theorist Phil

Agre illustrated this through his notion of "grammars of action,"[6] defined by the total set of moves needed to execute an activity, like moving through an entrance or exit on a freeway system. The total set of moves (signaling, merging, slowing down, and turning) is compiled and encoded into a single unit, which then becomes de facto for effectively operating in that system, whether computational or otherwise.[7]

Similarly, each grammar discussed in parts I through III contains its own highly specific protocols. Most of us rely on related grammars to use color in similar, semiautomated ways, unknowingly engaging in protocological activities that push the minutiae and subtlety of our mediated habits into the shadows of experience. The way that filters and special effects are prefabricated on video and imaging software like Instagram, Zoom, and Snapchat, or the requirement to select from prefabricated hues in color standards like PANTONE or HEX, illustrates the encoding of color protocols into a grammar. There is only one correct way to get off the freeway, just as there is only one way to correctly use a filter in a social media app. Most color protocols, our essays demonstrate, result in a process of obfuscation where embodied experiences

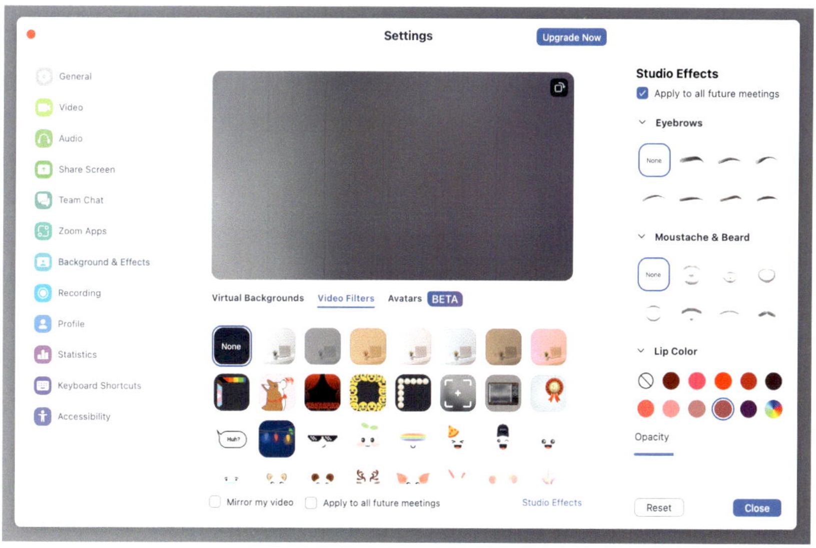

FIGURE I.3

"Studio effects" on Zoom's video interface (2023). Zoom offers a selection of preset filters to change the color of one's facial features. One can select only from these compressed set of color actions or "protocols."

and complex cultural narratives are boiled down into a singular, seemingly apolitical "choice" that is almost always coded by axiomatic whiteness.

DISCIPLINARY SCOPE

Before turning to an overview of the essays in the book, we offer a summary of the ways in which color has already been encoded through two general disciplinary protocols: (1) aesthetics and creative practice (art, design, architecture, and the history of science and technology); and (2) critical race studies. While our goal is ultimately to bring these two domains together, it is necessary to separate them before these foundational contours can be charted.

Despite the palpable ways in which scholars have begun linking visual culture and technology to race, color studies in the creative fields (such as architecture, fashion, design, and visual communications) and STEM disciplines have maintained a very traditional trajectory largely excluded from political acumen.[8] Accordingly, foundational color courses in these fields approach color through traditional Western models rooted in linear histories of art and science from the Renaissance through the Bauhaus, while applied fields in science, technology, and engineering have inherited equally uncritical legacies of color derived from nineteenth-century optical science, synthetic chemistry, and the perennially abstract world of mathematics. While scientists have long informed the work of artists and vice versa, for the most part, color research and its applications in these domains has been restricted to the Euro-American Enlightenment paradigm.[9]

The humanities and social sciences have more readily embraced such critiques. "Introduction to Part I: Color Encoding from the Enlightenment to 'Race Science'" expands on the formation of the modern concept of "race" commonly held to have emerged in the European Enlightenment, although differences of opinion continue to plague this genealogy. Here, we further provide a foundation for this discussion by addressing how recent scholarship in critical race studies intersects with notions of race as phenotypical color.

Over the last decade, best-selling and often controversial books such as Michelle Alexander's *The New Jim Crow* (2012), Robin DiAngelo's *White Fragility* (2019), Ibram X. Kendi's *How to Be an Anti-Racist* (2019), Iljeoma Oluo's *So You Want to Talk about Race* (2018), and Isabelle Wilkerson's *Caste*

(2020) have publicized critical approaches to race far beyond academia. These authors refute notions of "colorblindness" popular in the US during the 1990s and early 2000s, arguing that far from being a postracial society, systemic racism remains in the foundations of US institutions ranging from prisons to elementary schools.[10]

The roots of critical race theory (CRT) date back to the 1970s and 1980s, when groundbreaking scholars including Derrick Bell, Kimberlé Crenshaw, Richard Delgado, Cheryl Harris, Charles R. Lawrence, Mari Matsuda, and Patricia J. Williams sought to expand critical legal studies to address the role of racial bias in the legal system. By revealing the ways in which laws and policies have historically benefited white people by design, this first generation of CRT scholars revealed the ways in which race is shaped by distinct social, political, and economic forces. Beyond legal studies, fields such as, but not limited to, ethnic and diasporic studies; cultural studies; African American, Black, and Africana studies; Latinx and Chicanx studies; Asian American studies; Native and Indigenous studies; and critical refugee studies have likewise troubled the notion that race is fixed or natural, highlighting the instability and historical fluidity of color-based racial categories such as "White," "Black," "Brown," "Red," and "Yellow." Together, these disciplines frame race as a social construct that is nevertheless real, with oppressive consequences involving actual bodies and lived experiences.

Although there remains significant controversy as to whether race is purely constructed or can be scientifically justified by demonstrable genetic differences, most scholars of race and ethnicity agree that race, as it is construed today in Euro-American contexts, is a distinctly modern phenomenon.[11] The so-called science of race emerged alongside Enlightenment-era rationality and narratives of social progress that were eventually used to support and develop the eugenics movement in the late nineteenth century.[12] Natural scientists like Carl Linnaeus (1706–1778) and German naturalist Johann Friedrich Blumenbach (1752–1840) divided human beings into a set of discrete categories based on a set of physical, biological, and/or behavioral traits that, crucially, were then used to measure different degrees of "civilization" both within and between cultures, thus establishing a predetermined set of values undergirding power relations cloaked in the guise of objectivity. Although these phenotypical traits often included bone structure, hair texture and color, eye shape and color, and skin color, skin color and its linguistic equivalents ("White," "Black," "Red," "Yellow," and so on) became the most pervasive mode of describing racial categories (these ideas are expanded on in our essay introducing part I). There is now a wealth of scholarship on race in culture and media (including film and media studies, cultural studies, gender and sexuality studies, and communications), albeit mostly focusing on issues of representation rather than the ways in which racist ideologies are embedded in the materiality of media infrastructures.[13] *Color Protocols* amends this by moving beyond discourses of racial representation and inclusion and diving into the material substrate of the color technologies themselves. Color technologies and racialization are two sides of the same coin, seemingly distinct yet ultimately welded together.

OVERVIEW OF THE ESSAYS

The essays in part I, "Grammars," question the formation and validity of racialized color systems in primarily verbal, linguistic, and text-based media at the turn of the twentieth century. They consider how lines, words, and language systems have been shaped by early and late modern modes of engaging color, from the fraught concept of the "color line" to issues of deception and ephemerality that shroud any attempt to classify color as a fixed facet of surface and skin. The part I authors—Frederick Douglass, W. E. B. Du Bois, Nicholas Gaskill, Michael Keevak, Anne Anlin Cheng, and Michael Rossi[14]—question the supposed neutrality of color notation systems in the late nineteenth and early twentieth centuries. They ask the following questions: What role did the transatlantic slave trade and European colonial expansion play in promoting and solidifying color as a racial technology in the history of science, anthropology, and literature? How do nuanced color vocabularies succeed or fail as metaphors of race, and who benefits in the process? And how do these systems collectively undergird an explicitly "American grammar," as Hortense Spillers describes it?[15]

Part I is unique in that it contains writing by both living and deceased authors, with Douglass and Du Bois reflecting on the United States in the Reconstruction era as they experienced it in real time and Gaskill, Keevak, Cheng, and Rossi reflecting on the legacy of color-coded racism at the turn of the century and its implications for the present day. All the essays, however, look to the history of media, medicine, literature, anthropology, and technology to understand how these systems naturalized color as an index of racial difference by presenting it as objective or scientific. Specifically, drawing on textual, linguistic, and/or medical discourse, these authors reflect on the historical shift from colonial Europe as a once-central global power to the United States as the new locus of racial capitalism, where a complex history of color-based racial divisions have become deeply entrenched in the formation of the United States as a nation.[16]

The essays in part II, "Images," address the ways in which the racial grammars of part I hardened into the visual technologies of theater lighting, celluloid film, and analog photography. Despite a widespread cultural shift toward polychromatic screen media and the rise of official antiracist political orders following World War II, these media and technologies bolstered idealized images of archetypical whiteness.[17] These authors—Aileen Robinson, Richard Dyer, Lorna Roth, Genevieve Yue, Ali Feser, and Tina Post[18]—scrutinize the overwhelming conflation of skin color as the primary site of the production of meaning and justification for racism in twentieth-century visual culture. They interrogate image-based optical hierarchies and the ways in which they have been used to regulate skin tone along a continuum or schema, from medical reports to the calibration of film stock and the monitoring of bodies in time and space. Western culture's assumption that surface has some intrinsic relation to what lies beneath dates back centuries, as evidenced by Johann Wolfgang von Goethe's 1810 proclamation that "colour, in its infinite variety, exhibits itself on the surface of living beings . . . becom[ing] an important part of the outward indications . . . [to] discover what passes underneath."[19] However, this essentialist conflation of surface and content acquires new meaning in an age of visual-based, optic media when "color" on the surface refers to both bodies and skin. The authors in part II deconstruct this fallacy while critiquing how equating skin color with race often obscures other forms of less visible racial marginalization.

The essays in part III, "Algorithms," move from the visual to the computational. Written by Lisa Nakamura, Ruha Benjamin, Jianqing Chen, Amber Sweat, and Quran M. Karriem,[20] they address how informatic tools like search engines, digital filters, and facial recognition technologies exhibit racial bias akin to the bias for light skin in prior media, but now through algorithmic encoding.[21] Like the color protocols of their optical predecessors—but also with distinctly digital affordances—these technologies are unable to adequately "see" and recognize darker faces and non-Western features, having been programmed with prototypically white skin tones and facial features as their default. Racial inequity is once again reinforced by design.[22] In this sense, part III represents the culmination of *Color Protocols* not simply because it is the final section of the book, but because of the ways in which the protocols in earlier sections here become realized as networked infrastructure. These essays challenge the artificial nature of automated algorithms and the ways in which biometric technologies perpetuate preexisting racist structures of light skin bias and colorism, while also advocating for alternatives that escape this automated matrix. The authors ask: Will it ever be possible for these algorithms to reflect the plurality and variety of lived experiences and bodies, or are they forever tethered to dominant narratives of digital racial capitalism?

The parameters of this book reveal a capacious historical and theoretical scope that simultaneously attends to the medium-specific operations of race and/as color. However, despite the wide-ranging conversations and disciplines represented, one thing that we cannot claim is coverage. While critiques of racial bias are certainly global in scope, we also acknowledge that we cannot do justice to the unique racial politics of all cultural contexts in a single volume. The primary focus of the essays in *Color Protocols* is on the US context, which we are most familiar with and qualified to critique.[23] At the same time, a handful of essays do address color media and its racialization as a global phenomenon, as a consequence of East Asia and the Global South being pulled into the orbit of "American" corporate media empires.[24]

In this sense, we view *Color Protocols* as less a comprehensive survey than a conversational catalyst for focused discussions of chromatic media and race in the years to come. The essays represented here make a concerted effort to grapple with the double valence of color across modern media platforms, from language to biometrics. They reveal how a given medium can disrupt, entrench, or offer new possibilities for working with and understanding color as a racial technology. And in the process, they respond to—and propose new ways to combat—the legacies of violence and marginalization that seem to follow color's multifaceted histories.

NOTES

1. While Crayola changed "Flesh" to "Peach" in 1962 during the civil rights movement, popular accounts running through the 1980s and 1990s depict the "flesh-colored" crayon. During the latter decades of the twenty-first century, Crayola

released a "multicultural" pack intended to represent a more diverse array of skin tones. The multicultural pack is discussed in more detail in chapter 18.

2. For example, see Dalle Vache and Price, eds., *Color: The Film Reader,* and Watkins et al., eds. *Color and the Moving Image: History, Aesthetics, Archive.*

3. See Harvard University Press, *Racism in America: A Reader*; Herman et al., eds., *Living Racism: Through the Barrel of the Book*; White et al., eds. *Relating Worlds of Racism: Dehumanisation, Belonging, and the Normativity of European Whiteness*; and Bonilla-Silva et al., eds. *State of White Supremacy: Racism, Governance, and the United States.*

4. Cheng et al., *Race and Modern Architecture*, 18.

5. For more on this, see Galloway, *Protocol.*

6. Agre, "Surveillance and Capture," 746.

7. Agre, "Surveillance and Capture," 746.

8. For example, see Braham, *Modern Color/Modern Architecture*; Gatz and Wallenfang, *Color in Architecture;* Porter and Mikellides, *Color for Architecture*; Riley, *Color Codes*; Barnstone, *The Color of Modernism*; Riley, *Color Codes: Modern Theories of Color in Philosophy, Painting and Architecture, Literature, Music, and Psychology*; Kane, *Chromatic Algorithms*; Murray, *Bright Signals: A History of Color Television;* and Taussig, *What Color Is the Sacred?*

9. For more on this, see Kalba, *Color in the Age of Impressionism*; Brücke's *Die Physiologie der Farben für die Zwecke der Kunstgewerbe* (1887); Laugel's "Optics and the Arts" (1869); Henry's *Introduction to Scientific Aesthetics* (1885); Blanc's *Grammar of the Arts and Design* (1867); and Rood's *Modern Chromatics* (1879). Noteworthy exceptions include Dootson, *The Rainbow's Gravity*; Beyes, *Organizing Color*; Eaton, *Colour, Art, and Empire*; Batchelor, *Chromophobia*; Taussig, *What Color Is the Sacred?* and Wigley, *White Walls.*

10. Simultaneously, the far right's weaponization of "critical race theory" as a scare tactic has also obscured the field's history and the muddled working definitions of the term.

11. Classics scholar Benjamin Isaac has made a case for the existence of color-based racism in antiquity, which was subsequently adapted by Europeans until the Enlightenment (see his *The Invention of Racism in Classical Antiquity*). However, in a direct response to Isaac, David Goldenberg takes care to distinguish physical distaste toward darker skin tones based on aesthetic judgment and color symbolism from the *systemic* generalization that all people with certain physical traits will have the same emotional/intellectual characteristics (see his "Racism, Color Symbolism, and Color Prejudice"). Racism must be a form of group prejudice that is justified by rationalization and scientific knowledge, which historians largely agree did not solidify into an ideology until the eighteenth century.

12. Projit Bihari Mukharji's 2022 *Brown Coats, White Skins* provides a fascinating and troubling history of how European race science was imported to twentieth-century India.

13. See hooks, *Black Looks;* Hall et al., *Representation*; and Campbell, ed., *The Routledge Companion to Media and Race*. Even fewer scholars explicitly address racial bias in relation to color technologies. Key texts addressing the history of race science and its contemporary manifestations that do not discuss color explicitly include Nelson, *The Social Life of DNA*; Roberts: *Fatal Invention*; and Saini, *Superior*. Two recent monographs also pinpoint how racialized warfare and carceral capitalism make the skin a primary site of surveillance and exploitation: Tu, *Experiments in Skin*; and Visperas, *Skin Theory*.

14. Essays from this part printed here for the first time include the contributions from Gaskill and Rossi.

15. See Spillers, "Mama's Baby, Papa's Maybe: An American Grammar Book."

16. Unlike in the United States, racism in Europe has not been as historically bound to visible physical differences. Because targeted ethnic minorities (i.e., Jews) were often not physically distinguishable from the majority, they were often forced to label their Otherness with their clothing, badges, and other marks (Eliav-Feldon et al., *The Origins of Racism in the West*, 2). Even in other societies with a history of institutionalized systems of color-based racial segregation, these color categories do not neatly map onto those used in the United States. For example, during South African apartheid, the term "Coloured" referred to nonwhites without aboriginal African ancestry, which was treated as a separate racial category from "Black." In contrast, "colored" in the Jim Crow American South was a catch-all term for anyone with African ancestry (the "one-drop rule," or the racist notion that having a single Black ancestor would "contaminate" a person's bloodline, making them Black in the eyes of the law). Isabel Wilkerson also observes that many West Africans arriving in North America found that they did not conceive of themselves as "Black" until they entered the United States (*Caste*, 22).

17. See Melamed, *Represent and Destroy*, for a meticulous account of antiracist liberal orders in the United States after World War II.

18. Essays from this part printed here for the first time include the contributions from Robinson, Feser, and Post.

19. Goethe, *Theory of Colours*, 292.

20. Essays from this part printed here for the first time include the contributions from Chen, Sweat, Karriem, and the conversation between Zeitlin-Wu and Nakamura.

21. For more on this, see Noble, *Algorithms of Oppression: How Search Engines Reinforce Racism*; Katz, *Artificial Whiteness: Politics and Ideology in Artificial Intelligence*; and Chun, *Discriminating Data: Correlation, Neighborhoods, and the New Politics of Recognition*.

22. For example, see Joyrich's "Preface: Bringing Race and Media Technologies into Focus."

23. Many fascinating studies beyond the U.S. context may be used to expand this study. For example, in the New Zealand film *Once Were Warriors,* cinematographer Stuart Dryburgh engages with brown (Maori) skin tones through filters as an innovative strategy and as a seminal case study from the Global South.

24. Thomas, *Beneath the Surface: A Transnational History of Skin Lighteners,* provides an extensive discussion of colorism as it relates to postcolonial consumer culture and whiteness as a global beauty ideal.

BIBLIOGRAPHY

Agre, Philip E. "Surveillance and Capture. Two Models of Privacy." *Information Society* 10, no. 2 (1994): 101–127.

Alexander, Michelle. *The New Jim Crow: Mass Incarceration in the Age of Colorblindness.* New York: New Press, 2012.

Barnstone, Deborah Ascher. *The Color of Modernism: Paints, Pigments, and the Transformation of Modern Architecture in 1920s Germany.* London: Bloomsbury Visual Arts, 2022.

Batchelor, David. *Chromophobia.* London: Reaktion Books, 2000.

Beyes, Timon. *Organizing Color: Towards a Chromatics of the Social.* Palo Alto, CA: Stanford University Press, 2024.

Blanc, Charles. *Grammaire des Arts du Dessin.* Paris: Renouard, 1876.

Bonilla-Silva, Eduardo, Moon-Kie Jung, and João Helion Costa Vargas, eds. *State of White Supremacy: Racism, Governance, and the United States.* Palo Alto, CA: Stanford University Press, 2011.

Braham, William W. *Modern Color/Modern Architecture: Amédée Ozenfant and the Genealogy of Color in Modern Architecture.* London: Taylor and Francis, 2017.

Brücke, Ernst Wilhelm von. *Die Physiologie der Farben für die Zwecke der Kunstgewerbe.* Leipzig, Germany: Verlag von S. Hirzel, 1887.

Campbell, Christopher P., ed. *The Routledge Companion to Media and Race.* London: Taylor & Francis, 2016.

Cheng, Irene, Charles L. Davis II, and Mabel O. Wilson, eds. *Race and Modern Architecture: A Critical History from the Enlightenment to the Present.* Pittsburgh: University of Pittsburgh Press, 2020.

Chun, Wendy Hui Kyong. *Discriminating Data: Correlation, Neighborhoods, and the New Politics of Recognition.* Cambridge, MA: MIT Press, 2021.

Dalle Vache, Angela, and Brian Price, eds. *Color: The Film Reader.* New York: Routledge, 2006.

DiAngelo, Robin. *White Fragility: Why It's So Hard for White People to Talk about Racism.* Boston: Beacon Press, 2018.

Dootson, Kirsty Sinclair. *The Rainbow's Gravity: Colour, Materiality and British Modernity.* London: Paul Mellon Centre for Studies in British Art, 2023.

Dyer, Richard. *White: Essays on Race and Culture*. London: Routledge, 1997.

Eaton, Natasha. *Colour, Art, and Empire: Visual Culture and the Nomadism of Representation*. London: Bloomsbury, 2013.

Eliav-Feldon, Miriam, Benjamin Isaac, and Joseph Ziegler, eds. *The Origins of Racism in the West*. Cambridge: Cambridge University Press, 2009.

Galloway, Alexander R. *Protocol: How Control Exists After Decentralization*. Cambridge, MA: MIT Press, 2004.

Gatz, Konrad, and Wilhelm Otto Wallenfang. *Color in Architecture; A Guide to Exterior Design*. New York: Reinhold Pub. Corp, 1961.

Goethe, Johann Wolfgang von. *Theory of Colours*. Translated by Charles Lock Eastlake. Cambridge, MA: MIT Press, 1970.

Goldenberg, David. "Racism, Color Symbolism, and Color Prejudice." In *The Origins of Racism in the West,* edited by Miriam Eliav-Feldon, Benjamin Isaac, and Joseph Ziegler, 88–108. Cambridge: Cambridge University Press, 2009.

Hall, Stuart, Jessica Evans, and Sean Nixon. *Representation: Cultural Representations and Signifying Practices*. 2nd ed. Thousand Oaks, CA: SAGE Publications, 2013.

Harvard University Press. *Racism in America: A Reader*. Cambridge, MA: Harvard University Press, 2020.

Henry, Charles. *Introduction a une Esthétique Scientifique* [Introduction to scientific aesthetics].

Paris: La Revue contemporaine, 1885.

Herman, Cameron Khalfani, Eric A. Jordan, Thomas J. Mowen, Clarence R. Talley, Oliver Rollins, and Willie Jamaal Wright, eds. *Living Racism: Through the Barrel of the Book*. Lexington, KY: Lexington Books, 2017.

hooks, bell. *Black Looks: Race and Representation*. Boston: South End Press, 1992.

Isaac, Benjamin. *The Invention of Racism in Classical Antiquity*. Princeton, NJ: Princeton University Press, 2006.

Joyrich, Lynn. "Preface: Bringing Race and Media Technologies into Focus." *Camera Obscura* 24, no. 1 (Spring 2009): 1–5.

Laugel, Auguste. "L'Optique et les Arts." *Revue Des Deux Mondes (1829–1971)* 79, no. 4 (1869): 880–905.

Kalba, Laura Anne. *Color in the Age of Impressionism: Commerce, Technology, and Art*. University Park: Penn State University Press, 2017.

Kane, Carolyn L. *Chromatic Algorithms: Synthetic Color, Computer Art, and Aesthetics after Code*. Chicago: University of Chicago Press, 2014.

Katz, Yarden. *Artificial Whiteness: Politics and Ideology in Artificial Intelligence*. New York: Columbia University Press, 2020.

Kendi, Ibram X. *How to Be an Anti-Racist*. New York: One World, 2019.

Melamed, Jodi. *Represent and Destroy: Rationalizing Violence in the New Racial Capitalism*. Minneapolis: University of Minnesota Press, 2011.

Mukharji, Projit Bihari. *Brown Skins, White Coats: Race Science in India, 1920–66*. Chicago: University of Chicago Press, 2022.

Murray, Susan. *Bright Signals: A History of Color Television*. Durham, NC: Duke University Press, 2018.

Nelson, Alondra. *The Social Life of DNA: Race, Reparations, and Reconciliation after the Genome*. Boston: Beacon Press, 2016.

Noble, Safiya. *Algorithms of Oppression: How Search Engines Reinforce Racism*. New York: New York University Press, 2018.

Oluo, Ijeoma. *So You Want to Talk about Race*. Berkeley, CA: Seal Press, 2018.

Porter, Tom, and Byron Mikellides. *Color for Architecture*. New York: Van Nostrand Reinhold, 1976.

Riley, Charles A. *Color Codes: Modern Theories of Color in Philosophy, Painting and Architecture, Literature, Music, and Psychology*. Hanover, NH: University Press of New England, 1995.

Roberts, Dorothy. *Fatal Invention: How Science, Politics, and Big Business Re-create Race in the Twenty-first Century*. New York: New Press, 2012.

Rood, Ogden. *Modern Chromatics: With Application to Art and Industry*. London: C. Kegan Paul, 1878,

Saini, Angela. *Superior: The Return of Race Science*. Boston: Beacon Press, 2019.

Spillers, Hortense J. "Mama's Baby, Papa's Maybe: An American Grammar Book." *Diacritics* 17, no. 2 (Summer 1987): 64–81.

Taussig, Michael. *What Color Is the Sacred?* Chicago: University of Chicago Press, 2011.

Thomas, Lynn M. *Beneath the Surface: A Transnational History of Skin Lighteners*. Durham, NC: Duke University Press, 2020.

Tu, Thuy Linh Nguyen. *Experiments in Skin*. Durham, NC: Duke University Press, 2021.

Visperas, Cristina Meija. *Skin Theory: Visual Culture and the Postwar Prison Laboratory*. New York: New York University Press, 2022.

Watkins, Liz, Sarah Street, and Simon Brown, eds. *Color and the Moving Image: History, Aesthetics, Archive*. New York: Taylor & Francis, 2013.

White, Elisa Joy, Philomena Essed, Karen Farquharson, and Kathryn Pillay, eds. *Relating Worlds of Racism: Dehumanisation, Belonging, and the Normativity of European Whiteness*. Cham, Switzerland: Springer International Publishing, 2018.

Wigley, Mark. *White Walls, Designer Dresses: The Fashioning of Modern Architecture*. Cambridge, MA: MIT Press, 2001.

Wigley, Mark. "Chronic Whiteness" e-flux (November, 2020). https://www.e-flux.com/architecture/sick-architecture/360099/chronic-whiteness/.

Wilkerson, Isabel. *Caste: The Origins of Our Discontents*. New York: Random House, 2020.

I GRAMMARS

INTRODUCTION TO PART I: COLOR ENCODING FROM THE ENLIGHTENMENT TO "RACE SCIENCE"

In recent decades, a rich body of work has emerged from fields ranging from literary studies to the history of science and American studies, analyzing how the cross-pollination of biology and culture in the European Enlightenment paved the path for the explicitly racist scientific subfields of eugenics, phrenology, and physiognomy. Crucially, all these subfields relied on visible, physical differences as the basis for classifying human beings into arbitrary visual grammars to determine (ostensibly) moral character.[1] Although the presumption that race was an innate human characteristic was key to the formation of what became known as "race science" at the turn of the nineteenth century, contemporary scholars writing on this topic rarely explain why skin color, above all, has remained the most pervasive and defining trait in systemic racist legacies.[2] Why did natural scientists, anthropologists, and eugenicists rely so heavily on grammars and indexes of sensory color as fixed standards of measurement and classification? And why do such color taxonomies remain such powerful rhetorical stand-ins for race today?

The essays in part I navigate this capacious matrix of color and race by revisiting its origins in Western modernity, scrutinizing how various color grammars—rule-based systems of classification—in the late nineteenth and early twentieth centuries functioned as ideological, political, and moralistic ordering systems for human beings according to race. They reveal how skin color–based hierarchies proliferated through Enlightenment-era epistemologies and modernity's unwavering faith in social and technological progress, both of which mapped color onto Eurocentric narratives of "civility" versus "barbarism."[3]

While no single account could offer a comprehensive view of all color-based racial grammars, this introductory essay highlights some of the most formative—and pernicious—paradigms in the development of "race science" to provide readers with a necessary contextual overview from which the following essays can be better understood. Our intent is not to reify these pseudoscientific ideas or to canonize the historical figures involved, but rather to map out how the sorting and ranking of humans by race as color came to be justified in the name of "science."

Present-day race and ethnicity scholars typically concur that race is a distinctly modern phenomenon stemming from the grammars constructed in and around the Enlightenment,[4] although there are exceptions.[5] For one, it is difficult to deny the existence of what we might call protoracism prior to the Enlightenment, as David Bindman argues, these ingrained prejudices are not synonymous with racism proper, as they lack a comprehensive "grammar," or fully articulated *theory* of race, that would justify their prejudicial acts.[6] Furthermore, prior to the eighteenth century, the term "race" (*rasse* in German) was associated in countries like Germany, France, and England with a family, clan, or dynasty more than with larger communities or ethnic groups.[7] It wasn't until the late seventeenth century, in 1684, that French physician François Bernier (1625–1688) published *Nouvelle division de la terre par les différentes espèces ou races qui l'habitent* (New Division of the Earth by the Different Species or Races of Man That Inhabit It),[8] contouring a classification of humans into four "espèces ou races" by way of their physical traits and geographical habitats.[9] *Nouvelle division* is commonly taken as the first of its kind, establishing the beginnings of what Noémie Ndiaye calls "the racial matrix": an ever-growing cultural lexicon where, in the wake of the Age of Discovery and European colonization, skin color came to stand for race.[10]

In the eighteenth century, "race science" reemerged as a new category of "human variety," connected to the Latin *gens* (or *gentes*), and eventually *genus*. Swedish biologist Carl Linnaeus (1707–1787) drew on this concept in *Systema Naturae* (The Animal Kingdom), in 1735, a protogenetic set of taxonomic rankings (domain, kingdom, phylum, class, order, family, genus, and species) often considered the first comprehensive theory of race relative to skin color.[11]

Linnaeus's text builds on Aristotle's claim in the *Nicomachean Ethics* that man exists as both animal and human (*animal rationale,* or λόγον ἔχον),

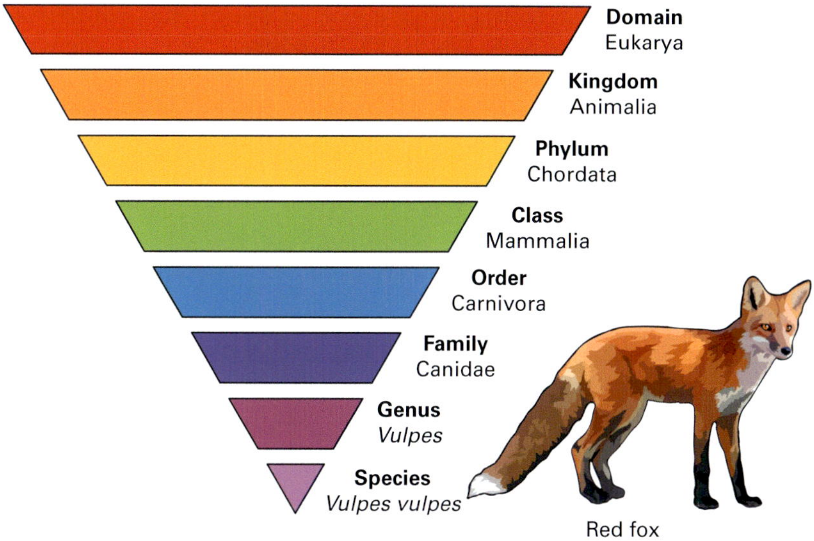

FIGURE PI.1
Taxonomic rank graph. The image shows the main taxonomic ranks as developed by Swedish biologist Carl Linnaeus in 1735. The hierarchy demonstrates how taxonomic ranking was used to designate related animals; the example used here is the red fox (*Vulpes vulpes*). Image by Annina Breen, 2015; courtesy of Creative Commons.

and can therefore be analyzed through physical taxonomies comparable to those used to analyze plants and animals.[12] Man could claim his own natural history, Linnaeus maintained, as "the genus of *Homo sapiens*, united by the ability to mate with . . . other kinds of humanity."[13] Post-1758 editions of *Systema Naturae* integrate material supporting Bernier's earlier, region-based racial claims that humans could also be sorted into four "varieties of man" [*Homo variat*] corresponding to the four continents then known to Europeans: *Europaeus albus* ("European White"), *Americanus rubescens* ("American Reddish"), *Asiatus fuscus* ("Asian tawny"), and *Africanus niger* ("African black"), with *Europaeus albus* and *Africanus niger* at the top and bottom of the hierarchy, respectively.[14] These four continents or "Four Quarters of the Earth," Linnaeus reasoned, corresponded with the "Four Temperaments" of the people who inhabited each one. It was only the white European (*sapiens Europaeus albus*) who, in his view, benefited from the proper epithet "sapiens."[15] *Systema Naturae* therein codified what came to be a pinnacle of systemic racism: a codified rationale for who gets to be considered "human" and who does not.

After Linnaeus's systematic assignment of moral and intellectual traits to each color variant of the world's people was readily adopted by Enlightenment-era scientists and philosophers, skin color became seamlessly folded into mainstream ethical claims surrounding relative human value, moral worth, and cultural supremacy. The placement of white European cultures at the top of this list was, not surprisingly, rooted in religious doctrines that extended beyond skin color—particularly Judeo-Christian mythologies of a prelapsarian unity when the "earth was populated by one people descended from one original couple."[16] The exploitation of non-European people "of color" thus appeared justified, if not divined.

A SCIENCE OF AESTHETICS

In the second half of the eighteenth century, investigations into the intersections of race and aesthetics fell to the academic sciences, primarily at European universities. The hope was that under the domain of "hard" science, aesthetics would be less susceptible to the whims of armchair philosophers and self-declared natural scientists.[17] By the same token, however, notions of "race" and "aesthetics," once under the aegis of the academy, became prime fodder for rigid scientific edifices further abstracted from embodied experience and tradition. Treatises such as German philosopher Alexander Gottlieb Baumgarten's (1714–1762) *Aesthetica* (1751) placed perception and sensation beneath the presumed superiority of rational abstraction (as Plato had proposed many centuries prior), and in doing so concretized a hierarchy of the human sensorium in which experiences that could not be explained by reason were denounced as culturally "uncivilized" and "vulgar."[18]

The elaborate philosophical edifices introduced by German philosopher Immanuel Kant (1724–1804) further concretized this drive to sort the "uncivilized" riffraff from the (presumably) intellectually gifted and morally superior. When Kant published *Observations on the Feeling of the Beautiful and the Sublime* (1764) only a few short years after Baumgarten's text, he unsurprisingly positioned "European man" as socially, morally, and intellectually superior to other races.[19]

We can trace the racial edifices in Kant's work to section 4 of his 1764 *Observations on the Feeling of the Beautiful and Sublime,* on "Of National Characteristics."[20] In this classification system (which built on the four-continent system of Linnaeus and the work of the Comte de Buffon), skin color serves not just as evidence of "racial" class, but also doubles as an indication of humankind's moral classifications.[21] Kant's aesthetic-qua-racial scheme classified humans into white (Europeans), yellow (Asians), Black (Africans), and red (Native Americans), and, as Emmanuel Chukwudi Eze (2019) shows, was further developed in his 1775 essay "On the Varieties of the Different Races of Man," in which Kant explicitly binds climate to skin color as follows::

STEM GENUS: white brunette

First race, very blond (northern Europe), of damp cold

Second race, Copper-Red (America), of dry cold

Third race, Black (Senegambia), of dry heat

Fourth race, Olive-Yellow (Indians), of dry heat[22]

Aside from the obviously reductive color-coding schema and designation of the "white brunette" as the default pinnacle of his "stem genus" from which all other races presumably deviate, another key problem arises (as it does with all sweeping theories of race): color always fails to fix objects, whether groups of people or inanimate entities, no matter how precise the terminology.

In Kant's *Anthropology from a Pragmatic Point of View* (1796), a compendium of the annual lecture series that he taught from 1756 through 1796,[23] he further adopted methods from anthropology and physical geography to jointly theorize the inner and outer aspects of human nature. Where the former dealt with moral conditioning, the latter addressed external bodily characteristics like skin, hair, and eye color, height, and facial characteristics.[24] Across both, the common thread was "Man," who was part of both culture (anthropology) and nature (geography). Kant's next step was classifying race and racial characteristics (namely, skin color) along the physical register. By mapping the visible (color) onto the invisible (morality), he concluded that "the difference in natural gifts between the various nations cannot be completely explained by means of causal [external, physical, or climatic] causes but rather must lie in the [moral] nature of Man himself."[25] In other words, in his view, racial differences among groups were innate but could only be observed externally, in the physical world. This logical contradiction, intrinsic to many theories of race, was now safely embedded in Kant's "moral law," and henceforth covertly became the cornerstone of outward racial bias and systemic stratification in the Enlightenment and beyond.[26]

Across the Atlantic, similar racially color-coded worldviews were endorsed through the natural rights philosophy of John Locke (1631–1704) and Charles Montesquieu (1689–1755), alongside several of the founders of the US. Thomas Jefferson, John Adams, and Benjamin Franklin (among others) proudly founded the country on principles of equality and freedom that were leveraged through a settler colonial economic system rooted in the enslavement and oppression of Black Africans and their descendants.

To justify the government's reliance on the transatlantic slave trade, American revolutionary leaders reverted to the theories of racial evolution noted here, which—as with so many of the racist doctrines outlined thus far—further naturalized monolithic progress narratives from the "primitive" to the "civilized."[27] The degree to which each race needed to "advance" to be treated as a group of "liberal subjects capable of self-governance" was foreclosed in advance to all who were not blessed with light, white skin. Much of this legislative twisting and turning in the American context fell to the work of "race science."[28]

Race science, as the name would suggest, presented itself as a progressive and forward-thinking enterprise aligned with democratic ideals and scientific innovation. Circa 1828, Blumenbach published his five-part race-based naming system, *The Elements of Physiology*, which linked skull morphology to human skin color.[29] Dividing humans into "Caucasians," "Mongolians," "Ethiopians," "Americans," and "Malays," he argued that the Caucasian skull clearly rose to the top, as the most "aesthetically pleasing." Blumenbach's so-called scientifically substantiated evidence of white superiority was adopted by many, including American anthropologist Louis R. Sullivan (1892–1925; discussed in chapter 5) and Francis Galton (1822–1911), Charles Darwin's cousin, who coined the term "eugenics" in 1883.[30] Blumenbach's skull morphology would eventually form the academic basis for the state-sponsored genocide of the Nazi regime.

Nineteenth-century attitudes about race continued to exhibit aspirations of Enlightenment-era rationality by maintaining related classification systems and racial grammars under the auspices of objective science. The use of increasingly more mathematically precise tools and techniques of measurement, such as spectroscopy, led to a scientific "metric" based on

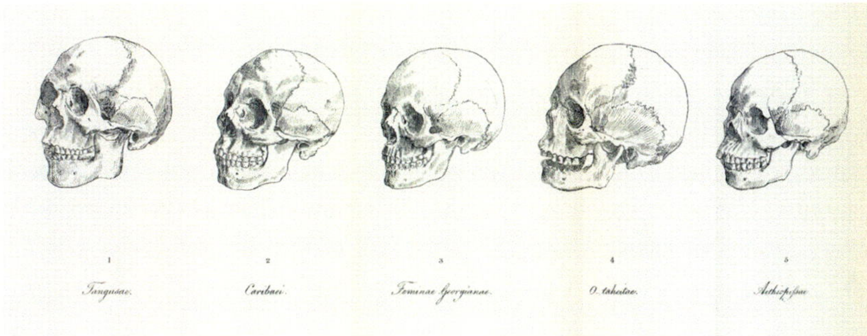

FIGURE PI.2
Blumenbach's five human races, organized by skull shape and skin color. From *De Generis Humani Varietate Native Liber* ("On the Natural Varieties of Mankind," 1795). The schema visualizes the "race theory" that Blumenbach expanded on in *The Elements of Physiology*.

color and pigmentation (discussed in chapter 4), lending even more legitimacy to race science by maintaining that "numbers are the ultimate test of objectivity."[31] Evolutionary theory also lent further credence to what we now take to be pseudoscientific racial frameworks like phrenology and eugenics. Both evolutionary biology and modern racial theories like social Darwinism endorsed beliefs that nature's laws had led to "natural" divisions between races and ethnic groups, reifying the notion of race as biologically determined in both academia and the public sphere.

While this account is far from exhaustive, we have nonetheless attempted to contextualize pivotal moments in a much broader deployment and development of racial taxonomies based on skin color in the Euro-American historical contexts. The essays that follow build on this foundation.

THE ESSAYS

The essays in part I reflect the historical shift from Europe to the United States as the superlative locus of racial capitalism in the late nineteenth and twentieth centuries and the ways that European color-based racial divisions took on new significance in a country which—even long after slavery had ended—continued to position "colored" Blackness and "unmarked" whiteness at the two poles of American life.

The first set of texts comes from intellectual, abolitionist, and statesman Frederick Douglass, who was born enslaved in Maryland circa 1817 (by his own estimate; he states in his first autobiography that he has no record of his exact age) and died in Washington, D.C., in 1895. After escaping slavery in 1839, Douglass and his first wife, Anna Murray, settled in Massachusetts, where they both became active members of the abolitionist movement.[32] In "The Color Line" (1881), originally published in the *North American Review*, Douglass argues that it was by conflating color with race that racism codified personhood on two sides of a Black-and-white binary. Even in Northern states without strident segregation laws, he writes, racial caste remained baked into institutions and everyday interactions, like an undetectable virus that "fill[ed] the air."[33] Douglass's usage of the term "color line" foreshadows Franz Fanon's later conception of racial epidermalization. Just as Fanon describes the experience of being unable to hide from his own Blackness in the famous "Look! A Negro!" passage from *Black Skin, White Masks* (1952),[34] Douglass observes that "the colored people of this country . . . carry in front [of them] the evidence which marks them for persecution."[35] At the same time, Douglass (who was of mixed African, Native American, and European ancestry) points to the contradictions inherent in this epidermal framework by calling out the absurdity of the "one-drop rule" codified as law in several states shortly after the essay's publication.[36] Although Douglass never quite gets to the bottom of why color is such a potent racial signifier, he does conclude that the only solution to dismantle what he calls "color-line feeling" is to decouple color from race.[37]

The second set of excerpts draw from American sociologist and civil rights activist W. E. B. Du Bois's (1868–1963) *The Souls of Black Folk: Essays and Sketches* (1903), including his well-known 1897 speech, "The Conservation of Races," and *Of Our Spiritual Strivings*, followed by an excerpt from "The Superior Race (An Essay)" (1923).[38] We also include a series of annotated images from Du Bois's innovative diagrams of African American life at the turn of the century, which he created with his students in Atlanta for the Exposition des Nègres d'Amérique at the Exposition Universelle in Paris in 1900. These color-coded charts, maps, and graphs—hidden in the shadow of Du Bois's scholarly legacy until recently[39]—offer some of the earliest examples of infographics in modern history.

For Du Bois, as stated at the opening of *The Souls of Black Folk*, "The problem of the Twentieth Century is the problem of the color-line."[40] Building on Douglass's work, he argues that this "color line" created an intrinsic division between Black and white people, a fixed boundary between two colors rather than a chromatic continuum or spectrum. At the same time, Du Bois maintains that race remains central to understanding what it means to be human. In his 1897 speech to the National Association for the Advancement of Colored People (NCAA), he writes: "The history of the world is the history, not of individuals, but of groups, not of nations, but of races, and he who ignores or seeks to override the race idea in human history ignores and overrides the central thought of all history."[41]

Du Bois develops a more nuanced take in his 1923 "The Superior Race (An Essay)."[42] In this work, a hypothetical conversation between himself and an imaginary white person unveils the contradictions underlying clear-cut assumptions of Black and white psychic and physical realities. Any hard-line demarcation of one color from another intrinsically fails to encompass holistic understandings of a shared humanity. Here, the rigid color line articulated in his early work becomes more of a grayscale (to continue the color metaphor), a succession of nuanced shades bridging the poles of Black and white. Du Bois would eventually abandon these binary oppositions altogether in favor of a more comprehensive color philosophy connected to sensual and aesthetic registers that could not be subsumed under any single system or grammar. In short, he came to conclude that no grammar—linguistic or otherwise—is sufficient to signify racial difference when race itself should instead be defined as "a vast family of human beings, generally of common blood and language, always of common history, traditions and impulses, who are both voluntarily and involuntarily striving together for the accomplishment of certain more or less vividly conceived ideals of life."[43]

The five remaining essays in part I—all from living authors—offer contemporary critiques of racialized color encoding in the arts and human sciences in the late nineteenth and early twentieth centuries. Nicholas Gaskill's "Like a Color: Chromatic Metaphors in Racial Discourse" further addresses the role of color and language by analyzing a series of racialized chromatic metaphors that include "the color line," "colorblindness," and "Black is beautiful." By looking at how these metaphors are taken up by both Black writers in the Harlem Renaissance and contemporary social movements to destabilize color as a referent of race, Gaskill asks: What does it mean to call people "Black," "white," "red," or "brown"? What ideas of color—and its material and psychological aspects in particular—underwrite such designations?

Two of the essays in part I interrogate the role that "yellowness" has played historically in constructing Asiatic Otherness. Selected excerpts from the third chapter of Michael Keevak's *Becoming Yellow: A Short History of Racial Thinking* (2011), titled "Measuring Skin Color" and "The Color Top," theorize how and why the term "yellow" came to describe people of East Asian descent in the Western imagination. In these excerpts, both of which are from that book's third chapter (titled "Nineteenth-Century Anthropology and the Measurement of 'Mongolian' Skin Color"), Keevak convincingly demonstrates how yellowness-as-Asianness is a Western construct resulting from the application of color science to the field of race science at the turn of the century. In "The Color Top," he reveals the sinister history of how Milton Bradley's color top, originally designed as a children's toy for the classroom, became an instrument of eugenicists and self-proclaimed race scientists, further illustrating how yellowness came to supplement whiteness and Blackness as a touchstone of color-coded systemic racism.

Historian of Science Michael Rossi's essay, "Pigment and Plaster: Anthropomorphizing Informatic Color in Hawai'i, 1921," focuses on the ways in which colored pigments became a crucial metonym and practical device for physical anthropologists as they struggled to define ideas of race and social order in the interwar period. Drawing on archival research conducted at the American Museum of Natural History, Rossi traces Sullivan's expeditions to Hawai'i in the 1920s, during which time he developed a practice of plaster-casting the heads of Native Hawaiians to classify racial "types." In this work, Rossi argues, both color and form played key roles in creating human taxonomies during a time when Hawaiian diversity posed a threat to social scientists who were attempting to navigate the consequences of so-called race mixing following the American annexation of several overseas territories in the Pacific. Using philosopher Gilbert Simondon's theory of information, Rossi concludes that the plaster-casting process was a way of codifying information, where painted skin became a special form of data. By way of a seemingly esoteric case study, "Pigments and Plaster" preempts discussions of algorithmic encoding through skin color that surface in part III of this book.

Finally, Anne Anlin Cheng's "Ornamentalism: A Feminist Theory for the Yellow Woman" (2018) offers the first serious undertaking of the construction of Asian female personhood in contemporary criticism by revealing the reverberations of nineteenth-century racial grammars in the present. This piece, which Cheng would develop into her monograph of the same name,[44] coins the term "ornamentalism" to describe a uniquely Asiatic form of femininity in the West, in which the Asian woman has been—and continues to be—simultaneously fetishized and decried as an aesthetic object. In this racial protocol, Cheng assertively reappropriates the culturally demeaning term "yellow" from the legacies of scientific racism that Keevak traces. In so doing, she points to a sustained tension between human skin and decoration ("thingness"), where "the yellow woman" blurs the lines between the human and nonhuman. As Cheng puts it, ornamentalism "relies on a decorative grammar, a fantasmatic corporeal syntax that is artificial and layered."[45] In this material grammar of Asiatic femininity, raced and gendered identity become inseparable from the color and luster of materials such as porcelain, gold, and silk. Ornamentalism, Cheng suggests, offers an alternative path to reassessing modernity's vast debates about subjecthood and agency, aesthetics, and politics. (Both Keevak's and Cheng's discussions of color-coded Asianness are addressed further in chapter 16, by Jianqing Chen, in part III of this book.)

By mapping some of these color-coded metrics and taxonomies from the early Enlightenment through the first quarter of the twentieth century, our goal here has been to chart the precarious road of "race science" that has paved the problems of the present. Once again, this essay is not meant to be an exhaustive or a comprehensive account of such grammars. Rather, we aim to highlight and expand on key moments in a longer history of color-based racial differentiation in Western thought that sets the stage for the radically different but related paradigms of racialized color protocols addressed in parts II and III of this book.

NOTES

1. For example, see Gould, *The Mismeasure of Man* (1981).

2. Keevak's *Becoming Yellow* and Hochman's *Savage Preservation* outline how standardized color charts (Keevak) and early color photographic processes (Hochman) played a key role justifying the emergence of race-based "science" in the late nineteenth and early twentieth centuries (see chapter 4 in this volume). More recently, Gaskill, *Chromographia* (2018), and Rossi, *The Republic of Color* (2019), have meticulously revealed the ways in which modern color science and language systems shaped racialized narratives of technological and social evolution in the United States in the progressive era.

3. Although racialized phenotypical traits also included bone structure, hair texture and color, and eye shape and color, it was skin color and its linguistic equivalents

(white, Black, red, and yellow) that became the most pervasive mode of describing (and exercising power and control over) racial categories in this era.

4. For further arguments claiming that the origin of racism is rooted in the European Enlightenment, see Horkheimer and Adorno, *Dialectic of the Enlightenment*; Mosse, *Toward the Final Solution*; and Bernal, *Black Athena*.

5. See Isaac, *The Invention of Racism in Classical Antiquity*; Wynter, "Unsettling the Coloniality"; and Wilson, quoted in Cheng et al., *Race and Modern Architecture*, 27. Both Isaac and Wynter argue, independently, that the European overdetermination of "human-ness" existed as early as Greco-Roman antiquity (Isaac) and Judeo-Christian cosmology in the fourteenth century, only later supplemented by the modern physical sciences (Wynter).

6. Bindman writes, "I would argue that cultural prejudice in the sense of believing, say, that certain non-European peoples are untrustworthy, is not the same as racism, which must have as a foundation a theory of race to justify the exercise of prejudice" (*Ape to Apollo*, 13).

7. Bindman, *Ape to Apollo*, 17.

8. Bernier, *Nouvelle division de la terre*, as cited in Bindman, *Ape to Apollo*, 17. Also see Hannaford, *Race: The History of an Idea in the West*, 191.

9. Bernier's pacification of humankind into four distinct races was contingent on a bizarre climatic-geographical delineation that attempted to explain degrees of diversity in humankind: (1) Europe, North Africa, the Middle East, India, Southeast Asia, and the Americas; (2) sub-Saharan Africans; (3) East and Northeast Asians; and (4) Sámi people (from Norway, Sweden, Finland, and parts of northern Russia).

10. Ndiaye, *Scripts of Blackness*, 4–5. See also Ndiaye's co-edited volume with Markey, *Seeing Race before Race*.

11. See von Linné et al. *Carolus Linnaeus Systema Naturae*, 17.

12. Aristotle, *Nicomachean Ethics*, Book I, 3, 6–7.

13. Linnaeus, paraphrased in Bindman, *Ape to Apollo*, 17.

14. Linnaeus, *Systema Natura* (1758) 34–35, 44–45; Blakey, "Scientific Racism," 32.

15. In these later editions, he added a number of other categories, including the Wild Man and various forms of monsters. See Linnaeus, *Systema Naturae* (1758), 44, 71; Bindman, *Ape to Apollo*, 17.

16. See Bindman, *Ape to Apollo*, 17. He further suggests that Europe's identification with Christendom began in the fifteenth century, and by the 1700s, it had become virtually synonymous with the word "Europe" (24).

17. Bindman, *Ape to Apollo*, 17.

18. Bindman, *Ape to Apollo*, 19

19. According to Kant, the former is associated with the truly beautiful (the transcendental aesthetic) and the latter with the merely agreeable (i.e., women, race, and color), all conveniently lumped together.

20. As noted, it is remarkable that accounts critiquing the racist implications of this work have been in circulation for over three-quarters of a century but are consistently dampened by "scholarly forgetfulness" (as Eze puts it), in line with hegemonic efforts to preserve Kant's work as pure philosophy, which concerns itself only with color-blind philosophical themes in the presumed sanctity of the Western tradition (Eze, "The Color of Reason," 200); see also Count, *This Is Race*, 704.

21. For Kant, Linnaeus's system constituted a problem because it was transcendentally ungrounded. Scientific knowledge, for Kant, was to be based on a universal and permanent (i.e., rational) foundation. Eze, "The Color of Reason," 220.

22. Kant, "On the Different Races of Man" [*Über die verschiedenen Rassen der Menschen*], 48; Eze, "The Color of Reason," in *Postcolonial African Philosophy*, 117–118. Also see Kant; *Anthropology from a Pragmatic Point of View* (1796), also cited in Eze, "The Color of Reason," 214.

23. Although the text was not published until late in Kant's lifetime, much of its material chronologically predates the three well-known *Critiques* and the previously noted essay. It includes *Critique of Pure Reason* (1781), *Critique of Practical Reason* (1788), and *The Critique of Judgement* (1790).

24. This may come as a surprise, given that Kant remains almost exclusively tethered to so-called continental philosophy. However, closer inspection reveals that anthropology and geography are very much part of philosophy due to the unique way in which Kant interpreted these terms as (1) a branch of metaphysics and (2) inextricably bound together as "twin sciences." In this division between practical anthropology and physical geography, we can already detect the long-standing hierarchical separation forthcoming between the abstract functions of the mind and the lower base functions of the body and its sensorial realities. See Eze, "The Color of Reason," 201.

25. See *Kant's philosophische Anthropologie*, ed. Starke, 352, Eze's translation; and Eze, "The Color of Reason," 215.

26. According to Earl W. Count, this text laid the foundation for Kant's "raciology," the most profound "raciological thought of the eighteenth century." *This Is Race*, 704.

27. Wilson, quoted in Cheng et al., *Race and Modern Architecture*, 36.

28. Bindman, *Ape to Apollo*, 11.

29. Blumenbach, *The Elements of Physiology*.

30. Gould, *The Mismeasure of Man*, 75.

31. Gould, *The Mismeasure of Man*, 26.

32. Unlike many formerly enslaved people, Douglass was not only literate, but also an eloquent and prolific writer. In fact, many did not believe initially that he had really written his first autobiography, *Narrative of the Life of Frederick Douglass* (1845), but its content overlapped substantially with the rousing speeches that he gave as an abolitionist orator. He also published two other autobiographies: *My Bondage and My Freedom* (1855) and *Life and Times of Frederick Douglass* (1881). In addition, he founded multiple abolitionist newspapers, delivered oratories around the US and abroad, and in 1877, was the first person of color to be named a US Marshal for the District of Columbia.

33. Douglass, "The Color Line," 568.

34. One key difference between the French-born Fanon and Douglass is that Douglass, writing half a century earlier, remarked on the fact that he experienced little to no racism in his travels to Europe, thus concluding that color-based racism was a uniquely American phenomenon. Fanon's writing marks a different period in the history of European (and particularly French) colonialism, and the differences as well as similarities between these two thinkers point to the historic entanglements among race, culture, and empire.

35. Douglass, "The Color Line," 568.

36. Douglass, "The Color Line," 569. The "one-drop rule" was a classification that emerged after the Civil War, where regardless of how "white-passing" a person was, if that person had a single Black ancestor, he or she was considered Black in the eyes of the law. The one-drop rule, which was used to police undesired "race-mixing" and justify segregation, shows the limitations of treating race as a purely visual phenomenon. As Nicholas Gaskill observes in chapter 3, this particular metaphor of race likens Blackness to dye: "For the metaphor to work, blackness has to be conceived not as one color clashing with another, but as a staining substance—like a dye—coming into contact with an uncolored solution, like water, or with an undyed cloth" (note 11 on page 83).

37. Douglass, "The Color Line," 575.

38. Du Bois was also the first African American to earn a doctorate from Harvard University, one of the founding members of the National Association for the Advancement of Colored People (NAACP), and, in his lifetime, published twenty-two single-author works.

39. The diagrams are compiled in *W. E. B. Du Bois's Data Portraits Visualizing Black America: The Color Line at the Turn of the Century* (2018). In 2023, the Cooper Hewitt Smithsonian Design Museum curated an exhibition of the original drawings called *Deconstructing Power: W. E. B. Du Bois at the 1900 World's Fair*. Autumn Womack further explores DU Bois's experimentation with data-driven technologies in *The Matter of Black Living*.

40. *The Souls of Black Folk*, 3.

41. "The Conservation of Races," 180.

42. This was first published in 1923 in *The Smart Set: A Magazine of Cleverness*, 70, no. 4 (April): 55–60. A similar, later version appears in *Dusk of Dawn: An Essay toward an Autobiography of a Race Concept* (1940).

43. Near the end of his life, Du Bois abandoned much of his earlier thinking and "immersed himself in the struggle for Pan Africanism and decolonization against the European colonial powers, and an emergent postcolonial 'African' or 'Pan-Negro' social and political identity culminating in his own life in his assumption of Ghanaian citizenship in 1963" (Gates, "The Black Letters on the Sign," xv). See also Gilroy, *The Black Atlantic*.

44. Cheng, *Ornamentalism* (2018)

45. Cheng, "Ornamentalism," 416.

BIBLIOGRAPHY

Aristotle. *Nicomachean Ethics*. 3rd ed. Translated by Terence Irwin. Indianapolis: Hackett Publishing Company, 2019. Originally published in 350 B.C.E.

Battle-Baptiste, Whitney, and Britt Rusert, eds. *W. E. B. Du Bois's Data Portraits Visualizing Black America: The Color Line at the Turn of the Century*. Princeton, NJ: Princeton Architectural Press, 2018.

Baumgarten, Alexander Gottlieb. *Aesthetica* (1751); cited in H. R. Schweizer, ed., *Alexander*

Gottlieb Baumgarten: Theoretische Ästhetik (Hamburg, 1980).

Bernal, Martin. *Black Athena: The Afroasiatic Roots of Classical Civilization, The Linguistic Evidence*. Piscataway, NJ: Rutgers University Press, 2006.

Bernier, François. *Nouvelle division de la terre par les différentes espèces ou races qui l'habitent* [Translated as "A New Division of the Earth," published in *Journal des Scavans*, April 24, 1684]. Translated by T. Bendyshe in *Memoirs Read before the Anthropological Society of London* 1 (1863–1864): 360–364.

Blakey, Michael L. "Scientific Racism and the Biological Concept of Race." *Literature and Psychology* 45, no. 1/2 (1999): 29–43.

Blumenbach, Johann Friedrich. *The Elements of Physiology*. 4th ed. Translated by John Elliotson. London: Longman, Rees, Orme, Brown, and Green, 1828.

Bindman, David. *Ape to Apollo: Aesthetics and the Idea of Race in the 18th Century*. Ithaca, NY: Cornell University Press, 2002.

Cheng, Anne Anlin. *Ornamentalism*. Oxford: Oxford University Press, 2018.

Cheng, Anne Anlin. "Ornamentalism: A Feminist Theory for the Yellow Woman." *Critical Inquiry* 44, no. 3 (Spring 2018): 415–446.

Cheng, Irene, Charles L. Davis II, and Mabel O. Wilson, eds. *Race and Modern Architecture: A Critical History from the Enlightenment to the Present*. Pittsburgh: University of Pittsburgh Press, 2020.

Count, Earl W. *This Is Race: An Anthology Selected from the International Literature on the Races of Man*. New York: Schuman, 1950.

Douglass, Frederick. "The Color Line." *North American Review* 132, no. 295 (June 1881): 567–577.

Douglass, Frederick. *Life and Times of Frederick Douglass*. Mineola, NY: Dover Publications, 2003.

Douglass, Frederick. *My Bondage and My Freedom*. London: Penguin Classics, 2003.

Douglass, Frederick. *Narrative of the Life of Frederick Douglass, an American Slave*. Cambridge, MA: Belknap Press of Harvard University Press, 1988.

Du Bois, W. E. B. "The Conservation of Races." In *The Souls of Black Folk*, edited by Brent Hayes Edwards, 179–188. Oxford: Oxford University Press, 2007.

Du Bois, W. E. B. *Dusk of Dawn: An Essay toward an Autobiography of a Race Concept*. New York: Harcourt Brace, 1940.

Du Bois, W. E. B. *The Souls of Black Folk: Essays and Sketches*. New York: Fawcett Publications, 1961.

Du Bois, W. E. B. "The Superior Race (An Essay)." *The Smart Set: A Magazine of Cleverness* 70, no. 4 (April 1923): 55–60.

Du Bois, W. E. B., Aldon D. Morris, and Mabel Wilson. *W. E. B. Du Bois's Data Portraits: Visualizing Black America: The Color Line at the Turn of the Twentieth Century*. Edited by Whitney Battle-Baptiste and Britt Rusert. Amherst, MA: The W. E. B. Du Bois Center at the University of Massachusetts Amherst and Princeton Architectural Press, 2018.

Eze, Emmanuel Chukwudi. "The Color of Reason: The Idea of 'Race' in Kant's Anthropology." *Bucknell Review* 38, no. 2 (January 1, 1995): 200–241.

Eze, Emmanuel Chukwudi, "The Color of Reason: The Idea of 'Race' in Kant's Anthropology." In *Postcolonial African Philosophy: A Critical Reader*, edited by Emmanuel Chukwudi Eze, 103–131. Cambridge, MA: Blackwell, 2019.

Fox, C., R. Porter, and R. Wokler, eds. *Inventing Human Science: Eighteenth-Century Domains*. Berkeley: University of California Press, 1995.

Gaskill, Nicholas. *Chromographia: American Literature and the Modernization of Color*. Minneapolis: University of Minnesota Press, 2018.

Gates, Henry Louis. "The Black Letters on the Sign: W. E. B. Du Bois and the Canon," in *Dusk of Dawn: An Essay toward an Autobiography of a Race Concept* [orig. 1940], xi–xxiv. Oxford: Oxford University Press, 2007.

Gilroy, Paul. *The Black Atlantic: Modernity and Double Consciousness*. Cambridge, MA: Harvard University Press, 1993.

Gould, Stephen J. *The Mismeasure of Man*. New York: W. W. Norton & Co., 1981.

Hannaford, Ivan. *Race: The History of an Idea in the West*. Washington, DC: Woodrow Wilson Center Press, 1996.

Hochman, Brian. *Savage Preservation: The Ethnographic Origins of Modern Media*. Minneapolis: University of Minnesota Press, 2014.

Horkheimer, Max, and Theodor Adorno. *Dialectic of the Enlightenment*. London: Verso, 1997.

Isaac, Benjamin. *The Invention of Racism in Classical Antiquity*. Princeton, NJ: Princeton University Press, 2004.

Jordan, Winthrop D. *White over Black: American Attitudes toward the Negro, 1550–1812*. Chapel Hill: University of North Carolina Press, 1968.

Kant, Immanuel. *Anthropology from a Pragmatic Point of View*. Translated by Victor Lyle Dowdell. Carbondale: Southern Illinois University Press, [1772–1796] 1978.

Kant, Immanuel. *Critique* of Judgment. Translated by Werner S. Pluhar. Indianapolis: Hackett Publishing, 1987.

Kant, Immanuel. Critique of Practical *Reason*. Translated by Mary J. Gregor. Revised edition. Cambridge: Cambridge University Press, 2015.

Kant, Immanuel. Critique of Pure *Reason*. Translated by Norman Kemp Smith. New York: St. Martin's Press, 1965.

Kant, Immanuel. "On the Different Races of Man." In Race and the Enlightenment: A *Reader*, edited by Emmanuel Chukwudi Eze, 38–48. Malden, MA: Wiley-Blackwell, 1997.

Keevak, Michael. "Taxonomies of Yellow: Linnaeus, Blumenbach, and the Making of a 'Mongolian' Race in the Eighteenth Century," in *Becoming Yellow: A Short History of Racial Thinking*, 43–69. Princeton, NJ: Princeton, University Press, 2011.

Linnaeus, Carl. *Systema naturae per regna tria naturae, secundum classes, ordines, genera, species, cum characteribus, differentiis, synonymis, locis*. 10th ed. Stockholm: Impensis Direct. Laurentii Salvii, 1758.

Mosse, George. *Toward the Final Solution: A History of European Racism*. Madison: University of Wisconsin Press, 2020.

Ndiaye, Noémie. *Scripts of Blackness: Early Modern Performance Culture and the Making of Race*. Philadelphia: University of Pennsylvania Press, 2022.

Ndiaye, Noémie, and Lia Markey, eds. *Seeing Race before Race: Visual Culture and the Racial Matrix in the Pre-modern World*. Tempe, AZ: ACMRS Press, 2023.

Rossi, Michael. *The Republic of Color: Science, Perception, and the Making of Modern America*. Chicago: University of Chicago Press, 2019.

von Linné, Carl, Maria Sara Johanna Engel-Ledeboer, and Hendrik Engel. *Carolus Linnaeus Systema Naturae, 1735*. Facsimile of the 1st Ed. With an Introd. and a First English Translation of the 'Observationes.'" *Dutch Classics on History of Science*, 8. Nieuwkoop: Utrecht, Hes & De Graaf Publishers, 2003.

Womack, Autumn. *The Matter of Black Living: The Aesthetic Experiment of Racial Data, 1880–1930*. Chicago: University of Chicago Press, 2022.

Wynter, Sylvia. "Unsettling the Coloniality of Being/Power/Truth/Freedom: Towards the Human, after Man, Its Overrepresentation—An Argument." *CR: The New Centennial Review* 3, no. 3 (Fall 1993): 263–264.

1 "THE COLOR LINE" (1881)

FREDERICK DOUGLASS (1818–1895)

Few evils are less accessible to the force of reason, or more tenacious of life and power, than a long-standing prejudice. It is a moral disorder, which creates the conditions necessary to its own existence, and fortifies itself by refusing all contradiction. It paints a hateful picture according to its own diseased imagination, and distorts the features of the fancied original to suit the portrait. As those who believe in the visibility of ghosts can easily see them, so it is always easy to see repulsive qualities in those we despise and hate.

Prejudice of race has at some time in their history afflicted all nations. "I am more holy than thou" is the boast of races, as well as that of the Pharisee. Long after the Norman invasion and the decline of Norman power, long after the sturdy Saxon had shaken off the dust of his humiliation and was grandly asserting his great qualities in all directions, the descendants of the invaders continued to regard their Saxon brothers as made of coarser clay than themselves, and were not well pleased when one of the former subject race came between the sun and their nobility. Having seen the Saxon a menial, a hostler, and a common drudge, oppressed and dejected for centuries, it was easy to invest him with all sorts of odious peculiarities, and to deny him all manly predicates. Though eight hundred years have passed away since Norman power entered England, and the Saxon has for centuries been giving his learning, his literature, his language, and his laws to the world more successfully than any other people on the globe, men in that country still boast their Norman origin and Norman perfections. This superstition of former greatness serves to fill out the shriveled sides of a meaningless race-pride which holds over after its power has vanished. With a very different lesson from the one this paper is designed to impress, the great Daniel Webster once told the people of Massachusetts (whose prejudices in the particular instance referred to were right) that they "had conquered the sea, and had conquered the land," but that "it remained for them to conquer their prejudices." At one time we are told that the people in some of the towns of Yorkshire cherished a prejudice so strong and violent against strangers and foreigners that one who ventured to pass through their streets would be pelted with stones.

Of all the races and varieties of men which have suffered from this feeling, the colored people of this country have endured most. They can resort to no disguises which will enable them to escape its deadly aim. They carry in front the evidence which marks them for persecution. They stand at the extreme point of difference from the Caucasian race, and their African origin can be instantly recognized, though they may be several removes from the typical African race. They may remonstrate like Shylock—"Hath not a Jew eyes'? hath not a Jew hands, organs, dimensions, senses, affections, passions? fed with the same food, hurt with the same weapons, subject to the same diseases, healed by the same means, warmed and cooled by the same summer and winter, as a Christian is?"—but such eloquence is unavailing. They are negroes—and that is enough, in the eye of this unreasoning prejudice, to justify indignity and violence. In nearly every department of American life they are confronted by this insidious influence. It fills the air. It meets them at the workshop and factory, when they apply for work. It meets them at the church, at the hotel, at the ballot-box, and worst of all, it meets them in the jury-box. Without crime or offense against law or gospel, the colored man is the Jean Valjean of American society. He has escaped from the galleys, and hence all presumptions are against him. The workshop denies him work, and the inn denies him shelter; the ballot-box a fair vote, and the jury-box a fair trial. He has ceased to be the slave of an individual, but has in some sense become the slave of society. He may not now be bought and sold like a beast in the market, but he is the trammeled victim of a prejudice, well calculated to repress his manly ambition, paralyze his energies, and make him a dejected and spiritless man, if not a sullen enemy to society, fit to prey upon life and property and to make trouble generally. When this evil spirit is judge, jury, and prosecutor, nothing less than overwhelming evidence is sufficient to overcome the force of unfavorable presumptions.

Everything against the person with the hated color is promptly taken for granted; while everything in his favor is received with suspicion and doubt.

A boy of this color is found in his bed tied, mutilated, and bleeding, when forthwith all ordinary experience is set aside, and he is presumed to have been guilty of the outrage upon himself; weeks and months he is kept on trial for the offense, and every effort is made to entangle the poor fellow in the confused meshes of expert testimony (the least trustworthy of all evidence). This same spirit, which promptly assumes everything against us, just as readily denies or explains away everything in our favor. We are not, as a race, even permitted to appropriate the virtues and achievements of our individual representatives. Manliness, capacity, learning, laudable ambition, heroic service, by any of our number, are easily placed to the credit of the superior race. One drop of Teutonic blood is enough to account for all good and great qualities occasionally coupled with a colored skin; and on the other hand, one drop of negro blood, though in the veins of a man of Teutonic whiteness, is enough of which to predicate all offensive and ignoble qualities. In presence of this spirit, if a crime is committed, and the criminal is not positively known, a suspicious-looking colored man is sure to have been seen in the neighborhood. If an unarmed colored man is shot down and dies in his tracks, a jury, under the influence of this spirit, does not hesitate to find the murdered man the real criminal, and the murderer innocent. Now let us examine this subject a little more closely. It is claimed that this wonder-working prejudice—this moral magic that can change virtue into vice, and innocence to crime; which makes the dead man the murderer, and holds the living homicide harmless—is a natural, instinctive, and invincible attribute of the white race, and one that cannot be eradicated; that even evolution itself cannot carry us beyond or above it. Alas for this poor suffering world (for four-fifths of mankind are colored), if this claim be true! In that case men are forever doomed to injustice, oppression, hate, and strife; and the religious sentiment of the world, with its grand idea of human brotherhood, its "peace on earth and good-will to men," and its golden rule, must be voted a dream, a delusion, and a snare.

But is this color prejudice the natural and inevitable thing it claims to be? If it is so, then it is utterly idle to write against it, preach, pray, or legislate against it, or pass constitutional amendments against it. Nature will have her course, and one might as well preach and pray to a horse against running, to a fish against swimming, or to a bird against flying. Fortunately, however, there is good ground for calling in question this high pretension of a vulgar and wicked prepossession. If I could talk with all my white fellow-countrymen on this subject, I would say to them, in the language of Scripture: "Come and let us reason together." Now, without being too elementary and formal, it may be stated here that there are at least seven points which candid men will be likely to admit, but which, if admitted, will prove fatal to the popular thought and practice of the times.

First. If what we call prejudice against color be natural, i.e., a part of human nature itself, it follows that it must be co-extensive with human nature, and will and must manifest itself whenever and wherever the two races are brought into contact. It would not vary with either latitude, longitude, or altitude; but like fire and gunpowder, whenever brought together, there would be an explosion of contempt, aversion, and hatred.

Secondly. If it can be shown that there is anywhere on the globe any considerable country where the contact of the African and the Caucasian is not distinguished by this explosion of race wrath, there is reason to doubt that the prejudice is an ineradicable part of human nature.

Thirdly. If this so-called natural, instinctive prejudice can be satisfactorily accounted for by facts and considerations wholly apart from the color features of the respective races, thus placing it among the things subject to human volition and control, we may venture to deny the claim set up for it in the name of human nature.

Fourthly. If any considerable number of white people have overcome this prejudice in themselves, have cast it out as an unworthy sentiment, and have survived the operation, the fact shows that this prejudice is not at any rate a vital part of human nature, and may be eliminated from the race without harm.

Fifthly. If this prejudice shall, after all, prove to be, in its essence and in its natural manifestation, simply a prejudice against condition, and not against race or color, and that it disappears when this or that condition is absent, then the argument drawn from the nature of the Caucasian race falls to the ground.

Sixthly. If prejudice of race and color is only natural in the sense that ignorance, superstition, bigotry, and vice are natural, then it has no better defense than they, and should be despised and put away from human relations as an enemy to the peace, good order, and happiness of human society.

Seventhly. If, still further, this aversion to the negro arises out of the fact that he is as we see him, poor, spiritless, ignorant, and degraded, then whatever is humane, noble, and superior, in the mind of the superior and more fortunate race, will desire that all arbitrary barriers against his manhood, intelligence, and elevation shall be removed, and a fair chance in the race of life be given him.

The first of these propositions does not require discussion. It commends itself to the understanding at once. Natural qualities are common and universal, and do not change essentially on the mountain or in the valley. I come therefore to the second point—the existence of countries where this malignant prejudice, as we know it in America, does not prevail; where character, not color, is the passport to consideration; where the right of the black man to be a man, and a man among men, is not questioned; where he may, without offense, even presume to be a gentleman. That there are such countries in the world there is ample evidence. Intelligent and observing travelers, having no theory to support, men whose testimony would be received without question in respect of any other matter, and should not be questioned in this, tell us that they find no color prejudice in Europe, except among Americans who reside there. In England and on the Continent, the colored man is no more an object of hate than any other person. He mingles with the multitude unquestioned, without offense given or received. During the two years which the writer spent abroad, though he was much in society, and was sometimes in the company of lords and ladies, he does not remember one word, look, or gesture that indicated the slightest aversion to him on account of color. His experience was not in this respect exceptional or singular. Messrs. Remond, Ward, Garnet, Brown, Pennington, Crummell, and Bruce, all of them colored, and some of them black, bear the same testimony. If what these gentlemen say (and it can be corroborated by a thousand witnesses) is true there is no prejudice against color in England, save as it is carried there by Americans—carried there as a moral disease from an infected country. It is American, not European; local, not general; limited, not universal, and must be ascribed to artificial conditions, and not to any fixed and universal law of nature.

The third point is: Can this prejudice against color, as it is called, be accounted for by circumstances outside and independent of race or color? If it can be thus explained, an incubus may be removed from the breasts of both the white and the black people of this country, as well as from that large inter mediate population which has sprung up between these alleged irreconcilable extremes. It will help us to see that it is not necessary that the Ethiopian shall change his skin, nor needful that the white man shall change the essential elements of his nature, in order that mutual respect and consideration may exist between the two races.

Now it is easy to explain the conditions outside of race or color from which may spring feelings akin to those which we call prejudice. A man without the ability or the disposition to pay a just debt does not feel at ease in the presence of his creditor. He does not want to meet him on the street, or in the market-place. Such meeting makes him uncomfortable. He would rather find fault with the bill than pay the debt, and the creditor himself will soon develop in the eyes of the debtor qualities not altogether to his taste.

Someone has well said, we may easily forgive those who injure us, but it is hard to forgive those whom we injure. The greatest injury this side of death, which one human being can inflict on another, is to enslave him, to blot out his personality, degrade his manhood, and sink him to the condition of a beast of burden; and just this has been done here during more than two centuries. No other people under heaven, of whatever type or endowments, could have been so enslaved without failing into contempt and scorn on the part of those enslaving them. Their slavery would itself stamp them with odious features, and give their oppressors arguments in favor of oppression. Besides the long years of wrong and injury inflicted upon the colored race in this country, and the effect of these wrongs upon that race, morally, intellectually, and physically, corrupting their morals, darkening their minds, and twisting their bodies and limbs out of all approach to symmetry, there has been a mountain of gold—uncounted millions of dollars—resting upon them with crushing weight. During all the years of their bondage, the slave master had a direct interest in discrediting the personality of those he held as property. Every man who had a thousand dollars so invested had a thousand reasons for painting the black man as fit only for slavery. Having made him the companion of horses and mules, he naturally sought to justify himself by assuming that the negro was not much better than a mule. The holders of twenty hundred million dollars' worth of property in human chattels procured the means of influencing press, pulpit, and politician, and through these instrumentalities they belittled our virtues and magnified our vices, and have made us odious in the eyes of the world. Slavery had the power at one time to make and unmake Presidents, to construe the law, dictate the policy, set the fashion in national manners and customs, interpret the Bible, and control the church; and, naturally enough, the old masters set themselves up as much too high as they set the manhood of the negro too low. Out of the depths of slavery has come this prejudice and this color line. It is broad enough and black enough to explain all the malign influences which assail the newly emancipated millions to-day. In reply to this argument it will perhaps be said that the negro has no slavery now to contend with, and that having been free during the last sixteen years, he ought by this time to have contradicted the degrading qualities which slavery formerly ascribed to him. All very true as to the letter, but utterly false as to the spirit. Slavery is indeed gone, but its shadow still lingers over the country and poisons more or less the moral atmosphere of all sections of the republic. The money motive for assailing the negro which slavery represented is indeed absent, but love of power and dominion, strengthened by two centuries of irresponsible power, still remains.

Having now shown how slavery created and sustained this prejudice against race and color, and the powerful motive for its creation, the other four points made against it need not be discussed in detail and at length, but may only be referred to in a general way.

If what is called the instinctive aversion of the white race for the colored, when analyzed, is seen to be the same as that which men feel or have felt toward other objects wholly apart from color; if it should be the same as that sometimes exhibited by the haughty and rich to the humble and poor, the same as the Brahmin feels toward the lower caste, the same as the Norman felt toward the Saxon, the same as that cherished by the Turk against Christians, the same as Christians have felt toward the Jews, the same as that which murders a Christian in Wallachia, calls him a "dog" in Constantinople, oppresses and persecutes a Jew in Berlin, hunts down a socialist in St. Petersburg, drives a Hebrew from an hotel at Saratoga, that scorns the Irishman in London, the same as Catholics once felt for Protestants, the same as that which insults, abuses, and kills the Chinaman on the Pacific slope—then may we well enough affirm that this prejudice really has nothing whatever to do with race or color, and that it has its motive and mainspring in some other source with which the mere facts of color and race have nothing to do.

After all, some very well informed and very well meaning people will read what I have now said, and what seems to me so just and reasonable, and will still insist that the color of the negro has something to do with the feeling entertained toward him; that the white man naturally shudders at the thought of contact with one who is black—that the impulse is one which he can neither resist nor control. Let us see if this conclusion is a sound one. An argument is unsound when it proves too little or too much, or when it proves nothing. If color is an offense, it is so, entirely apart from the manhood it envelops. There must be something in color of itself to kindle rage and inflame hate, and render the white man generally uncomfortable. If the white man were really so constituted that color were, in itself, a torment to him, this grand old earth of ours would be no place for him. Colored objects confront him here at every point of the compass. If he should shrink and shudder every time he sees anything dark, he would have little time for anything else. He would require a colorless world to live in—a world where flowers, fields, and floods should all be of snowy whiteness; where rivers, lakes, and oceans should all be white; where islands, capes, and continents should all be white; where all the men, and women, and children should be white; where all the fish of the sea, all the birds of the air, all the "cattle upon a thousand hills," should be white; where the heavens above and the earth beneath should be white, and where day and night should not be divided by light and darkness, but the world should be one eternal scene of light. In such a white world, the entrance of a black man would be hailed with joy by the inhabitants. Anybody or anything would be welcome that would break the oppressive and tormenting monotony of the all-prevailing white.

In the abstract, there is no prejudice against color. No man shrinks from another because he is clothed in a suit of black, nor offended with his boots because they are black. We are told by those who have resided there that a white man in Africa comes to think that ebony is about the proper color for man. Good old Thomas Whitson—a noble old Quaker—a man of rather odd appearance—used to say that even he would be handsome if he could change public opinion.

Aside from the curious contrast to himself, the white child feels nothing on the first sight of a colored man. Curiosity is the only feeling. The office of color in the color line is a very plain and subordinate one. It simply advertises the objects of oppression, insult, and persecution. It is not the maddening liquor, but the black letters on the sign telling the world where it may be had. It is not the hated Quaker, but the broad brim and the plain coat. It is not the hateful Cain, but the mark by which he is known. The color is innocent enough, but things with which it is coupled make it hated. Slavery, ignorance, stupidity, servility, poverty, dependence, are undesirable conditions. When these shall cease to be coupled with color, there will be no color line drawn. It may help in this direction to observe a few of the inconsistencies of the color-line feeling, for it is neither uniform in its operations nor consistent in its principles. Its contradictions in the latter respect would be amusing if the feeling itself were not so deserving of unqualified abhorrence. Our Californian brothers, of Hibernian descent, hate the Chinaman, and kill him, and when asked why they do so, their answer is that a Chinaman is so industrious he will do all the work, and can live by wages upon which other people would starve. When the same people and others are asked why they hate the colored people, the answer is that they are indolent and wasteful, and cannot take care of themselves. Statesmen of the South will tell you that the negro is too ignorant and stupid properly to exercise the elective franchise, and yet his greatest offense is that he acts with the only party intelligent enough in the eyes of the nation to legislate for the country. In one breath they tell us that the negro is so weak in intellect, and so destitute of manhood, that he is but the echo of designing white men, and yet in another they will virtually tell you that the negro is so clear in his moral perceptions, so firm in purpose, so steadfast in his convictions, that he cannot be persuaded by arguments or intimidated by threats, and that nothing but the shot-gun can restrain him from voting for the men and measures he approves. They shrink back in horror from contact with the negro as a man and a gentleman, but like him very well as a barber, waiter, coachman, or cook. As a slave, he could ride anywhere, side by side with his white master, but as a freeman, he must be thrust into the smoking-car. As a slave, he could go into the first cabin; as a freeman, he was not allowed abaft the wheel. Formerly it was said he was incapable of learning, and at the same time it was a crime against the State for any man to teach him to read. To-day he is said to be originally and permanently inferior to the white race, and yet wild apprehensions are expressed lest six millions of this inferior race will somehow or other manage to rule over thirty-five millions of the superior race. If inconsistency can prove the hollowness of anything, certainly the emptiness of this pretense that color has any terrors is easily shown. The trouble is that most men, and especially mean men, want to have something under them. The rich man would have the poor man, the white would have the black, the Irish would have the negro, and the negro must have a dog, if he can get nothing higher in the scale of intelligence to dominate. This feeling is one of the vanities which enlightenment will dispel. A good but simple-minded Abolitionist said to me that he was not ashamed to walk with me down Broadway arm in-arm, in open daylight, and evidently thought he was saying something that must be very pleasing to my self-importance, but it occurred to me, at the moment, this man does not dream of any reason why I might be ashamed to walk arm-in-arm with him through Broadway in open daylight. Riding in a stage-coach from Concord, New Hampshire, to Yergennes, Vermont, many years ago, I found myself on very pleasant terms with all the passengers through the night, but the morning light came to me as it comes to the stars; I was as Dr. Beecher says he was at the first fire he witnessed, when a bucket of cold water was poured down his back—"the fire was not put out, but he was." The fact is, the higher the colored man rises in the scale of society, the less prejudice does he meet.

The writer has met and mingled freely with the leading great men of his time,—at home and abroad, in public halls and private houses, on the platform and at the fireside,—and can remember no instance when among such men has he been made to feel himself an object of aversion. Men who are really great are too great to be small. This was gloriously true of the late Abraham Lincoln, William H. Seward, Salmon P. Chase, Henry Wilson, John P. Hale, Lewis Tappan, Edmund Quincy, Joshua R. Giddings, Gerrit Smith, and Charles Sumner, and many others among the dead. Good taste will not permit me now to speak of the living, except to say that the number of those who rise superior to prejudice is great and increasing. Let those who wish to see what is to be the future of America, as relates to races and race relations, attend, as I have attended, during the administration of President Hayes, the grand diplomatic receptions at the executive mansion, and see there, as I have seen, in its splendid east room the wealth, culture, refinement, and beauty of the nation assembled, and with it the eminent representatives of other nations,—the swarthy Turk with his "fez," the Englishman shining with gold, the German, the Frenchman, the Spaniard, the Japanese, the Chinaman, the Caucasian, the Mongolian, the Sandwich

Islander, and the negro,—all moving about freely, each respecting the rights and dignity of the other, and neither receiving nor giving offense.

"Then let us pray that come it may,
As come it will for a' that,
That sense and worth, o'er a' the earth,
May bear the gree, and a;' that;
That man to man, the world o'er,
Shall brothers be, for a' that."

Frederick Douglass.

NOTE

The North American Review 132, no. 295 (June 1881): 567577.

2 *THE SOULS OF BLACK FOLK: ESSAYS AND SKETCHES* (1903) AND "THE SUPERIOR RACE (AN ESSAY)" (1923)

W. E. B. DU BOIS (1868–1963)

W. E. B. Du Bois gave a speech called "The Conservation of Races" before the first conference of the American Negro Academy, the scholarly society founded by Alexander Crummell. In it, Du Bois discusses a number of themes elaborated on in the first chapter of The Souls of Black Folk, *"Of Our Spiritual Strivings." The speech was subsequently published by the academy in 1897 as the second of the organization's Occasional Papers.*

EXCERPTS FROM APPENDIX I, "THE CONSERVATION OF RACES," IN *THE SOULS OF BLACK FOLK: ESSAYS AND SKETCHES* (1903)

The American Negro has always felt an intense personal interest in discussions as to the origins and destinies of races: primarily because back of most discussions of race with which he is familiar, have lurked certain assumptions as to his natural abilities, as to his political, intellectual and moral status, which he felt were wrong. He has, consequently, been led to deprecate and minimize race distinctions, to believe intensely that out of one blood God created all nations, and to speak of human brotherhood as though it were the possibility of an already dawning to-morrow.

Nevertheless, in our calmer moments we must acknowledge that human beings are divided into races; that in this country the two most extreme types of the world's races have met, and the resulting problem as to the future relations of these types is not only of intense and living interest to us, but forms an epoch in the history of mankind.

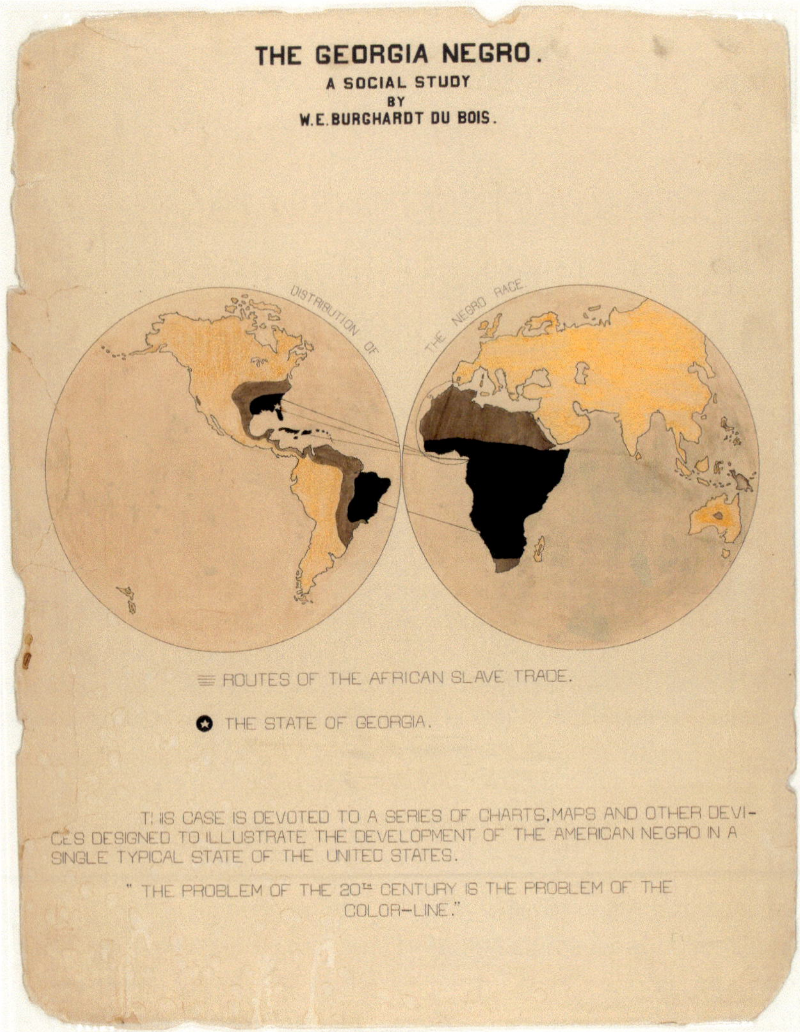

FIGURE 2.1

W. E. B. Du Bois and Atlanta University, Plate 1: "The Georgia Negro: A Social Study." Circa 1900, Du Bois and his students created a series of sixty innovative data visualizations for the Exposition des Nègres d'Amérique at the Exposition Universelle in Paris. In this plate, the first in the set, two circles are connected by a thin black line denoting the historical routes and development of the transatlantic slave trade. Image courtesy of the Daniel Murray collection at the Library of Congress, Prints and Photographs Division.

It is necessary, therefore, in planning our movements, in guiding our future development, that at times we rise above the pressing, but smaller questions of separate schools and cars, wage-discrimination and lynch law, to survey the whole question of race in human philosophy and to lay, on a basis of broad knowledge and careful insight, those large lines of policy and higher ideals which may form our guiding lines and boundaries in the practical difficulties of every day. For it is certain that all human striving must recognize the hard limits of natural law, and that any striving, no matter how intense and earnest, which is against the constitution of the world, is vain. The question, then, which we must seriously consider is this: What is the real meaning of Race; what has, in the past, been the law of race development, and what lessons has the past history of race development to teach the rising Negro people?

When we thus come to inquire into the essential difference of races we find it hard to come at once to any definite conclusion. Many criteria of race differences have in the past been proposed, as color, hair, cranial measurements and language. And manifestly, in each of these respects, human beings differ widely. They vary in color, for instance, from the marble-like pallor of the Scandinavian to the rich, dark brown of the Zulu, passing by the creamy Slav, the yellow Chinese, the light brown Sicilian and the brown Egyptian. Men vary, too, in the texture of hair from the obstinately straight hair of the Chinese to the obstinately tufted and frizzled hair of the Bushman. In measurement of heads, again, men vary; from the broad-headed Tartar to the medium-headed European and the narrow-headed Hottentot; or, again in language, from the highly-inflected Roman tongue to the monosyllabic Chinese. All these physical characteristics are patent enough, and if they agreed with each other it would be very easy to classify mankind. Unfortunately for scientists, however, these criteria of race are most exasperatingly intermingled. Color does not agree with texture of hair, for many of the dark races have straight hair; nor does color agree with the breadth of the head, for the yellow Tartar has a broader head than the German; nor, again, has the science of language as yet succeeded in clearing up the relative authority of these various and contradictory criteria. The final word of science, so far, is that we have at least two, perhaps three, great families of human beings—the whites and Negroes, possibly the yellow race. That other races have arisen from the intermingling of the blood of these two. This broad division of the world's races which men like Huxley and Raetzel have introduced as more nearly true than the old five-race scheme of Blumenbach, is nothing more than an acknowledgment that, so far as purely physical characteristics are concerned, the differences between men do not explain all the differences of their history. It declares, as Darwin himself said, that great as is the physical unlikeness of the various races of men their likenesses are greater, and upon this rests the whole scientific doctrine of Human Brotherhood.

Although the wonderful developments of human history teach that the grosser physical differences of color, hair and bone go but a short way toward explaining the different roles which groups of men have played in Human Progress, yet there are differences—subtle, delicate and elusive, though they may be—which have silently but definitely separated men into groups. While these subtle forces have generally followed the natural cleavage of common blood, descent and physical peculiarities, they have at other times swept across and ignored these. At all times, however, they have divided human beings into races, which, while they perhaps transcend scientific definition, nevertheless, are clearly defined to the eye of the Historian and Sociologist. If this be true, then the history of the world is the history, not of individuals, but of groups, not of nations, but of races, and he who ignores or seeks to override the race idea in human history ignores and overrides the central thought of all history. What, then, is a race? It is a vast family of human beings, generally of common blood and language, always of common history, traditions and impulses, who are both voluntarily and involuntarily striving together for the accomplishment of certain more or less vividly conceived ideals of life.

Turning to real history, there can be no doubt, first, as to the widespread, nay, universal, prevalence of the race idea, the race spirit, the race ideal, and as to its efficiency as the vastest and most ingenious invention for human progress. We, who have been reared and trained under the individualistic philosophy of the Declaration of Independence and the laisser-faire philosophy of Adam Smith, are loath to see and loath to acknowledge this patent fact of human history. We see the Pharaohs, Caesars, Toussaints and Napoleons of history and forget the vast races of which they were but epitomized expressions. We are apt to think in our American impatience, that while it may have been true in the past that closed race groups made history, that here in conglomerate America *nous avons changer tout cela*[1]—we have changed all that, and have no need of this ancient instrument of progress. This assumption of which the Negro people are especially fond, cannot be established by a careful consideration of history.

We find upon the world's stage today eight distinctly differentiated races, in the sense in which History tells us the word must be used. They are, the Slavs of eastern Europe, the Teutons of middle Europe, the English of Great Britain and America, the Romance nations of Southern and Western Europe, the Negroes of Africa and America, the Semitic people of Western Asia and Northern Africa, the Hindoos of Central Asia and the Mongolians of Eastern Asia. There are, of course, other minor race groups, as the American Indians, the Esquimaux and the South Sea Islanders; these larger races, too, are far from homogeneous; the Slav includes the Czech, the Magyar, the Pole and the Russian; the Teuton includes the German, the Scandinavian and the Dutch; the English include the Scotch, the Irish and the conglomerate American. Under Romance nations the widely-differing Frenchman, Italian, Sicilian and Spaniard are comprehended. The term Negro is, perhaps, the most indefinite of all, combining the Mulattoes and Zamboes of America[2] and the Egyptians, Bantus and Bushmen of Africa. Among the Hindoos are traces of widely differing nations, while the great Chinese, Tartar, Corean and Japanese families fall under the one designation—Mongolian.

The question now is: What is the real distinction between these nations? Is it the physical differences of blood, color and cranial measurements? Certainly we must all acknowledge that physical differences play a great part, and that, with wide exceptions and qualifications, these eight great races of to-day follow the cleavage of physical race distinctions; the English and Teuton represent the white variety of mankind; the Mongolian, the yellow; the Negroes, the black. Between these are many crosses and mixtures, where Mongolian and Teuton have blended into the Slav, and other mixtures have produced the Romance nations and the Semites. But while race differences have followed mainly physical race lines, yet no mere physical distinctions would really define or explain the deeper differences—the cohesiveness and continuity of these groups. The deeper differences are spiritual, psychical, differences—undoubtedly based on the physical, but infinitely transcending them. The forces that bind together the Teuton nations are, then, first, their race identity and common blood; secondly, and more important, a common history, common laws and religion, similar habits of thought and a conscious striving together for certain ideals of life. The whole process which has brought about these race differentiations has been a growth, and the great characteristic of this growth has been the differentiation of spiritual and mental differences between great races of mankind and the integration of physical differences.[3]

EXCERPTS FROM *THE SOULS OF BLACK FOLK: ESSAYS AND SKETCHES* (1903)

THE FORETHOUGHT

Herein lie buried many things which if read with patience may show the strange meaning of being black here in the dawning of the Twentieth Century. This meaning is not without interest to you, Gentle Reader; for the problem of the Twentieth Century is the problem of the color-line.

Atlanta, Ga., Feb. 1, 1903

OF OUR SPIRITUAL STRIVINGS

Between me and the other world there is ever an unasked question: unasked by some through feelings of delicacy; by others through the difficulty of rightly framing it. All, nevertheless, flutter round it. They approach me in a half-hesitant sort of way, eye me curiously or compassionately, and then, instead of saying directly, How does it feel to be a problem? they say, I know an excellent colored man in my town; or, I fought at Mechanicsville;[4] or, Do not these Southern outrages make your blood boil? At these I smile, or am interested, or reduce the boiling to a simmer, as the occasion may require. To the real question, How does it feel to be a problem? I answer seldom a word. And yet, being a problem is a strange experience,—peculiar even for one who has never been anything else, save perhaps in babyhood and in Europe. It is in the early days of rollicking boyhood that the revelation first bursts upon one, all in a day, as it were. I remember well when the shadow swept across me. I was a little thing, away up in the hills of New England, where the dark Housatonic winds between Hoosac and Taghkanic to the sea. In a wee wooden schoolhouse, something put it into the boys' and girls' heads to buy gorgeous visiting-cards—ten cents a package—and exchange. The exchange was merry, till one girl, a tall newcomer, refused my card,—refused it peremptorily, with a glance. Then it dawned upon me with a certain suddenness that I was different from the others; or like, mayhap, in heart and life and longing, but shut out from their world by a vast veil. I had thereafter no desire to tear down that veil, to creep through; I held all beyond it in common contempt, and lived above it in a region of blue sky and great wandering shadows. That sky was bluest when I could beat my mates at examination time, or beat them at a foot-race, or even beat their stringy heads. Alas, with the years all this fine contempt began to fade; for the worlds I longed for, and all their dazzling opportunities, were theirs, not mine. But they should not keep these prizes, I said; some, all, I would wrest from them. Just how I would do it I could never decide: by reading law, by healing the sick, by telling the wonderful tales that swam in my head,—some way. With other black boys the strife was not so fiercely sunny: their youth shrunk into tasteless sycophancy, or into silent hatred of the pale world about them and mocking distrust of everything white; or wasted itself in a bitter cry, Why did God make me an outcast and a stranger in mine own house? The shades of the prison-house closed round about us all: walls strait and stubborn to the whitest, but relentlessly narrow, tall, and unscalable to sons of night who must plod darkly on in resignation, or beat unavailing palms against the stone, or steadily, half hopelessly, watch the streak of blue above.

After the Egyptian and Indian, the Greek and Roman, the Teuton and Mongolian, the Negro is a sort of seventh son, born with a veil, and gifted with second-sight[5] in this American world,—a world which yields him no true self-consciousness, but only lets him see himself through the revelation of the other world. It is a peculiar sensation, this double consciousness, this sense of always looking at one's self through the eyes of others, of measuring one's soul by the tape of a world that looks on in amused contempt and pity. One ever feels his two-ness,—an American, a Negro; two souls, two thoughts, two unreconciled strivings; two warring ideals in one dark body, whose dogged strength alone keeps it from being torn asunder.[6]

EXCERPT FROM "THE SUPERIOR RACE (AN ESSAY)" (1923)

I. WHEN the obsession of his race consciousness leaves him, my white friend is quite companionable; otherwise he is impossible. He has a way of putting an excessive amount of pity in his look and of stating as a general and incontrovertible fact that it is "horrible" to be an Exception. By this he means me. He is more than certain that I prove the rule. He is not a bright person, but of that famous average, standardized and astonished at anything that even seems original. His thesis is simple: The world is composed of Race superimposed on Race; classes superimposed on classes; beneath the whole thing is "Our Family" in capitals, and under that is God. God seems to be a cousin, or at least a blood relative of the Van Diemans.

"Of course," he says, "you know Negroes are inferior."

I admit nothing of the sort, I maintain. In fact, having known with considerable intimacy, both male and female, the people of the British Isles, of Scandinavia, of Russia, of Germany, north and south, of the three ends of France and the two ends of Italy; specimens from the Balkans and black and white Spain; the three great races of Asia and the melange of Africa, without mentioning America, I sit here and maintain that black folk are much the superior of white.

"You are either joking or mad," he says.

Both and neither. This race talk is, of course, a joke, and frequently it has driven me insane and probably will permanently in the future; and yet, seriously and soberly, we black folk are the salvation of mankind.

He regards me with puzzled astonishment and says confidentially:

"Do you know that sometimes I am half afraid that you really believe this? At other times I see clearly the inferiority complex."

The former after lunch, I reply, and the latter before.

"Very well," he says, "let's lunch."

Where? I ask quizzically, we being at the time in the roaring Forties.

"Why—oh, well!—their refusal to serve you lunch at least does not prove your superiority."

Nor yet theirs, I answer; but never mind, come with me to Second Avenue.

We start again with the salad.

"Now, superiority consists of what?" he argues.

Life is, I remark, (1) Beauty and health of body, (2) Mental clearness and creative genius, (3) Spiritual goodness and receptivity, (4) Social adaptability and constructiveness.

"Not bad," he answers. "Not bad at all. Now, I contend that the white race conspicuously excels in one, two and four and is well abreast even in three."

And I maintain that the black race excels in one, three and four and is well abreast in two.

"Sheer nonsense and pure balderdash! Compare the Venus of Milo and the Apollo Belvedere with a Harlem or Beale Street couple."

With a Fifth Avenue Easter parade or a Newport Dance. In short, compare humanity at its best or worst with the Ideal, and humanity suffers. But black folk in most attributes of physical beauty, in line and height and curve, have the same norms as whites and differ only in small details of color, hair and curve of countenance. Now can there be any question but that as colors bronze, mahogany, coffee and gold are far lovelier than pink, gray and marble? Hair is a matter of taste. Some will have it drab and stringy and others in a gray, woven, unmoving mass. Most of us like it somewhere between, in tiny tendrils, smoking curls and sweeping curves. I have loved all these varieties in my day. I prefer the crinkly kind, almost wavy, in black brown and glistening. In faces I hate straight features; needles and razors may be sharp—but beautiful, never.

"All that is personal opinion. I prefer the colors of heaven and day: sunlight hair and blue eyes, and straight noses and thin lips, and that incomparable air of haughty aloofness and aristocracy."

And I, on the contrary, am the child of twilight and night, and choose intricately curly hair, black eyes, full and luscious features, and that air of humility and wonder which streams from moonlight. Add to this, voices that caress instead of rasp, glances that appeal rather than repel, and a sinuous litheness of movement to replace Anglo-Saxon stalking—there you have my ideal. Of course you can bury any human body in dirt and misery and make it horrible. I have seen the East End of London.

"Beauty seems to be simply opinion, if you put it that way."

To be sure. But whose opinion?

"Bother beauty. Here we shall never agree. But, after all, I doubt if it makes much difference. The real point is Brains: clear thinking, pure reason, mathematical precision and creative genius. Now, without blague, stand and acknowledge that here the white race is supreme."

Quite the contrary. I know no attribute in which the white race has more conspicuously failed. This is white and European civilization; and as a system of culture it is idiotic, addle-brained, unreasoning, topsy-turvy, without precision, and its genius chiefly runs to marvelous contrivances for enslaving the many and enriching the few. I see absolutely no proof that the average ability of the white man's brain to think clearly is any greater than that of the yellow man or of the black man. If we take even that doubtful but widely heralded test, the frequency of individual creative genius (when a real racial test should be the frequency of ordinary common sense)—if we take the Genius as the savior of mankind, it is only possible for the white race to prove its own incontestable superiority by appointing both judge and jury and summoning only its own witnesses.

. . .

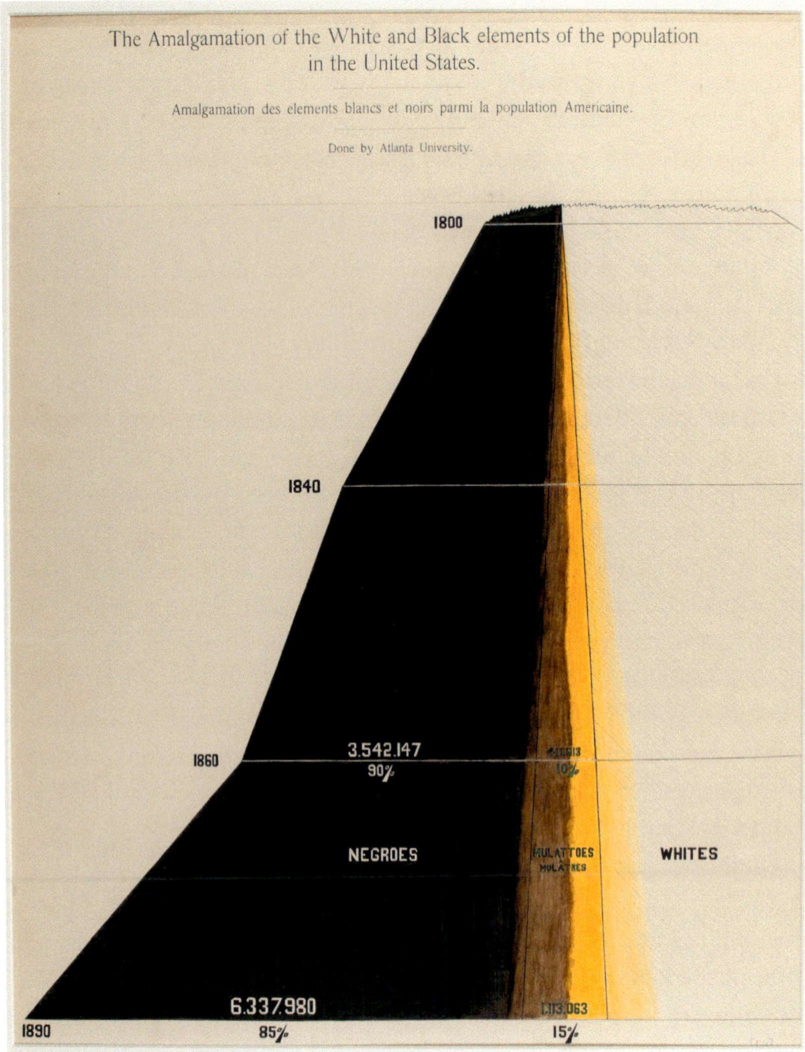

FIGURE 2.2
W. E. B. Du Bois and Atlanta University: "The Amalgamation of the White and Black Elements of the Population in the United States." Circa 1900. Image courtesy of Library of Congress Prints and Photographs Division Washington, D.C.

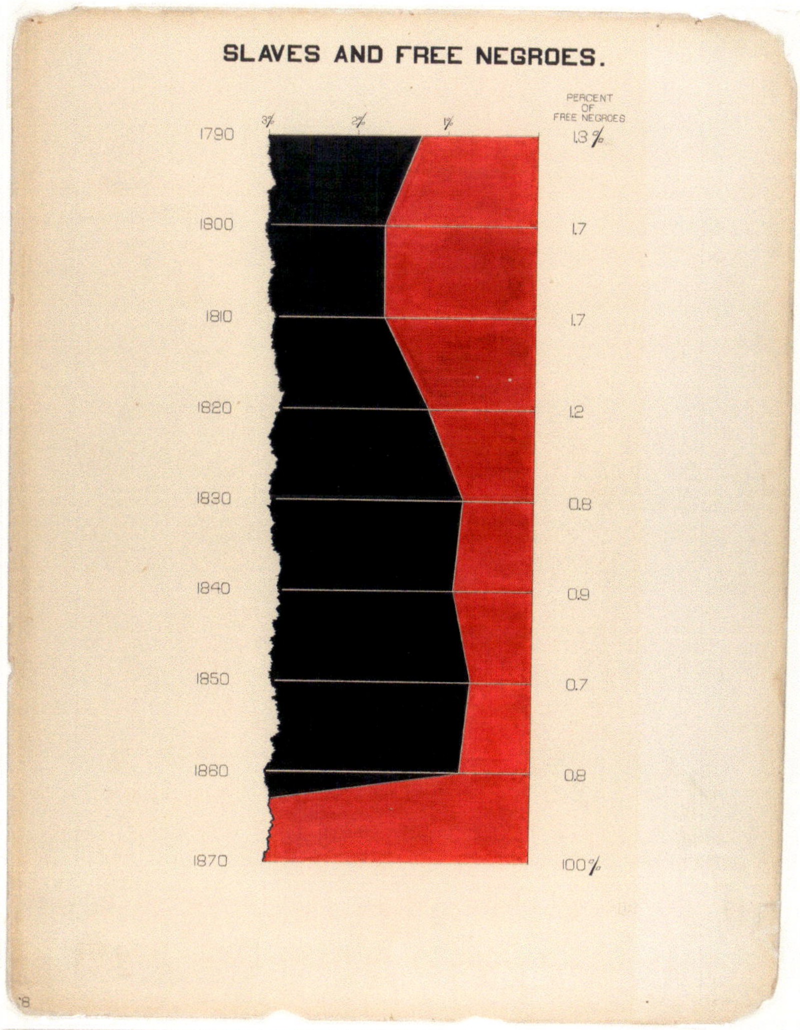

FIGURE 2.3

W. E. B. Du Bois and Atlanta University, Plate 12: "Slaves and Free Negroes." The visual tension in this graphic derives from the jagged edge separating right from left. The latter denotes enslaved Georgians between 1790 to 1890, while the right (red) denotes the transition from enslaved to "free" Georgians during this same period. Image courtesy of the Daniel Murray collection at the Library of Congress, Prints and Photographs Division.

FIGURE 2.4
W. E. B. Du Bois and Atlanta University, Plate 25: "Assessed Value of Household and Kitchen Furniture Owned by Georgia Negroes." In this experimental data visualization, each of the six colored lines indicates the average value of the household and kitchen furniture owned by Black Georgians in five-year increments between 1875 and 1899. Image courtesy of the Daniel Murray collection at the Library of Congress, Prints and Photographs Division.

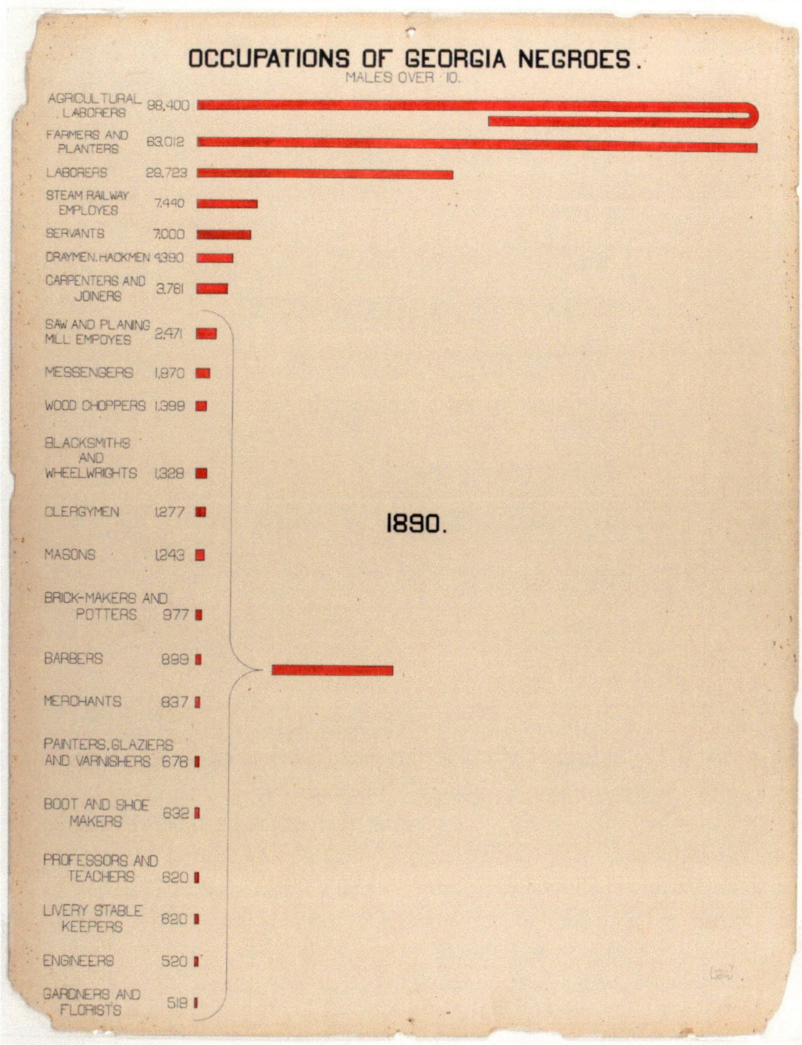

FIGURE 2.5

W. E. B. Du Bois and Atlanta University, Plate 26: "Occupations of Georgia Negroes. Males over 10." This plate documents the changing professions of Black Georgians in 1890. The overwhelming number of agriculture laborers, farmers, and planters points to the historical and social realities of life in the immediate aftermath of the Civil War (1861–1865). Image courtesy of the Daniel Murray collection at the Library of Congress, Prints and Photographs Division.

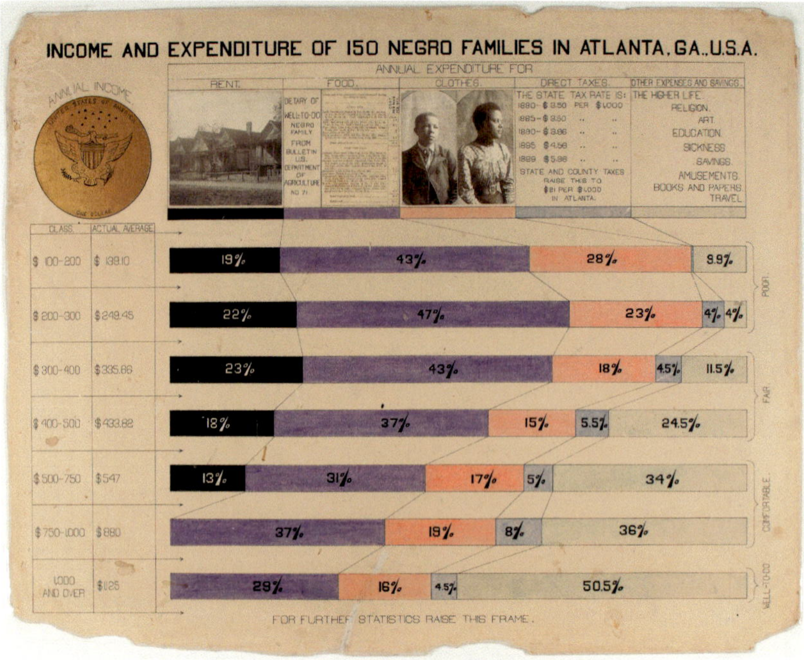

FIGURE 2.6

W. E. B. Du Bois and Atlanta University, Plate 31: "Income and Expenditure of 150 Negro Families in Atlanta, GA, U.S.A." The mixed-media collage (including inked typography, charts colored in gouache, photographic prints, photomechanical reproductions of type, and gold-leaf painting in a seal in the upper-left corner) presents a visually engaging and easy-to-read infographic representing 150 Black Atlanta families, with their budgets divided into categories of food, clothing, rent, taxes, and other expenses. Image courtesy of the Daniel Murray collection at the Library of Congress, Prints and Photographs Division.

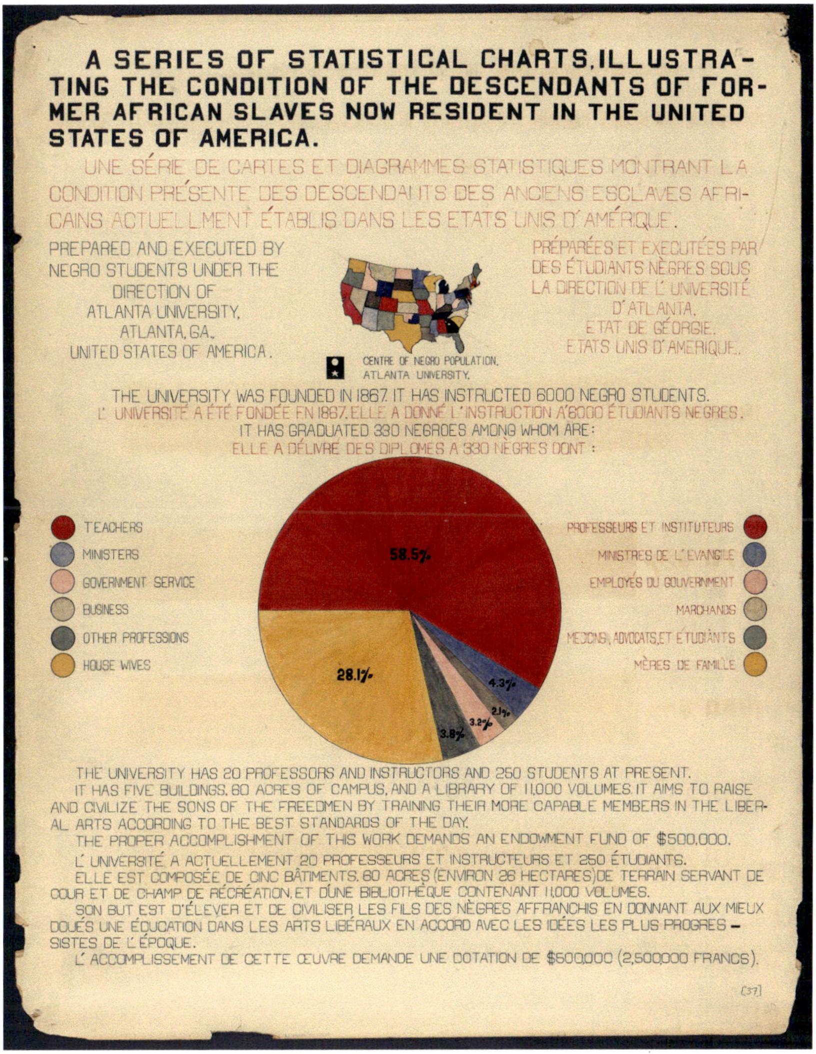

FIGURE 2.7
W. E. B. Du Bois and Atlanta University, Plate 37: "A Series of Statistical Charts Illustrating the Condition of the Descendants of Former African Slaves Now in Residence in the United States of America." The plate is designed with gouache and ink to communicate the historical trajectory of formerly enslaved African Americans residing in the United States circa 1900. Image courtesy of the Daniel Murray collection at the Library of Congress, Prints and Photographs Division.

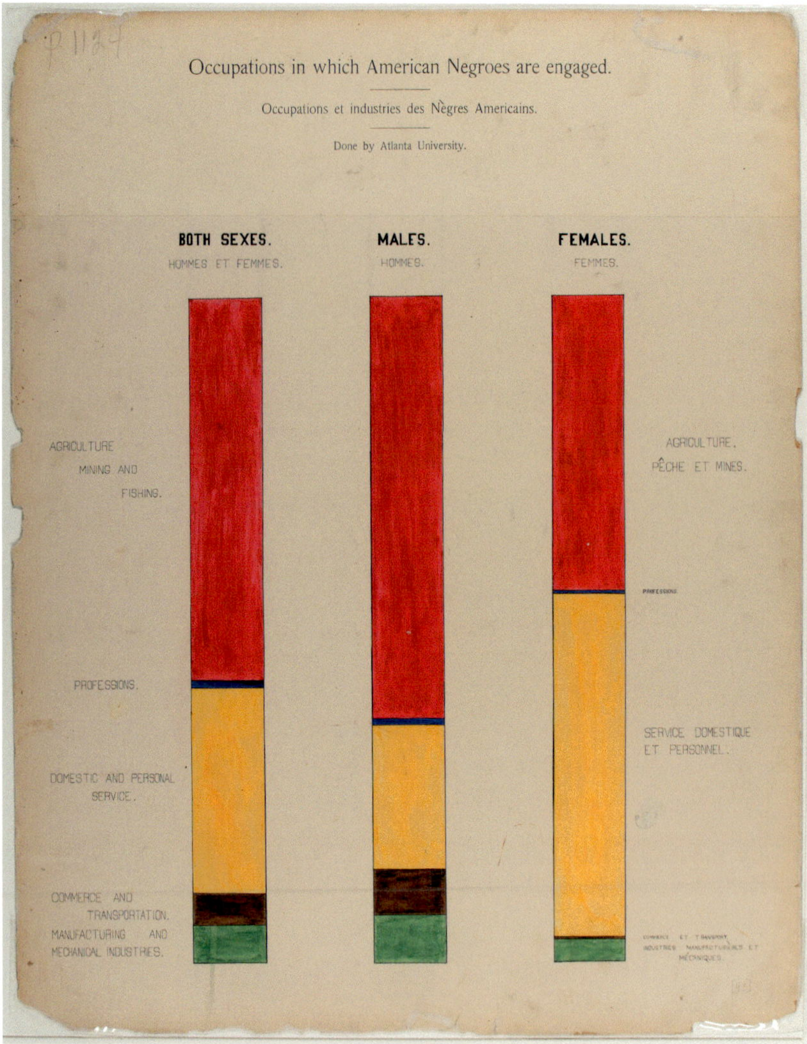

FIGURE 2.8

W. E. B. Du Bois and Atlanta University, Plate 43: "Occupations in Which American Negroes Are Engaged." Here, we see three vertically aligned bars conveying information about the occupations of Black Georgians, classified by gender. Despite the gender distinction, it is clear (once again) that the majority of Black Georgians circa 1900 were engaged in agriculture, mining, and fishing. Image courtesy of the Daniel Murray collection at the Library of Congress, Prints and Photographs Division.

NOTES

1. *Nous avons changer tout cela* (French): a slightly incorrect version of *nous avons changé tout cela* ("we have changed all that").

2. Peoples of mixed-race descent.

3. W. E. B. Du Bois, "The Conservation of Races" (1897), 179–182.

4. The location in Virginia of a crucial Civil War battle (also known as the Battle of Beaver Dam Creek) of June 1862.

5. Born with a veil . . . second-sight: in African American folk culture, it is believed that children born with a caul (a membrane from the placenta covering the infant's face at birth) are gifted with prophetic and psychic abilities.

6. *Of Our Spiritual Strivings*, 7–8.

3 LIKE A COLOR: CHROMATIC METAPHORS IN RACIAL DISCOURSE

NICHOLAS GASKILL

You can't talk about race without talking about color. I mean that literally, though I've got metaphors in mind. Our most common terms for designating racial difference—Black, white, brown, people of color—are chromatic, and they carry the connotation of visual qualities, even when those qualities are largely beside the point. The modern grammar of race relies on chromatic metaphors. What consequences do those metaphors have? What happens when we speak of humans in terms of colors, especially when those colors are meant not in their visual senses but as something more figurative?

I will approach these questions by looking at how African American writers have contested the chromatic discourse of modern race. My working assumption is that the language of color in racial discourse—because it is metaphorical—is always up for grabs to some extent, but also that any attempt to reconfigure the values embedded in the discourse are subject to broader material conditions. There's room for troping, especially in literature, but the liberatory effects can extend only so far, given the entrenched economic and social barriers to Black equality. I should note that a full analysis would encompass not just the Black-white dyad but also the figuration of Asians as "yellow," Native Americans as "red," and so on.[1] Even so, I hope that my more circumscribed approach, focused as it is on a structurally fundamental racial distinction in the U.S., will illuminate important aspects of the metaphorics of color in racial discourse. I'll look in particular at three tropes roughly corresponding to three successive moments in the history of the American grammar of race: the color line, "Black is Beautiful," and colorblindness. In the first section, on the color line, I'll consider authors who criticized attempts to define race through color, often by pointing out the illogic of assigning people to color categories that don't even match their visual hues. In the second, I'll discuss writers who took a different tack by foregrounding the aesthetic value of actual skin colors, as if to bend the metaphoric colors back to literal perception. This is the "Black is beautiful" strategy. And in the third section, I'll show how writers and artists have responded to the most recent twist on the chromatic metaphor—the claim that one could be "colorblind" and so not see race—by embracing color's capacity to make visual distinctions. I'll conclude by suggesting how we might give the color trope yet another turn by using it to understand the slippery ontology of race.

THE COLOR LINE

The entanglement of color and race goes back as far as the modern attempts to concoct a system of racial differences. Building on even earlier remarks about "dark" or "pale" faces from colonial encounters, eighteenth- and nineteenth-century race science and ethnography attempted to tie race to the colors of skin, hair, and eyes. All of those efforts failed. None of the specific systems proposed, from Carl Linnaeus's division of the genus *Homo* into "whitish" Europeans, "reddish" Americans, "dusky" and later "yellow" Asians, and "black" Africans, to the anthropometrics of Paul Broca and Francis Galton, proved empirically useful for classifying individuals.[2] But rhetorically, the trick of using color as a shorthand for race stuck, especially as colonial enterprise, the slave trade, and immigration moved vast numbers of people around the globe and so made geographic designators like "European" and "African" far less useful for racial thinking than "white" or "Black."

Color, then, was available as a racial classifier early on, but in the U.S. it was the Civil War—or, more precisely, the aftermath of Emancipation and Reconstruction—that brought color to the foreground of racial discourse. If, before the war, the salient distinctions had been between *enslaved* and *free* or, in the conflicts between European settlers and Indigenous peoples, *Christian* and *heathen*, after 1865 it was between *Black* and *white*. This is the context for the emergence of the *color line* as a dominant metaphor in American racial discourse, one that lives on in W. E. B. Du Bois's oft-quoted opening to *The Souls of Black Folk* (1903): "The problem of the Twentieth Century is the problem of the color line."[3]

Before the 1870s, the phrase "color line" referred to the flag-bearers in an army regiment. And that martial meaning may have been in play in some of the early instances in descriptions of political battles in the South during Reconstruction, which tended to refer to a clash between groups or to a forcible "drawing" of the line. Certainly one of the most important features of the metaphor is this sense of division: a hard line separating two sides, rather than a smooth chromatic continuum.[4]

This apparent tension—between "color," which moves through gradations, and "line," which denotes a hard division—led many writers and artists to approach the injustice and logical absurdity of modern racial systems by taking the metaphor literally. The famous orator and author Frederick Douglass took this approach in his 1881 essay "The Color Line," in which he entertains the idea, offered by an imagined interlocutor, "that the color of the negro has something to do with the feeling entertained toward him; that the white man naturally shudders at the thought of contact with one who is black." Douglass pushes the logic of the metaphor to its absurd conclusion:

> If the white man were really so constituted that color were, in itself, a torment to him, this grand old earth of ours would be no place for him. Colored objects confront him here at every point of the compass. If he should shrink and shudder every time he sees anything dark, he would have little time for anything else.[5]

Color doesn't cause racial prejudice, Douglass continues; rather, it provides an "instantly recognized" quality through which racism is organized and deployed.[6] What he calls "the color-line feeling" has nothing to do with the properties of the actual colors and everything to do with "the things with which it is coupled." When the conditions of "[s]lavery, ignorance, stupidity, servility, poverty, [and] dependence [. . . .] shall cease to be coupled with color, there will be no color line drawn."[7]

Douglass's analysis singles out the all-important role of feeling and emotion in the chromatic metaphors of racial discourse. Colors have long been understood to stir an emotional response, especially when considered in isolation. Think of Wassily Kandinsky's early experiments in abstract art, which sought an aesthetic language of the emotions in arrangements of "pure" color. The hope was to find a way of correlating particular hues to particular feelings. Though never as crude as the spiritualist or psychological attempts to find one-to-one correlations between colors and feelings (e.g., red = energetic or angry), Kandinsky and other early abstract artists—including the Synchromist painter Stanton Macdonald-Wright and pioneers in "color-music," such as Alexander Wallace Rimington—attempted to forge a grammar of aesthetic feeling based on an emotional language of color.[8] That specific aspect of early modernist abstraction didn't fare any better than the project of classifying humans according to skin color: as filmmaker Sergei Eisenstein argued, there are no universal color-feelings. Rather, the tenor of the emotional responses that "pure" colors stimulate depends on the cultural, historical, and material factors influencing the color experience.[9] How redness makes you feel will change depending on the red object and how it's valued by the larger culture. Douglass makes this point for blackness, and in doing so he indicates how part of the power of the color-line is precisely the way that it enlists the emotional potency of abstract color to endow racial difference with a feeling of affective immediacy.

Indeed, the wicked genius of the color-line metaphor is that even though it seems to appeal to visual color and to a preexisting perceptual difference, it in fact introduced a semantic contrast in order to *intensify* the feeling of visual disparity. No one, as Richard Dyer and others have pointed out, is literally white or black (or red or yellow).[10] And so whenever someone like Douglass takes the metaphor literally, detractors can agree to the absurdity of thinking that there's something horrible about blackness as such and yet still continue to insist on a notion of racial difference which they signify through color. In other words, the function of color terms here is not to denote visual difference—not when visibly "white" people could nonetheless be "Black" according to the so-called "one-drop rule," the other dominant metaphor for thinking about race at the turn of the twentieth century.[11] Instead of pointing to a clearly-defined perceptual distinction, then, color terms in racial discourse simultaneously activate a *feeling* of visual difference and stretch that feeling across invisible realms, so that it encompasses not just skin but also intangible aspects of character, behavior, and mental capacity: everything that gets folded into "race." Thus the strange way that somebody might be said to "sound white," a phenomenon that has been satirized in Black art from Charles Chesnutt's *The Marrow of Tradition* (1901) to the film *Sorry to Bother You* (dir. Boots Riley, 2018).[12]

We can see these rhetorical gymnastics at work in the majority decision of *Plessy v. Ferguson*, the landmark Supreme Court case that, in 1896, gave legal grounds for segregation. The author of that decision, Justice Henry Billings Brown, argued that the "merely . . . legal distinction" between races imposed by segregation was "founded in the color of the two races" and that it "must always exist so long as white men are distinguished from the other race by color."[13] You would never guess from this reasoning that Homer Plessy, the plaintiff, looked so "white" that he had to inform the train conductor that he belonged in the "colored" car. The chromatic distinction Brown invokes as obvious didn't exist in the case at hand. Plessy's strategy—like Douglass's—was to trust the self-evidence of visible color to point up the injustice and absurdity of Jim Crow laws. Instead, the Court appealed to the self-evidence of "the color of the two races"—a distinction given a chromatic *name* but referring to a more nebulous constellation of traits, projections, fears, and styles—to support the idea that a person who looked white could in fact be black. It was the "color-line feeling," rather than any true perceptual hues, that Brown had in mind.

The "color" of the color line is, then, simultaneously like and unlike visual chroma. It's unlike visual color in the sense that it does not literally describe the visual tone of people's skin. But it nonetheless adopts one of the most crucial functions of color perception, which—as philosopher of color M. Chirimuuta explains—is to create a visual contrast.[14] From an evolutionary perspective, color perception helps us to make distinctions, to pick out the red fruit from the green leaves. Color puts that quality to work in racial discourse, ensuring that whenever we think of people as "Black," "white," or otherwise "colored," we activate a feeling of visual difference even in its absence.

BLACK IS BEAUTIFUL

The operations of the "color-line feeling" would seem to stack the deck against any invocation of color on the part of those interested in contesting the social order. And, indeed, many turned away from the visual, as if any invocation to *look* could only, within a perceptual field stratified by race and its metaphors, strengthen existing divisions. Du Bois, for instance, leaned on *sound* rather than vision when he invited his white readers to acknowledge the spiritual gifts of Black Americans. Song and music take precedent.

At least, that is, in *The Souls of Black Folk* from 1903. By the time of *Dusk of Dawn*, in 1940, Du Bois had changed his tune. In an imagined dialogue between personifications of the "white and colored environment," Du Bois replies to the white representative's claim (familiar from Thomas Jefferson's *Notes on the State of Virginia* [1785]) that white skin is more beautiful than black with an excited litany of hues: "Now can there be any question but that *as colors*, bronze, mahogany, coffee and gold are far lovelier than pink, gray, and marble?"[15]

With "the color line," Du Bois marked out color as a mechanism of division, and, accordingly, he asked his readers to *listen* rather than look. But in this passage, he embraces color as a sensuous and aesthetic quality, and he gambled that readers could perceive color outside of the simplifying frame of racial categories. The move is similar to that of contemporary visual artists who have created perceptual representations of *actual* skin tones as a criticism of the abstract color schemas of racial thinking. To take two prominent examples: both Byron Kim's *Synecdoche* (1991–present) and Angelíca Dass's *Humanæ* (2012–ongoing) use an array of human hues to show up the absurdity of drawing hard lines between people through color. *Synecdoche*

FIGURE 3.1
Byron Kim, *Synecdoche*, 1991–present. Oil and wax on lauan plywood, birch plywood, and plywood, each pane: 25.4 × 20.32 cm (10 × 8 in.); overall installed: 305.44 × 889.64 cm (120 ¼ × 350 ¼ in.). National Gallery of Art, Washington. Copyright Byron Kim. Courtesy the artist and James Cohan, New York. Photo: Dennis Cowley.

consists of a grid of 8 × 10-inch color panels, each of which presents the skin tone of an individual person whose name is listed in the signage. What looks at first blush like another entry into the aesthetics of abstract color turns out, in fact, to be a representation of color as charged with racial significance, such that individuals are represented solely by their color (the "synecdoche" of the title).

Similarly, Angelíca Dass's *Humanæ* presents a similar grid of photographs of people of all different skin tones, set against backgrounds that attempt to match the precise hue of their bodies using the Pantone system. The specific insights of these works differ—for instance, Dass's juxtaposition of person and hue more readily shows up the variations of skin tone on any one person, let alone each "race"—but each aims to counter the abstractions of racial color with visual presentations of the hues of actual bodies. The result is that the range of racial color is shown to be simultaneously too narrow, because it can't account for the wealth of particular variations, and too wide, since human shades are not as different as white and black but instead cluster in a small band of pinks and browns, with tinges of orange.[16]

FIGURE 3.2
Angélica Dass, Exhibition of *Humanae* project in Valencia, Spain, October 24, 2014. Photograph by LOLAOMI. Image courtesy of Wikimedia and Creative Commons CC0 1.0 Universal Public Domain Dedication.

Du Bois's rhetorical move is similar to these contemporary works, but with a crucial difference: where Kim and Dass lean on the scientific presentation of skin tone, as if to contest the terms of racial logic on empirical grounds, Du Bois appeals instead to *aesthetic* experience, inviting readers to celebrate and take pleasure in the many hues lumped together as "Black." He gives us colors "as colors" and then cloaks them in language that keeps them as close as possible to the material world of actual objects and people (not abstract "brown" but sturdy "Mahogany," brilliant "bronze," or invigorating "coffee").

Du Bois's strategy of contesting the impoverished schema of racial color through rhetorical appeals to the beauty and particularity of actual colors was almost never used in the U.S. in the nineteenth century, but it is a staple of racial discourse in the twentieth. I've argued elsewhere that the first flourishing of this way of handling racial colors came in the literature of the Harlem Renaissance.[17] During the 1920s, when the use of color in commercial industry reached spectacular new heights, African American writers such as Claude McKay, Zora Neale Hurston, and Nella Larsen attempted to take advantage of the culture-wide fascination with bright colors to position life in uptown New York as exciting, stimulating, and essentially *modern*.

Litanies of color abound in this literature. For the sake of space, I'll offer just two. In the first, Claude McKay, who would later pen a controversially color-heavy novel about Black life in New York, *Home to Harlem* (1928), complains about the white-washing of Black performers in the hit musical *Shuffle Along* (1921). "Instead of making up to achieve a uniform near-white complexion," he wrote, the all-Black chorus "might have made up to accentuate the diversity of shades among 'Afro Americans' and let the white audiences in on the secret of the color nomenclature of the Negro World," including "chocolate, chocolate-to-the-bone, brown, low-brown, teasing-brown," and so on.[18] Note how McKay immediately jumps from a visual presentation of chromatic variety to a "nomenclature" that can capture and hold that variety in mind, as if the colors simply won't be seen—and certainly won't be enjoyed or appreciated—without the accompanying language.

Second, in Nella Larsen's *Quicksand*, the protagonist Helga "marvel[s] at the gradations within this oppressed race of hers" while at a Harlem cabaret: "sooty black, shiny black, taupe, mahogany, bronze, copper, gold, orange, yellow, peach, ivory, pinky white, and pasty white."[19] Like McKay and other Black writers of the 1920s, Larsen used language to accentuate variety without reproducing the color prejudices that elevated light skin over dark. There are essentially two moves here: first, an insistence on the particular and specific hues that appear on human bodies, against the tendency to overshadow those nuances with the pseudo-visual terms "Black" and "white"; second, an arrangement of those proliferated hues in the form of a long litany or ongoing catalogue, one that resolutely dispenses with the hierarchical arrangements of racist color coding.

Such strategies have their advantages, but in the context of the longstanding libel about the sensuousness of non-white people, descriptions like those in *Quicksand* or *Home to Harlem* always ran the risk of being pulled back into the dichotomy of a "primitive" delight in bright colors and a "civilized" (read: white) restraint. The narrative of *Quicksand*, in which the mixed-race Helga shuttles between the two poles of the white-Black dichotomy, turns on precisely that threat. But my point here is that one move within the rhetorical battle around racial color has been to resist the abstraction of "Black" and "white" through an emotionally-charged description of hues that have texture, taste, and detail. In the Harlem Renaissance, such invocations sought to lure readers uptown, as it were, with the hopes of gaining equality through the cultural recognition of the indisputable beauty and creativity of African Americans. In important ways, that effort did not succeed: Jim Crow persisted. But the attempt by Larsen, McKay, and others to create a different rhetorical scrim through which to view race persisted into the 1960s, where it found its enduring slogan: "Black is beautiful."

The Black Arts and Black Power movements that ultimately enshrined that slogan are usually understood to have had a more radical political agenda than the most well-known writers of the Harlem Renaissance, but key figures like the poet Amiri Baraka (at that time, LeRoi Jones) and the photographer Kwame Brathwaite nonetheless built on the strategy of aesthetically reevaluating Blackness pioneered in the 1920s. Often that involved seeing skin color "as color." One complex example of this comes in James Baldwin's novel *Another Country* (1962), when Rufus gives his sister Ida an Indian shawl. When she tries it on, "something shook in him which had never been touched before. He had never seen the beauty of black people before." Newly adorned, "she became associated with the colours of the shawl, the colours of the sun, and with a splendor incalculably older than the grey stone of the island [of Manhattan] on which they had been born." Seeing his sister in this way, he realized anew that "Ida had not been merely the descendent of slaves."[20] Many of the tropes from the Harlem Renaissance are revived here: the contrast of a rich delight in colors with the frigidity of the white world, a solidarity among the formerly colonized and enslaved peoples of the world figured through a relation to color, and—most directly—a technique of setting skin tones amid chromatic arrangements that resist reductive racial grammars.

There is a difference, however, between the chromatic catalogues of the 1920s and the decentering of European notions of beauty in the 1960s: in short, the insistence that "Black is beautiful" accepts the broad racial categories (white vs. color) with the aim of shifting their relative values, rather than attempting to do away with the categories altogether. In the decades after Civil Rights, as inequalities persisted, that proved the more attractive strategy.

COLORBLINDNESS

One of the distinctive characteristics of color, as it has operated in racial discourse, is that it is superficial, inherently meaningless. In one context, green means "step on the gas pedal"; in another, it means "eliminate fossil fuels." We've already seen how Douglass pointed out the essential semantic poverty of colors in "The Color Line." Now we can hear this aspect put to different use—or given a different meaning—in the offhand way people remark that they don't care if someone is "white, black, brown, or purple." The color, such a gesture suggests, is beside the point. In this case, the superficiality of color underlies the rhetoric of "colorblindness," the variation of the chromatic metaphor in racial discourse that came to the fore in the 1980s and persisted, arguably, until the emergence of Black Lives Matter and the election of Donald Trump. As Eduardo Bonilla-Silva argues in *Racism without Racists*, colorblindness names a "racial ideology" which defines a nonracist stance as someone who "does not see color," but who also fails to see the "subtle, institutional, and apparently nonracial" means by which "contemporary racial inequality is produced."[21] The imagery of colorblindness goes hand-in-hand with assertions of a "post-racial" society in which, at long last, equal opportunity has been secured for all and any remaining inequalities must boil down to nonracial factors, such as market forces or cultural limitations. Within such a context, appeals to the superficiality of color or even the beautiful tonalities of dark skin feed into an ideology that's very happy to embrace difference (as a way of capturing a wider consumer market or a deeper pool of human capital) and to denounce crass racism but that is unable to recognize the less overt ways in which white supremacy operates in American life.

How have Black writers and artists responded to this changed rhetorical and political situation? The poet and critic Evie Shockley has argued that in the age of colorblindness, poets and artists have created works that "demand to be engaged both as text and as image" in order to experiment with ways of making blackness visible and legible against an ideology that claims to "not see color" and that resents any direct invocation of race as "playing the race card."[22] One example of what Shockley calls "Black graphics" is Robin Coste Lewis's "Voyage of the Sable Venus," a 2015 poem composed entirely of wall-texts, museum catalog descriptions, and titles of artworks depicting Black women. The conceptual task of the poem is to make visible the tropes, figures, and strategies by which Black women have been represented and visualized in Western art. Yet despite this close engagement with descriptions of visual art, color terms are in relatively short supply in the text. Unlike Larsen or McKay, Lewis doesn't want to create a new aesthetically-charged vocabulary for seeing Blackness; she wants instead to deform and rearrange the accumulated vocabulary in order to help us apprehend it as the relentlessly distorting and dehumanizing grammar that it is. She wants to help readers see how their *seeing* has been conditioned.[23]

Another example comes from Glenn Ligon: *Untitled (I Feel Most Colored When I Am Thrown Against a Sharp White Background)* (1990). To create this painting, Ligon repeatedly stenciled a sentence from Zora Neale Hurston's essay "How It Feels to Be Colored Me" (1928) onto a tall wooden door.

The sentence itself—"I feel most colored when I am thrown against a sharp white background"—draws attention to the relational nature of race, the way that Blackness implies whiteness, or, more generally, how each racial identity takes its shape in relation to the others. At first the sentence is stenciled cleanly. There's a sharp white background. But as the text moves downward, the black of the letters blurs and smudges into the surrounding area, to the point that the words themselves become illegible. Those confronting the painting begin as readers and end as viewers, and that transition from text to image corresponds to a shift from a clear black-and-white dyad to an illegible blur. Race (or how it feels to be "colored") exists in that imbroglio of language and perception, metaphor and sight, and it requires a

FIGURE 3.3

Glenn Ligon, *Untitled (I Feel Most Colored When I Am Thrown Against a Sharp White Background)*, 1990, Oil stick, gesso, and graphite on wood, 203.2 × 76.2 cm (80 × 30 in.). © Glenn Ligon. Courtesy the artist, Hauser & Wirth, New York, Regen Projects, Los Angeles, Thomas Dane Gallery, London and Galerie Chantal Crousel, Paris.

set of hierarchical contrasts to be seen or felt at all. Ligon, like Lewis, wants us to see the "envelop of discourse" that structures racial sight.[24]

Ligon's painting marks a changed tactic in the way African American artists have engaged the color terms of racial discourse, one increasingly familiar through the critical analyses and black-and-white signage of the Black Lives Matter movement. If the Harlem Renaissance writers used their color catalogues to implore their readers to drop abstract racial categories and look at the actual people in front of them, contemporary Black writers have focused instead on helping readers to look beyond individual people and towards the abstract structures that shape their lives. The *relational* aspect of color that before seemed dangerous—in that there was no way to think beyond racial divisions while sticking with color, especially the black/white dichotomy—becomes essential when the task is to convince people that racism is not simply a personal concern.

CONCLUSION

My aim in this essay has been to show how our reliance on color metaphors in racial discourse comes with its own consequences and affordances for how we think and talk about race. These consequences shift according to context, such that, for instance, the relational or superficial qualities of color can have different valences within the racial imaginaries of different periods. I've focused in particular on how African American writers and artists have responded to the color-race metaphor since, in the U.S., they have been most keenly aware of racial discourse as a contested field with real stakes for those involved.

From this perspective, it would be wrong to assume that color rhetoric in race is necessarily bad. In many ways "Black" and "white" seem to me less pernicious than alternative terms like "African American" or "Caucasian," since the latter rely on notions of natality and bloodlines and, in the case of "African American," make an insidious contrast with the unmarked or unqualified category of "American." The problem is the wide category of race more than any one of its metaphors, and as we've seen, color terms can be useful for making the operations of racial categories more visible.

In fact, I want to conclude with a suggestion for how the color metaphor might prove helpful for thinking about race. I think that the slippery ontology of color can help us to understand the equally slippery ontology of race. Philosophers have long worried over the reality of color, wondering if it inheres only in the mental operations of the observer or if it corresponds to objects in the world, and the best answers, in my opinion, are those that insist that color is a *relational* and *ecological* activity.[25] In order to understand color, that is, we must look to the wider context of organisms and environments within which color appears and functions, for it is only in the interaction of perceivers and the perceived that color comes into being. Race works similarly. This is how we might acknowledge both the way in which race is an obvious fiction—something that has been constructed historically and socially—and very real, and so not likely to disappear after being exposed as a fiction. Race and color alike are *made real* within specific contexts, and the critical task is to understand and ultimately intervene in how it is made. Here, then, is a way that the color rhetoric in racial discourse might become an asset in helping us to conceive of the ontological workings of racial categories. Race is real like color is real.

NOTES

1. For parts of this broader picture, see Keevak, *Becoming Yellow* (an excerpt of which appears in chapter 4 of this book), and Shoemaker, "How Indians Got to Be Red."

2. Linnaeus made his classifications in *Systema naturae*, first published in 1735. For discussions of his racial system as well as other attempts to tie racial difference to color, see Kane and Zeitlin-Wu's "Introduction to Part I: Color Encoding from the Enlightenment to 'Race Science'"; Kastan, with Farthing, *On Color*, especially chapter 3; Rossi, "Ten Short Notes on Color, Anthropology, and Makeup," and his *Republic of Color*, chapter 7; and Gossett, *Race: The History of an Idea in America*, especially chapter 1.

3. Du Bois, *Writings*, 359.

4. For a reflection on the meanings of the "color line," with special attention to Du Bois's work in visual culture, see Smith, *Photography on the Color Line*.

5. Douglass, "The Color Line."

6. Douglass, "The Color Line," 575.

7. Douglass, "The Color Line," 575.

8. I elaborate this account of Kandinsky and early justifications for abstract art, including those given by Rimington for his color-music, in Gaskill, "The Articulate Eye." See also Gage, *Color and Meaning*, chapter 19, and *Color in Art*, chapter 5.

9. For a discussion of Eisenstein and modernist color theory, see Gaskill et al., "Literature and the Performing Arts."

10. See Dyer, *White*, an excerpt of which is reprinted in part II of this book.

11. It's interesting to note that the "one drop of black blood" metaphor shifts the visual analogue from paint to dye. No one, as Ralph Ellison noted to satiric effect in the "Optic White" section of *Invisible Man*, can detect a single drop of black in a bucket of white paint. For the metaphor to work, blackness has to be conceived not as one color clashing with another, but as a staining substance—like a dye—coming into contact with an uncolored solution, like water, or with an undyed cloth.

12. For a discussion of Chesnutt's novel in this vein, as well as a more in-depth analysis of the rhetoric of color in racial discourse at the turn of the twentieth century, see Gaskill, *Chromographia*, especially chapter 4.

13. Thomas, ed. *Plessy v. Ferguson*, 43.

14. Chirimuuta, *Outside Color*, See also Thompson, *Colour Vision*.

15. Du Bois, *Writings*, 657 (emphasis added).

16. I've leaned here on Kastan's illuminating discussion of Kim's work in *Color*, 71–74, and Rossi's equally illuminating discussion of Dass in "Ten Short Notes," 6–8.

17. For more on the entangled aesthetics of color and race in the Harlem Renaissance, see Gaskill, *Chromographia*, chapter 5.

18. McKay, "A Negro Extravaganza," 64.

19. Larsen, *Quicksand*, 54.

20. Baldwin, *Another Country*, 17.

21. Bonilla-Silva, *Racism without Racists*, 2–3.

22. Shockley, "On Seeing and Reading the 'Nothing,'" 500.

23. Lewis, *Voyage of the Sable Venus and Other Poems*.

24. Kastan, *Color*, 78. Shockley also discusses Ligon's work in the context of color-blind racial ideology.

25. Chirimuuta, *Outside Color*, and Evans, *The Perception of Color*, offer examples of this approach.

BIBLIOGRAPHY

Baldwin, James. *Another Country*. London: Penguin, 2001.

Bonilla-Silva, Eduardo. *Racism without Racists: Color-Blind Racism and the Persistence of Inequality in America*. Lanham, MD: Rowman & Littlefield Publishers, 2014.

Chirimuuta, M. *Outside Color: Perceptual Science and the Puzzle of Color in Philosophy*. Cambridge, MA: MIT Press, 2015.

Douglass, Frederick. "The Color Line," *North American Review* 132 (June 1881): 567–577.

Du Bois, W. E. B. *Writings*. New York: Library of America, 1986.

Dyer, Richard. *White: Essays on Race and Culture*. London: Routledge, 1997.

Ellison, Ralph. *Invisible Man*. New York: Vintage, 1995.

Evans, Ralph M. *The Perception of Color*. New York: Wiley-Interscience, 1974.

Gage, John. *Color in Art*. London: Thames and Hudson, 2006.

Gage, John. *Color and Meaning: Art, Science, and Symbolism*. Berkeley: University of California Press, 1999.

Gaskill, Nicholas. "The Articulate Eye: Color-Music, the Color Sense, and the Language of Abstraction," *Configurations* 25, no. 4 (2017): 475–505.

Gaskill, Nicholas. *Chromographia: American Literature and the Modernization of Color*. Minneapolis: University of Minnesota Press, 2018.

Gaskill, Nicholas, Sarah Street, and Joshua Yumibe, "Literature and the Performing Arts." In *A Cultural History of Color in the Modern Age*, eds. Anders Steinvall and Sarah Street, vol. 6 of *A Cultural History of Color*, 135–154. London: Bloomsbury Academic, 2021.

Gossett, Thomas F. *Race: The History of an Idea in America*, new edition. Oxford: Oxford University Press, 1997.

Kastan, David Scott, and Stephen Farthing. *On Color*. New Haven, CT: Yale University Press, 2018.

Keevak, Michael. *Becoming Yellow: A Short History of Racial Thinking*. Princeton, NJ: Princeton University Press, 2011.

Larsen, Nella. *Quicksand*. Mineola, NY: Dover, 2006.

Lewis, Robin Coste. *Voyage of the Sable Venus and Other Poems*. New York: Alfred A. Knopf, 2020.

McKay, Claude. "A Negro Extravaganza." In *The Passion of Claude McKay: Selected Poetry and Prose, 1912–1948*, ed. Wayne F. Cooper, 63–64. New York: Schocken Books, 1973.

Rossi, Michael. *The Republic of Color: Science, Perception, and the Making of Modern America*. Chicago: University of Chicago Press, 2019.

Rossi, Michael. "Ten Short Notes on Color, Anthropology, and Makeup," *Grey Room* 76 (Summer 2019): 6–23.

Shockley, Evie. "On Seeing and Reading the 'Nothing': Poetry and Blackness Visualized," *New Literary History* 50, no. 4 (autumn 2019): 499–528.

Shoemaker, Nancy. "How Indians Got to Be Red," *American Historical Review* 102, no. 3 (June 1997): 625–44.

Smith, Shawn Michelle. *Photography on the Color Line: W. E. B. Du Bois, Race, and Visual Culture*. Durham, NC: Duke University Press, 2004.

Thomas, Brook, ed. *Plessy v. Ferguson: A Brief History with Documents*. Boston: Bedford Books, 1997.

Thompson, Evan. Colour Vision: A Study in Cognitive science and the Philosophy of Perception. London: Routledge, 1995.

4 "MEASURING SKIN COLOR" AND "THE COLOR TOP" (2011)

MICHAEL KEEVAK

MEASURING SKIN COLOR

It was only a matter of time before nineteenth-century science began to wonder the degree to which racial difference—including skin color—might be measured in greater detail. . . .[1] People's entire bodies were evaluated with hundreds of tiny measurements of every imaginable sort. Yet it was also typical of Western race preoccupations that this obsession almost immediately spread to the question of color, as if it were now the explicit aim (although, of course, for many it wasn't) to prove that some people were "more white" than others.

The year 1859 . . . marks the publication of Darwin's *The Origin of Species*, which far from squelching the polygenesis debate once and for all (although the book barely mentions the question of race), actually seemed to strengthen the idea of even more rigid and scientifically proven racial hierarchies. The question of evolution or transmutation had been hotly debated since the eighteenth century, and while in many ways Darwin seemed to solve these problems, he also gave credence to even more daring notions about natural selection, competition, and "survival of the fittest." The essential separateness of the races, for many readers, was now indisputable and supposedly supported by scientific inquiry.

For also in that year the French surgeon Paul Broca, best known for his pioneering researches on the human brain, founded the Société d'anthropologie in Paris, soon to be followed by the Anthropological Society of London in 1863.[2] His contributions to early anthropology were immense, and particularly what came to be known as physical anthropology. He was personally responsible for some two dozen measuring devices, including craniometers, goniometers, and micrometric compasses, all of which were standard tools by the time of his death in 1880.[3] Less well known today are his contributions to color measurement, even though Broca's *table chromatique* was regularly used well into the twentieth century, sometimes with minor variations. This table, looking much like a sheet of paint or makeup samples, featured twenty colored circles (for eyes) and thirty-four colored rectangles (for skin and hair) lithographed onto a piece of white paper or card stock. It was first published as part of the "Instructions générales pour les recherches et observations anthropologiques" (General Instructions for Anthropological Research and Observation), which appeared in the Société's *Mémoires* in 1865 as well as in monograph form that same year. A second revised edition appeared in 1879, where the text regarding color remained nearly identical, and the color table was now divided into three pages owing to the small pocket size of the volume.

That is to say, the book was designed as a practical guide for scientific (and armchair) travelers, who, it cannot be overstressed, no longer examined cadavers or skulls in laboratories but took measurements on the living. Tens of thousands of white European subjects were measured as well, it must be pointed out, but in 1874 the British Association for the Advancement of Science published its own revealingly titled version, which also included Broca's table: *Notes and Queries on Anthropology for the use of Travellers and Residents in uncivilized Lands*.[4] There was no question, in other words, that physical anthropology was intimately concerned with the measurement of racial difference, and that manuals such as these were geared toward evaluating those who were regularly referred to as "savages."

The main purpose of Broca's guide was to instruct observers to measure the head, face, torso, and limbs of their subjects, the results of which could then be recorded in a form that appeared at the end of the book in both "complete" and "abbreviated" versions . . . In addition to anatomical measurements, the form asked for a few general remarks about the weight of the subject, the pulse rate, the shape of the lips and nose, beard growth, teeth, as well as the color of the skin (both covered and exposed parts), hair, beard, and eyes. The observer was to record the numbers of the corresponding colors provided on the color table. The abbreviated version greatly reduced the number of measurements for each category but still requested color information.[5]

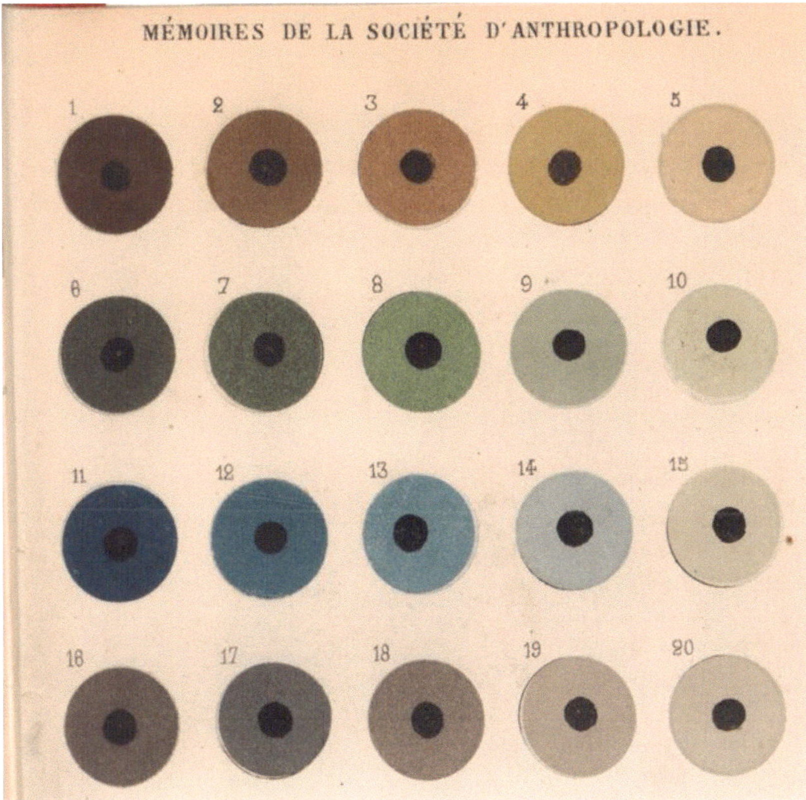

FIGURE 4.1
Paul Broca. Chromatic table of eye colors from *Instructions générales pour les recherches et observations anthropologiques* (1879). Each vertical row represents five shades of brown, green, blue, and gray that become lighter as one moves from left to right. Image courtesy of Princeton University Libraries.

The text went into some detail about how to match the color of the eye, but far less space was devoted to questions of skin and hair. In fact Broca claimed that the determination of skin color was much easier than the other two categories, since the eye was complicated by many factors such as the structure of the cornea, the pupil, and the iris, the effects of light, the fact that eyes were often not composed of one uniform color, and so on. Hair was affected by light and not generally composed of a single color either. Yet the color of the skin, he claimed, was always flat and uniform, and the only caveat offered was that certain skins embodied a shiny and oily reflection that was not a serious difficulty.[6] No mention was made

FIGURE 4.2
Paul Broca. Table of hair and skin colors from *Instructions générales pour les recherches et observations anthropologiques* (1879). Each horizontal row begins with a shade of brown or black that becomes lighter as one moves from left to right (although the colors have faded and are now difficult to discern). Broca was vague about how these colors were chosen except for number 48 in the lower-left corner, which he claimed to be "pure black." Image courtesy of Princeton University Libraries.

of developing controls for consistent color measurement, or even which part of the body should be utilized. Nor did anyone seem to be troubled by larger issues of color perception, in relation both to the physiology of the observer and to his or her language and culture.[7] Indeed the whole point of the color table was that it was supposed to provide international standards for colors by assigning them numerical values instead of conventionally vague and impressionistic descriptive terms such as "yellowish."

It turned out, however, that the many nuances of "yellow" and "red" skin were the most challenging. Once again it was the "in-between" shades that seemed to cause the greatest trouble, but this was not because yellow and red were really more difficult to judge. Rather, white and black were simply being taken for granted as self-evident entities. As one of Broca's most important successors put it, black and white were "primordial" and "incontestable," while red and yellow were "less evident, especially red." Little wonder that they were "less evident" since these were precisely the two colors that had been so arbitrary from the start, which is to say so much more than straightforward terms of objective description, even though "white" and "black" were equally fantasized notions. Centuries of observers had wrestled with the question of just what shade East Asians were, but by this time their Mongolian yellowness was a matter beyond question . . . Yet in the Instructions [Broca] admitted that when he first attempted to construct the table he was frustrated by the "multiplicity and proximity of the fundamental nuances relating to two colors only, the yellow and the red, as well as their exceedingly variable proportions when mixed together."[8]

In other words, yellow and red were not so simple after all, and it was clearly impossible to organize the skin and hair color samples into any sort of coherent order, as Broca claimed to have done with the twenty shades for eyes, which were arranged in five rows of four columns each, with each row representing five "degrees" of brown, green, blue, and gray. But how would one go about choosing basic skin and hair colors and then deciding upon their lighter and darker nuances? Perhaps one could start with black, brown, and blonde for hair (gray is already represented in the table for eyes), but what about skin? Obviously white, black, red, and yellow, the colors by which the races were almost universally known, were unworkable points of departure, since no one's skin was really any of these colors.

In fact this dilemma seemed to have caused considerable delay in the completion of the Instructions, which were announced as early as 1862, and Broca never fully explained how the resulting thirty-four colors had actually been chosen. He noted merely that the order in which they were laid out was not by chance, although it was not always a "regular" or "unbroken" sequence with respect to their "natural series." He did mention, however, that number 48 (in the lower left corner) was a pure black, and that while some observers claimed to have encountered people with completely black hair, the skin color of even the darkest Africans he had seen in Paris was always a shade mixed with red or yellow—colors that, he said, were represented at the beginning of the three rows above, numbers 27, 34, and 41.[9] In other words, each of the four rows at the foot of the table began with a shade of black (pure black, black mixed with yellow, black mixed with red, and black mixed with white or gray?) that became lighter as one moved to the right. This left only the row at the top, which seemed to contain shades with no black at all and were therefore positioned one space to the right. The effect of this arrangement was that when observers actually began to record their data it was only the white races whose skin tone matched the samples given in the top row. Although Broca certainly did not make a point of it, the table appeared to be laid out in such a way that, when it came to skin color, those with no trace of black in their skin were duly separated from the multitude of "colored" races that populated the four rows at the bottom.

Broca claimed to have overcome numerous material difficulties in the printing of the colors, including how to maintain consistency in each copy of the table, but the end result was a disturbingly confident tool for matching human skin, hair, and eye color in every part of the world . . . [10] [T]he sentiment was that Broca's table was still not practical or fully reliable, because the colors chosen were not useful for the particular group being tested, because color lithography was such an unstable media, or because it was difficult to isolate a particular color on the card as it was held up to the skin.[11]

Until these issues could be resolved and some sort of international agreement was in place, the guides recommended that an observer could resort to simpler and more impressionistic color judgments, which, as [Smithsonian curator Aleš Hrdlička] noted, "in most cases are quite sufficient." He suggested a number of terms for each of the four "classes of color," which he identified as white, yellow, brown, and black. Note that yellow was presumed from the outset, even as red, that other bugbear shade, seemed to have been dropped as an acceptable descriptor. In later editions yellow and brown were actually combined and the number of terms was slightly reduced, but it is noteworthy that in both versions "light brown" appeared as a shade for the white as well as the yellow racial categories, since it was precisely the problem that Europeans and East Asians were often indistinguishably "light brown." In 1900 Joseph Deniker proposed a different set of ten "principal nuances" in which the white and yellow were more securely separated ("brownish white" as opposed to "yellowish white"), a distinction that was accompanied by a number of comparisons from other registers that the white category did not seem to require: the lightest yellows were compared to grains of wheat and described as *terreux* (i.e., muddy, pasty; in the English version: "a sickly hue"). The middle yellow (*jaune épais,* dull yellow, also called olive) was likened to "the color of new portmanteau leather," and dark yellow (*jaune brun*) was clarified as the color of dead leaves.[12]

Meanwhile other researchers tried to improve upon Broca's lithographed card by introducing new media. The most well known examples were Felix von Luschan's thirty-six colors of opaque glass, first developed around 1904 and reissued in the 1920s, and Gustav Fritsch's forty-eight tints in oil, produced around 1916. Others tried to replicate people's complexions by using watercolors and mixing them on site. One physician suggested leather samples as they more closely resembled the texture of the skin, and it was even claimed that toward the end of his life Broca had relied upon a table of more than one hundred colors produced by a glove factory. Some researchers developed more limited tonal ranges for particular peoples or regions, and others tried cutting out individual colors from tables like Broca's and mounting them onto separate strips.[13]

Something of a climax was reached in 1927 with Arthur Hintze's table of 358 colors mounted on fourteen cardboard plates, attached in the form of three large flexible fans that could be opened out to find the desired tint. Next to each color was a hole so that the fans could be laid directly on the surface of the skin. An initial match made with the "general" fan could then be refined with the "normal" (sic) and "red" fans designed to measure lighter and "middle" tones with greater precision. This was probably in response to numerous complaints that color tables were much easier to use with respect to darker skin, which was after all their primary purpose. Both Fritsch and [Hrdlička] had remarked that tables lacked sufficient matches for "normal" complexions, and in the 1920s one researcher came up with a table exclusively for use on European skin rather than the "exotic races."[14]

THE COLOR TOP

But perhaps the oddest development regarding skin color during this period was an instrument known as the color top, which was actually a child's toy designed for the schoolroom to teach the fundamentals of color and color mixing. It was based on experiments in color and color perception conducted in the 1850s by James Clerk Maxwell, and color wheels had been around for centuries. Maxwell's great innovation was to attach colored disks to the wheel, with each disk having a slit cut from its circumference to the center that would allow the user to interlock and combine the colors in any desired proportion. When the wheel or top was spun the colors blended. "Maxwell disks" became a Victorian optical toy, and in the early 1890s, Milton Bradley, best known to later generations as the king of board games, decided that a simpler version would be beneficial for children's education. In 1895 the Bradley color top was patented and soon sold in Woolworth's at six cents each. It consisted of a wooden spindle attached to a thick cardboard disk, the outer rim of which was divided into units so that the percentage of each colored disk could be easily quantified. An assortment of paper disks were included: red, orange, yellow, green, blue, and violet, plus white and black.[15]

Many anthropologists, surprisingly enough, adopted this toy as an eminently pragmatic (if not absolutely scientifically sound) research tool. This was especially true in the United States, partly because the top was more readily available there, but even more so because, as we shall see presently, it was seen as the most effective way of measuring blackness as opposed to any other color. The advantage of its portability was obvious, and because each of the colors could actually be quantified according to the gradations printed on the base of the top, it seemed far more accurate and copious than color scales such as Broca's or Luschan's. It first came to the attention of the scientific world as a way of studying the color of small animals and birds, and the major proponent of the method was biologist (and later eugenicist) Charles Davenport . . . [16] Here is Davenport's description of the procedure involved:

> [The assistant] visited the homes of the colored people and obtained all the genealogical data that could be furnished. Then the sleeve was rolled up to above the elbow and a part of the skin that is usually covered from the sunlight was thus exposed. The arm was placed on the table by a good light and a Bradley color-top was spun close to the arm and the disks adjusted until they matched, when spun,

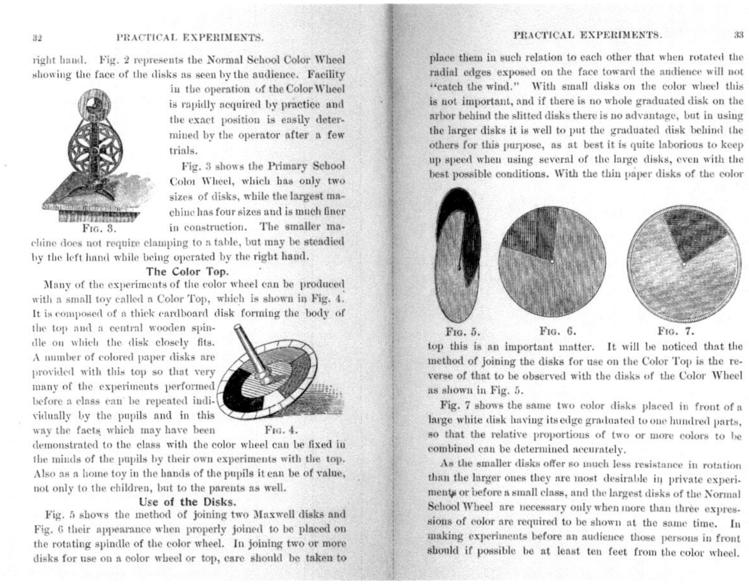

32 PRACTICAL EXPERIMENTS.

right hand. Fig. 2 represents the Normal School Color Wheel showing the face of the disks as seen by the audience. Facility in the operation of the Color Wheel is rapidly acquired by practice and the exact position is easily determined by the operator after a few trials.

Fig. 3 shows the Primary School Color Wheel, which has only two sizes of disks, while the largest machine has four sizes and is much finer in construction. The smaller machine does not require clamping to a table, but may be steadied by the left hand while being operated by the right hand.

FIG. 3.

The Color Top.

Many of the experiments of the color wheel can be produced with a small toy called a Color Top, which is shown in Fig. 4. It is composed of a thick cardboard disk forming the body of the top and a central wooden spindle on which the disk closely fits. A number of colored paper disks are provided with this top so that very many of the experiments performed before a class can be repeated individually by the pupils and in this way the facts which may have been demonstrated to the class with the color wheel can be fixed in the minds of the pupils by their own experiments with the top. Also as a home toy in the hands of the pupils it can be of value, not only to the children, but to the parents as well.

FIG. 4.

Use of the Disks.

Fig. 5 shows the method of joining two Maxwell disks and Fig. 6 their appearance when properly joined to be placed on the rotating spindle of the color wheel. In joining two or more disks for use on a color wheel or top, care should be taken to

PRACTICAL EXPERIMENTS. 33

place them in such relation to each other that when rotated the radial edges exposed on the face toward the audience will not "catch the wind." With small disks on the color wheel this is not important, and if there is no whole graduated disk on the arbor behind the slitted disks there is no advantage, but in using the larger disks it is well to put the graduated disk behind the others for this purpose, as at best it is quite laborious to keep up speed when using several of the large disks, even with the best possible conditions. With the thin paper disks of the color

FIG. 5. FIG. 6. FIG. 7.

top this is an important matter. It will be noticed that the method of joining the disks for use on the Color Top is the reverse of that to be observed with the disks of the Color Wheel as shown in Fig. 5.

Fig. 7 shows the same two color disks placed in front of a large white disk having its edge graduated to one hundred parts, so that the relative proportions of two or more colors to be combined can be determined accurately.

As the smaller disks offer so much less resistance in rotation than the larger ones they are most desirable in private experiments or before a small class, and the largest disks of the Normal School Wheel are necessary only when more than three expressions of color are required to be shown at the same time. In making experiments before an audience those persons in front should if possible be at least ten feet from the color wheel.

FIGURE 4.3

Milton Bradley. Color top from *Elementary Color* (1895). In this image, "Fig. 4" depicts the gradations on the top, while "Figs. 5–7" demonstrate how the disks were interlocked and adjusted to the desired proportion. The top was regularly used by anthropologists throughout the early twentieth century to measure skin color. Image courtesy of Princeton University Libraries.

the color of the skin. Various combinations of black (N), red (R), yellow (Y), and white (W) gave a close approximation to the skin color.

A recent commentary has referred to this form of measurement as an inadvertent self-parody, and indeed it is hard to imagine that a five-and-dime child's toy would become such an important research tool able to Broca's scale since the colors were arranged in a clear progression from white to black . . . [17]

But exactly what did it measure? It was soon recognized that the disks chosen (black, white, red, and yellow) were not necessarily the equivalents of the pigments present in human skin. The optical effect of mixing black and white to spectral colors makes them "broken," so all the color top really did was to produce a range or lighter or darker red-yellows, which is to say shades of brown and orange. And isn't all human skin one shade or another of those colors anyway? Yet it is nonetheless worth pondering why Davenport had selected these four disks in the first place. Basic color theory dictates that many different combinations could have been used to match

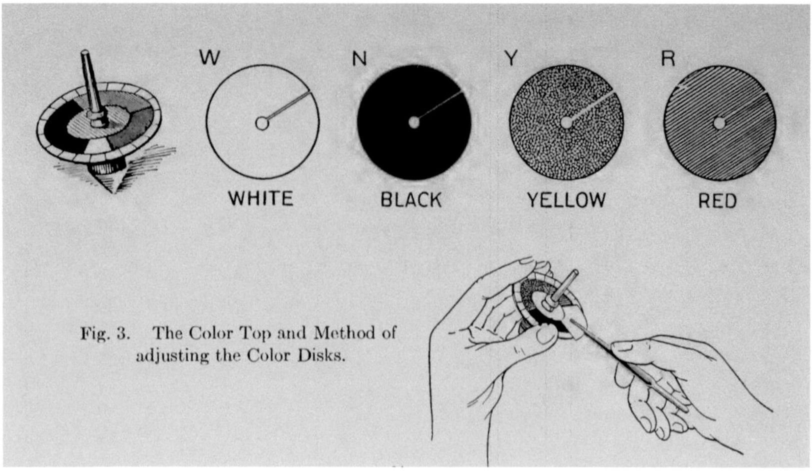

FIGURE 4.4
Louis R. Sullivan. Color top from *Essentials of Anthropometry* (1928). The illustration here was featured in a pocket-sized guidebook for the measurement of physical differences published by the American Museum of Natural History. Courtesy American Museum of Natural History and Princeton University Libraries.

the rather limited range that comprises the color of human skin, and there was also a confusion between the standard palette for mixing colored pigments (red, green, and blue) and for mixing colored light (magenta, yellow, and cyan). Davenport himself offered a rather half-baked and oddly retrospective explanation: that each color did represent an element in human pigmentation (black for melanin, yellow for "lipochrome" or carotene, red for blood, and white because it was "reflected from the opaque skin"). But he also claimed that the disks "indicated" that these were the pigments that compose skin color rather than the other way around . . . [18]

The real focus of these initial experiments, however, was the blackness of the N disk (N for *niger* or *nigrum*), just as in the beginning the only people who were actually measured were black and white "hybrids"—as if the centuries-old classification of mulattos, mestizos, quadroons, and so on, might actually be quantifiable. It was no longer necessary to refer to dark skin as café au lait or chocolate-colored, both of which had been used extensively throughout the nineteenth century. As Luschan wryly noted in 1916, any skin would be the color of *Milchkaffee* depending on how it was prepared.[19] From a eugenics point of view the aim was to prove that skin pigmentation "segregated" and that it was passed on according to Mendelian laws of heredity. Davenport thought he had demonstrated that there were two biotypes of black African, one full-blooded and the other a hybrid, with the latter having much lower levels of black in the skin. When two mulatto parents had children their pigmentation "Mendelized" and both very white and very black children could result, particularly in the second and third generations . . .

Yet the top's ability to measure nonwhiteness may have been more than just a matter of racial prejudice. It may have been built into the nature of the very colors that were selected . . . Milton Bradley himself noted that it was something of a mystery that the addition of only a little white changed the tone of a color drastically, while it required a great deal of black to produce a noticeable change (and that this is exactly the opposite of what occurs when mixing paint). It is perhaps for this reason that researchers noticed that the white disk was not really able to distinguish different Caucasian (and Mongolian) skin tones . . . Others complained that the yellow disk was just a mixture of red and green that merely produced a sensation of "yellow," or that the Milton Bradley company had changed the cardboard platform on which the disks rested from gray to brown, which threw everything considerably out of whack. Sometimes puzzling optical effects were recorded, as when a yellow object registered a higher percentage of red, or when a visibly lighter skin actually showed a larger amount of black than a darker one. Worst of all, there was as much as a twenty-point overlap between the amount of the N disk in darker Caucasians and lighter-colored blacks.[20]

It was then discovered in the early 1920s that the red disk wasn't even red at all (not spectral red, that is), since it was more of an oxblood shade that, when analyzed, turned out to contain 59% black.[21] The four colors on the top, ironically enough, were even blacker than Davenport had originally supposed, and all his early data needed to be recalculated if it was to be of any value at all. And yet the top continued to be used. In 1926 he became even bolder and published a popular essay in *Natural History* arguing that the skin color of all the races could be "describe[d] in exact quantitative terms" with the top. . . .[22]

What made this essay different, however, was that Davenport also ventured into the much more uncertain waters regarding the "yellow" and "red" races, although he gave far more attention to the former than the latter. Since there were no earlier studies of East Asians, he measured the upper arm of a Japanese colleague, only to find that "the Y element is not exceptionally great" and that the readings of each of the four disks were almost identical to those of a group of Caucasian children he had just examined. Obviously this came as something of a surprise, particularly in light of the fact that "some other observations [he] has made suggest that Eastern Chinese show a large per cent of yellow in their skins." And without any evidence at all he concluded on the next page that if "the eastern Asiatics were formerly classified as the yellow race" it was "a terminology justified by the large size of the yellow factor."[23]

What clearer example could there be of the assumption that East Asians simply "were" yellow, even if it was not actually borne out by the results of investigation? A 1930 essay by Beatrice Blackwood of Oxford soon came to the rescue, since it had actually carried out color top measurements on a substantial number of each of the "four races" . . . [24]

But at just this time the color top's star had also begun to fade, as new and more sophisticated devices for color measurement were being developed utilizing colored light (generally red, green, and violet) and with the aid of electricity. These devices went under a number of different names (the photometer, the spectrophotometer, the colorimeter, the photoelectric reflection meter), but in each case the principle was the same: they did not simply measure color in any simple sense but instead gauged the way that an object both reflected and absorbed light at various points in the visible color spectrum . . . [25] [I]n the 1920s and 1930s the device was seen to demonstrate with even greater scientific reliability the steady increase in pigmentation among the "dark races," from the lightest East Asian to the darkest African. And these researchers, too, regularly took the yellowness of East Asians for granted . . . [26]

Whether they were "actually" yellow was no longer the focus, perhaps, but at the same time there could be no mistake that they were not white. The point of so much measurement was to show that the "other" races really did differ from the "normal" one, the group to which of course the anthropologists themselves generally belonged . . . The threat of the "yellow" race, like all the other nonwhite ones, had become a problem . . .

NOTES

Chapter 4, "Measuring Skin Color" and "The Color Top," excerpted from ""Nineteenth-Century Anthropology and the Measurement of Mongolian Skin Color," originally published in *Becoming Yellow: A Short History of Racist Thinking*, by Michael

Keevak, © 2011, Princeton University Press. Used with permission of Princeton University Press; permission conveyed through Copyright Clearance Center, Inc.

1. The following passages were originally published in "Nineteenth-Century Anthropology and the Measurement of 'Mongolian' Skin Color," chapter 3 of Keevak's *Becoming Yellow* (Princeton, NJ: Princeton University Press, 2011).

2. The Société ethnologique de Paris had been founded in 1839, followed by the Ethnological Society of New York in 1842, the Ethnological Society of London in 1843, and the Koninklijk Instituut voor Taal, Land-en Volkenkunde in Holland in 1851.

3. Schiller, *Paul Broca*, 163.

4. This went through six editions and, beginning with the second edition of 1892, was published by the Royal Anthropological Institute in London as *Notes and Queries on Anthropology*. It was reprinted as late as 1971. Competing guides published in France and Germany include "Instruction générale adressée aux voyageurs," *Instructions générales aux voyageurs,* and *Anleitung zu wissenschaftlichen Beobachtungen auf Reisen.*

5. Broca, *Instructions générales pour les recherches et observations anthropologiques,* 134–136.

6. Broca, *Instructions générales pour les recherches et observations anthropologiques,* 54.

7. An exception was Yule, "On the Influence of Bias" (1906).

8. Topinard, *L'anthropologie*, 364 (*Anthropology*, 344); Broca, *Recherches sur l'hybridité animale*, 463; Broca, *Instructions générales pour les recherches et observations anthropologiques*, 55.

9. Broca, "Instructions générales pour les recherches anthropologiques," *Bulletins de la Société d'anthropologie de Paris* 3 (1862); Broca, *Instructions générales pour les recherches et observations anthropologiques*, 54–55.

10. Broca, "Échelle chromatique des yeux," 592–593; Broca, "Tableau chromatique des yeux," 768–773; Broca, *Instructions générales pour les recherches et observations anthropologiques*, 45.

11. Topinard, *Éléments d'anthropologie Générale* (Delahaye and Lecrosnier, 1885), 317; *Notes and Queries on Anthropology*, 64; Hrdlička, *The Old Americans*, 22; *Hrdlička's Practical Anthropometry*, 107.

12. Hrdlička, "Anthropometry," 46–67, 290, 314; *Hrdlička's Practical Anthropometry*, 4th ed., 107; Deniker, *Les races et les peuples de la terre*, 64 (*The Races of Man*, 47). See also Martin, *Lehrbuch der Anthropologie*, 206; and Comas, *Manual of Physical Anthropology*, 264–265.

13. Luschan, "Einige wesentliche Fortschritte in der Technik der physischen Anthropologie": 465–466; "Über Hautfarbentafeln": 402–406; Virchow, "Von Luschan'schen Farbentafel": 328; Fritsch, "Die Feststellung der menschlichen Hautfarben"; "Bemerkungen zuder Hautfarbentafel": 183–185; Martin, *Lehrbuch der Anthropologie*; Hrdlička, "Anthropometry," 290; Thomson, "On the Treatment and

Utilization of Anthropological Data," 27; Ranke, "Über die Hautfarbe der südamerikanischen Indianer": 61–73; Sarasin and Sarasin, *Ergebnisse naturwissenschaftlicher Forschungen auf Ceylon*, 91–96; Sarasin and Sarasin, *Materialien zur Naturgeschichte der Insel Celebes*, vol. 5, part 2, table 2; Gates, *Pedigrees of Negro Families*, 253–254.

14. Hintze, "Der Hautfarbenfächer und das Hautfarbendiagramm": 254–278; Fritsch, "Feststellung der menschlichen Hautfarben," 88; Hrdlička, "Anthropometry," 313; Herskovits, *The Anthropometry of the American Negro*, 39; Basler, "Über eine Hautfarbentafel für Europäer": 525–530.

15. Maxwell, *Scientific Papers*: 126–154; Bradley, *Color in the School-Room*, 52–59; Bradley, *Elementary Color*, 18–22; *Bradley's Kindergarten Material and School Aids*, 59. On Maxwell's experiments, see Sherman, *Colour Vision in the Nineteenth Century*, 153–183; and Harman, *The Natural Philosophy of James Clerk Maxwell*, 37–48.

16. Strong, "A Quantitative Study of Variation in the Smaller North-American Shrikes": 271–298.

17. Davenport, *Negro-White Crosses*, 2; Sollors, *Neither Black Nor White Yet Both*, 137; Bowman, "The Color-Top Method of Estimating Skin Pigmentation," 59; Herskovits, *Anthropometry*, 39.

18. Davenport, "Heredity of Skin Pigment," 664.

19. Luschan, "Über Hautfarbentafeln," 402. See Pillsbury, "A New Color Scheme": 114; Bradley, "The Color Question Again": 175–176; and *Elementary Color*, 19–20.

20. Bradley, *Elementary Color*, 35; Blackwood, "Racial Differences in Skin-Colour as Recorded by the Colour Top," 163; Herskovits, *Anthropometry*, 73–74; Eickstedt, *Rassenkunde und Rassengeschichte der Menschheit*, 49–50; Bowman, "Color-Top Method," 60–61; Todd et al., "Skin Pigmentation," 189; Sumner, "Linear and Colorimetric Measurements of Small Mammals," 203; Todd and Van Gorder, "The Quantitative Determination of Black Pigmentation in the Skin of the American Negro," 247.

21. Todd and Van Gorder, "Quantitative Determination," 243–244; Harris, "San Blas Indians," 47.

22. Davenport, "Skin Colors of Races," 44–46

23. Davenport, "Skin Colors of Races," 47–48; cf. Davenport, *Negro-White Crosses*, 28.

24. Blackwood, "Racial Differences," 144, 152, 160–165.

25. Sheard and Brown, "The Spectrophotometric Analysis of the Color of the Skin": 816–831; Shaxby and Bonnell, "On Skin Colour" (1928): 60–64; Hardy, "A New Recording Spectrophotometer": 305–319; Sheard and Brunsting, "The Color of the Skin as Analyzed by Spectrophotometric Methods": 559–613; Weiner, "A Spectrophotometer for Measurement of Skin Colour," 152.

26. Robins, *Biological Perspectives on Human Pigmentation*, 98–103; Dorno, "Beiträge zur Kenntnis," 74–75; Edwards and Duntley, "The Pigments and Color of Living Human Skin," 6.

BIBLIOGRAPHY

Basler, Adolf. "Über eine Hautfarbentafel für Europäer und mit vorgenommene Untersuchungen." *Zeitschrift für Morphologie und Anthropologie* 25 (1926): 525–530.

Blackwood, Beatrice. "Racial Differences in Skin-Colour as Recorded by the Colour Top." *Journal of the Royal Anthropological Institute of Great Britain and Ireland* 60 (1930): 137–168.

Bowman, H. A. "The Color-Top Method of Estimating Skin Pigmentation." *American Journal of Physical Anthropology* 14 (1930): 59–72.

Bradley, Milton. "The Color Question Again." *Science* 19 (1892): 175–176.

Bradley, Milton. *Color in the School-Room: A Manual for Teachers*. Springfield, MA: Milton Bradley Co., 1890.

Bradley, Milton. *Elementary Color*. Springfield, MA: Milton Bradley Co., 1895.

Bradley's Kindergarten Material and School Aids. Springfield, MA: Milton Bradley Co., 1904.

Broca, Paul. "Échelle chromatique des yeux." *Bulletins de la Sociétéé d'anthropologie de Paris* 4 (1863): 592–605.

Broca, Paul. "Instructions générales pour les recherches anthropologiques." *Bulletins de la Société d'anthropologie de Paris* 3 (1862): 411–414.

Broca, Paul. *Instructions générales pour les recherches anthropologiques à faire sur le vivant*. 2nd ed. Paris: Masson, 1879.

Broca, Paul. *Instructions générales pour les recherches et observations anthropologiques (anatomie et physiologie)*. Paris: Masson, 1865 [also in *Mémoires de la Société d'anthropologie* 2 (1865): 69–204; reprinted in *Archives de médecine navale* 3 (1865): 369–504].

Broca, Paul. *Recherches sur l'hybridité animale en général et sur l'hybridité humaine en particulier*. Paris: Claye, 1860.

Broca, Paul. "Tableau chromatique des yeux, de la peau et des cheveux pour les observations anthropologiques." *Bulletins de la Société d'anthropologie de Paris* 5 (1864): 767–73.

Comas, Juan. *Manual of Physical Anthropology*, Rev. ed. Springfield, IL: Thomas, 1960.

Davenport, Charles B. *The Heredity of Skin Color in Negro-White Crosses*. Washington, DC: Carnegie Institution, 1913.

Davenport, Charles B. "Heredity of Skin Pigment in Man." *American Naturalist* 44 (1910): 641–72, 705–31.

Davenport, Charles B. *Heredity of Skin Color in Negro-White Crosses*. Washington, DC: Carnegie Institute of Washington, 1913.

Davenport, Charles B. "The Skin Colors of the Races of Mankind." *Natural History* 26 (1926): 44–49.

Deniker, Joseph. *Les races et les peuples de la terre* (1900). 2nd ed. Paris: Masson, 1926.

Deniker, Joseph. *The Races of Man: An Outline of Anthropology and Ethnography*. London: Scott, 1900.

Dorno, Carl. "Beiträge zur Kenntnis des Sonnen-und Quarzlicht-Erythems und-Pigmentes." *Strahlentherapie* 22 (1926), 74–75.

Edwards, Edward A., and S. Quimby Duntley. "The Pigments and Color of Living Human Skin." *American Journal of Anatomy* 65 (1939): 1–33.

Eickstedt, Egon Freiherr von. *Rassenkunde und Rassengeschichte der Menschheit.* Stuttgart: Enke, 1934.

Fritsch, Gustav. "Bemerkungen zu der Hautfarbentafel." *Mitteilungen der Anthro- pologischen Gesellschaft in Wien* 46 (1916): 183–85.

Fritsch, Gustav. "Die Feststellung der menschlichen Hautfarben." *Zeitschrift für Ethnologie* 48 (1916): 86–89.

Gates, R. Ruggles. *Pedigrees of Negro Families.* Philadelphia: Blakiston, 1949.

Hardy, Arthur C. "A New Recording Spectrophotometer." *Journal of the Optical Society of America* 25 (1935): 305–19.

Harman, P. M. *The Natural Philosophy of James Clerk Maxwell.* Cambridge: Cambridge University Press, 1998.

Harris, Reginald G. "The San Blas Indians." *American Journal of Physical Anthropology* 9 (1926): 17–63.

Herskovits, Melville J. *The Anthropometry of the American Negro: A Study in Racial Crossing.* New York: Columbia University Press, 1930.

Hintze, Arthur. "Der Hautfarbenfächer und das Hautfarbendiagramm." *Zeitschrift für Ethnologie* 59 (1927): 254–78.

Hrdlička, Aleš. "Anthropometry." *American Journal of Physical Anthropology* 2 (1919): 43–67, 175–94, 283–319, 401–28; 3 (1920): 147–73 [reproduced as *Anthropometry.* Philadelphia: Wistar Institute of Anatomy and Biology, 1920].

Hrdlička, Aleš. *Hrdlička's Practical Anthropometry.* 4th ed. Ed. T. D. Stewart. Philadelphia: Wistar Institute of Anatomy and Biology, 1952.

Hrdlička, Aleš. *The Old Americans.* Baltimore: Williams & Wilkins, 1925.

Martin, Rudolf. *Lehrbuch der Anthropologie in systematischer Darstellung.* 2nd ed. 3 vols. Jena, Germany: Fischer, 1928.

Maxwell, James Clerk. *Scientific Papers.* Edited by W. D. Niven. 2 vols. Cambridge: Cambridge University Press, 1890.

Notes and Queries on Anthropology. 2nd ed. London: Anthropological Institute, 1892.

Pillsbury, J. H. "A New Color Scheme." *Science* 19 (1892): 114.

Ranke, Karl Ernst. "Über die Hautfarbe der südamerikanischen Indianer." *Zeitschrift für Ethnologie* 30 (1898): 61–73.

Robins, Ashley H. *Biological Perspectives on Human Pigmentation.* Cambridge: Cambridge University Press, 1991.

Sarasin, Paul, and Fritz Sarasin. *Ergebnisse naturwissenschaftlicher Forschungen auf Ceylon.* 4 vols. Wiesbaden: Kreidel, 1887–1908.

Sarasin, Paul, and Fritz Sarasin. *Materialien zur Naturgeschichte der Insel Celebes*. 5 vols. Wiesbaden, Germany: Kreidel, 1898–1906.

Schiller, Francis. *Paul Broca: Founder of French Anthropology, Explorer of the Brain*. Berkeley: University of California Press, 1979.

Shaxby, J. H., and H. E. Bonnell. "On Skin Colour." *Man* 28 (1928): 60–64.

Sheard, Charles, and George E. Brown. "The Spectrophotometric Analysis of the Color of the Skin." *Archives of Internal Medicine* 38 (1926): 816–831.

Sheard, Charles, and Louis A. Brunsting. "The Color of the Skin as Analyzed by Spectrophotometric Methods." *Journal of Clinical Investigation* 7 (1929): 559–613.

Sherman, Paul D. *Colour Vision in the Nineteenth Century*. Bristol, UK: Hilger, 1981.

Sollors, Werner. *Neither Black nor White Yet Both: Thematic Explorations of Interracial Literature*. Oxford: Oxford University Press, 1997.

Strong, R. M. "A Quantitative Study of Variation in the Smaller North-American Shrikes." *American Naturalist* 35 (1901): 271–298.

Sumner, Francis B. "Linear and Colorimetric Measurements of Small Mammals." *Journal of Mammalogy* 8 (1927): 177–206.

Thomson, Arthur. "On the Treatment and Utilization of Anthropological Data." *Knowledge* (1 February 1899): 25–27.

Todd, T. Wingate, Beatrice Blackwood, and Harry Beecher. "Skin Pigmentation: The Color Top Method of Recording." *American Journal of Physical Anthropology* 11 (1928): 187–204.

Todd, T. Wingate, and Leona Van Gorder. "The Quantitative Determination of Black Pigmentation in the Skin of the American Negro." *American Journal of Physical Anthropology* 4 (1921): 239–260.

Topinard, Paul. *L'anthropologie*. Paris: Reinwald, 1876.

Topinard, Paul. *Anthropology*. Translated by Robert T. H. Bartley. London: Chapman and Hall, 1878.

Topinard, Paul. *Éléments d'anthropologie Générale*. Paris: Delahaye and Lecrosnier, 1885.

Virchow, Hans. "Von Luschan'schen Farbentafel zur Bestimmung der Hautfarbe." *Zeitschrift für Ethnologie* 58 (1926): 328.

von Luschan, Felix. "Einige wesentliche Fortschritte in der Technik der physischen Anthropologie." *Zeitschrift für Ethnologie* 36 (1904): 465–66.

von Luschan, Felix. "Über Hautfarbentafeln." *Zeitschrift für Ethnologie* 48 (1916): 402–406.

Weiner, J. S. "A Spectrophotometer for Measurement of Skin Colour." *Man* 51 (1951): 152–153.

Yule, G. Udny. "On the Influence of Bias and of Personal Equation in Statistics of Ill-Defined Qualities." *Journal of the Anthropological Institute of Great Britain and Ireland* 36 (1906): 325–381.

5 PIGMENT AND PLASTER: ANTHROPOMORPHIZING INFORMATIC COLOR IN HAWAI'I (1921)

MICHAEL ROSSI

PROLEGOMENON

When we ask about the role color technologies have played in encoding racial hierarchies over the course of the twentieth century, the question we're posing is really one that has to do with the relationship between color and information. This is firstly because, as several contributors to this volume have demonstrated, for more than a century, a diverse array of actors have understood the problem of color as a matter of translating abstract, perceptual data into systematic, mathematicized forms—what would come to be understood as "information" in the specialized, cybernetic sense.[1] In the second place, technologies for managing color are "informational" in the more traditional sense of information-as-pedagogy: they are used not simply to structure color per se, but also to bring sensory percepts into alignment with a wider array of philosophical, ideological, and social structures. To put it another way, they make meaning out of data. Finally, in the specific case of race, researchers from various scientific disciplines have depended on color to inform (to literally and figuratively *form*) the nebulous but powerful fantasy of classifying human "types." A specific instance makes this proposition especially clear.

In the early twentieth century, the problem of race—and especially "race-mixing"—weighed heavy on the minds of American social scientists as they contemplated the recent annexation of a number of overseas territories in the wake of the Spanish American War.[2] Although the Kingdom of Hawai'i was not, strictly speaking, one of these wartime acquisitions, it nevertheless served as a touchstone for American scientific (and political) thinking about the place of a multiracial population in American governance. Hawai'i was home to a diverse population of human beings: indigenous Hawaiian people, people of European descent, and people from Asia. These people lived close to one another, worked side by side with one another, and—to the intense concern of many American scientists—had children with one another. The fact that Hawai'i appeared to be a peaceful, prosperous, and multiracial extension of the territorial holdings of the United States was therefore a cause for anthropological investigation.

It was for this reason that Louis R. Sullivan set out in 1920 to investigate the racial "types" of the Hawaiian Islands. Sullivan was a physical anthropologist at the American Museum of Natural History in New York. The "problem" of race in Hawai'i was, as the Museum's director Henry Fairfield Osborn put it, "one of concern to every American."[3] Any scientific investigation would be both a political instrument for the consolidation of American power, and an addition to the store of knowledge contributed by anthropology to the study of humankind. Osborn, along with many of his peers in elite American society, was a eugenicist—a believer in the idea that the racial demography of a nation was directly tied to that nation's political harmony and industrial productivity. Given the island's diverse population, Osborn felt that Hawai'i was an ideal "natural laboratory" for the study of "race-mixing." Understanding the racial "types" of Hawai'i would therefore lend understanding to the project of harmony and productivity in the United States more broadly.

How, then, did anthropologists study the racial types of Hawai'i? In the arsenal of early twentieth-century anthropology, there were a number of possible technologies for turning ill-defined ideas of "race" into precise, scientific "information," but the most powerful technology of all was the plaster cast: covering an individual's face—or occasionally whole body—in plaster in order to make a mold, then filing that mold again with plaster to create a positive replica. The resulting form would capture the precise, facial structure of the individual, which could then be colored in order to re-create the individual with (ostensible) exactitude. The final model would then be displayed in a museum to edify onlookers interested in the "races

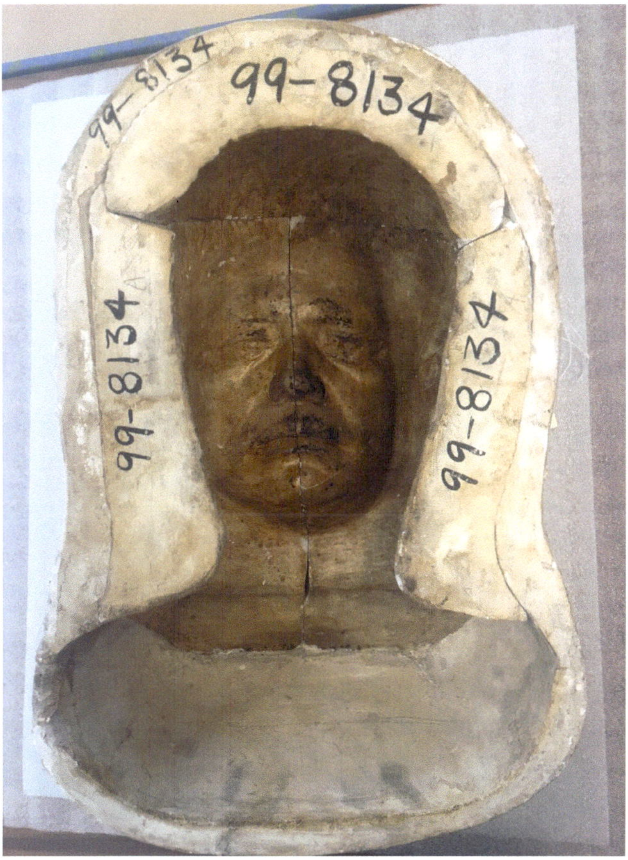

FIGURE 5.1
The mold for cast #52. Courtesy of Anthropology Archives, American Museum of Natural History.

of man," or else could be stored and studied by anthropologists. In this act of re-creation and reproduction, the racial type of the individual would be reified—made (almost literally) concrete through the action of plaster on skin, paint on plaster.

By looking to this seemingly esoteric case study—Sullivan's plaster casting project in Hawai'i, circa 1921—this chapter illustrates how color and form, as instantiated by pigment and plaster, were the necessary components of racial informatics for early twentieth-century physical anthropology. In demonstrating how early twentieth-century anthropometry was an information-collecting act, I reveal how the function of the casts was to simultaneously sort color data, measurement data, and physical data into racial order. Ultimately, Sullivan and his peers would fail to arrive at decisive way of classifying color by way of race, and vice versa. Color proved too elusive, human morphology too varied, the concept of "race" too nebulous to sustain a research program with an output of pigment on plaster. Yet I argue that it was nevertheless the very ambiguity of the informational relationship between color and race that made color such a necessary part of Sullivan's project.

We speak in the twenty-first century about genetic "information" about race. But what is this information, precisely? A century on, it is worth considering the ways in which color can become transformed into genetic information, precisely as we are in an age in which, on the one hand, we fret about genetic information, and on the other hand, in which scientists and their clients assiduously work to separate genes from scientific fact. In the case of plaster-casting in early twentieth-century Hawai'i, stringing together the tiny islands of meaning insinuated by color ordering technologies across the vast, white expanse of plaster casts made race—or rather the promise of race as a scientific fact—momentarily intelligible, significant, and (for a short time) *informative*. Ironically, then, pigment and plaster were therefore the most precise genetic technologies of their time, even as the information that they yielded was ambivalent at best.

PHILOSOPHY AND PSYCHOLOGY

It is worth pausing for a moment to consider the profound strangeness of Sullivan's task as he set off for Hawai'i. His job was to classify humans—to render them "stereotypical"—but his method was paradoxically to try to reproduce individuals in their most minute particularities. He was dedicated to classifying the very innermost essences of human beings, but his taxonomic technologies came down mostly to the play of pigments on a white, plaster surface. Indeed, the entirety of anthropological casting was a musing on the most infinitesimal possible point of contact—that is to say, the point at which skin meets plaster, and the point at which light reflects off of pigment, hits the observer's eye, and is interpreted (by the observer and their cultural communities) as significant. To say that Sullivan's work was superficial (or, indeed "skin deep") is not pejorative, but merely descriptive.[4] Race—for Sullivan, Osborn, and their colleagues who held this view—often came down to the thinnest of possible criteria.

What, then, was the theory behind Sullivan's work? There are two ways of looking at this. In the one sense, molding and casting gives us a clue about how we might think of casting and coloring as modes of producing information. In this instance, we might follow philosopher of technology Gilbert Simondon in imagining the casting process itself as an information technology.[5] That is to say, for Simondon, the ways in which matter and form interact to become an individual something—rather than a nothing or an everything—can be thought of as an information process. He gives the simple example the formation of a brick. One might say that a brick is simply clay that's been forced into a square mold. But Simondon suggests instead that it is neither the clay which makes the brick, nor is it the mold that forces the clay into brick shape—rather, it is the interplay between clay and mold which *is* the brick, and "information" is the tension that forms—or informs—the possibility of the brick.[6] As Simondon puts it "information is not a thing; rather it is the operation of a thing arriving in a system in which it produces a transformation."[7] On the one hand, he notes, "[i]nformation cannot be defined outside of this incidence of transformation and the operation of reception."[8] On the other hand, it is noteworthy that in this formulation, the reverse must also be true: the act of transformation is precisely that which constitutes "information."

Thought of in this way, the racial information that Sullivan produced existed in the world as a kind of tension. It lived in the first place in the tension between plaster and human skin when Sullivan and his assistants made a mold. It was then re-produced, revitalized, in the contact between the hard plaster of the mold, and the liquid plaster which flowed into the mold and took the form of a face. This *information*—this way of ordering simultaneously perceptual and morphological data—was that which instructed scientists and their audiences about the racial typology of the person who had been cast. To put it another way, as sociologist of science James Greismer proposes, certain forms of museum display—taxidermy, for instance—should be thought of as "models of data constructed out of the physical entities being modeled."[9] For its part, the technology of casting simply replaced the "data" constituted by taxidermied skin with an infinitesimally small point of contact between skin and plaster medium.

At the same time, however, a plaster cast itself comes out of its mold a uniform, monochromatic white. One might not expect this to be a problem for anthropology, if all that is necessary for pronouncements about racial typology as form. But for Sullivan, Osborn, and their colleagues, the whiteness of the plaster had to be painted in order for the cast to achieve its fullest effect. It was the color of the cast—insofar as the cast represented the individual—that made race explicit. One is here reminded of Franz Fanon's concept of the "epidermalization" of race: the way in which, as Fanon puts it, racial identity, historicity, indeed, individual subjectivity is reduced—or rather reproduced, or even projected—entirely onto the skin of non-white colonial subjects.[10] If, in Simondon's scheme, the "information" of the cast is co-constitutive of the cast itself; and in Greismer's telling, skin can be a special form of data; by thinking with Fanon we realize that—in the instance of Sullivan and Osborn's physical anthropology—information and data inhered not singularly in the act of obtaining a copy of the individual from which to catalog typological judgements, but also in the re-coloring of the pure white cast to reflect trained judgments about what a true racial type should look like.[11]

PRACTICE AND PRACTICALITY

Few people would have understood the problems of color as an information technology better than Sullivan. After a stint measuring soldiers in the U.S. Army during the First World War, and then another stint doing a PhD in physical anthropology with Franz Boas, he went to work for the American Museum of Natural History (hereafter AMNH), where he was promptly dispatched to Hawaii. After he returned to New York, he wrote a short manual based on his experiences entitled *Essentials of Anthropometry: A Handbook for Explorers and Museum Collectors* (1923). In *Essentials of Anthropometry,* Sullivan discussed the finer points of how to "collect" human beings for anthropological study. By way of introductory remarks, Sullivan noted that "[t]here is practically no limit to the number of measurements and observations which may be taken on the living human body." Different anthropologists could measure different parts of their human subjects. In order that data could be compared between scientific observers—and scientific information about human beings generated—Sullivan advised that the number of measurements be reduced to seven: "stature, head length, head breadth, face breadth, anatomical face height, nasal height, and nasal width." "These measures and indices," Sullivan continued, "should be supplemented by descriptive observations" including observations of eye color, hair color, and skin color among other lesser qualitative characteristics such as nose and eye shape, and patterns of body hair. Through this metrical reduction of human morphology to numerical information, the careful anthropometrist could capture the "form" of any human "type."

In order to achieve the first orders of measurement, Sullivan emphasized the use of carefully-chosen instruments. A "stature rod," for instance, would measure the height of subjects with precision; Sullivan advised that the best stature rod on the market was made by one P. Herman of Zurich. "Spreading calipers," for their parts, could be used for measuring "head length" and "head breadth." P. Herman also made the best model of these, but in a pinch, Sullivan advised consulting "the nearest large police station," since police also used the calipers for measuring the heads of criminals.[12] And "sliding calipers" (P. Herman, again) would come in handy for measuring nasal height and nasal width. If would-be anthropologists made sure that the instruments were well calibrated, and followed the instructions in Sullivan's book to the letter, even the most harried amateur anthropologist could contribute useful data for the study of humankind.

On the subject of color, however, Sullivan was more equivocal. It was true, he noted, that precision was important. The best color scale available for anthropological work, Sullivan felt, was a set of set of thirty-six small, opaque glass bricks developed by German anthropologist Felix von Luschan. Each brick measured roughly two by three centimeters, and ranged in color from an almost-chromatically-white brick labeled number "1," to an almost-chromatically black brick labeled number "36." The shades in between varied from pinkish-white ("7"), to lemony-yellow ("20"), to deep, cinnamon brown ("30"). By holding a tile from the scale next to a subject's skin, the anthropologist could assign a standardized numerical "grade" to the subject's skin color, and thereby group the skin colors of different notional races.

Sullivan was clear that even if the anthropometric investigator couldn't get their hands on a von Luschan scale, it was desirable to use *some* sort of color scale for measuring skin tone. French anthropologist Paul Broca, for example, had published a less-reliable paper series of skin color scales in 1864, which the Government Printing Office in Washington D.C. had reissued. Or, one might turn to still-more unwieldy color systems like the one

FIGURE 5.2
Felix von Luschan. *Hautfarben-Tafel* (Table of skin colors), ca. 1900. Museum Collection. Von Luschan's scales for measuring skin color. © President and Fellows of Harvard College, Peabody Museum of Archaeology and Ethnology, PM2005.1.168.

designed by Smithsonian ornithologist Robert Ridgeway in 1888. Used most frequently by scientists classifying birds, mammals, mushrooms, and other non-human living things, the color system could nevertheless be employed in a pinch for anthropometry, and the results later recalibrated to Von Luschan's index. Indeed, Sullivan emphasized, the anthropometrist should if necessary devise their own scale rather than go without, insofar as "almost any scale is an improvement on attempting to estimate color off-hand."[13]

Nevertheless, when it came to generating actual data from color descriptions, Sullivan admitted that it was "impossible to measure these . . . accurately," and defaulted to a plea for his readers simply to describe colors as uniformly as possible. For example, although Broca had also published a chart of eye color and hair colors in 1864, Sullivan nevertheless suggested that anthropometrists in the field should limit their descriptive vocabulary for hair color to "black, dark brown, reddish brown, light brown, blond, golden, red." Red hair, he continued, "should be further qualified as light brick or auburn."[14] In a similar way, he urged anthropometrists to "describe the eye color as nearly as possible the following terms: black, dark brown, medium brown, light brown, gray green, blue-brown, gray-brown, dark blue, light blue." This was complicated, however, by the fact that gray ought to be used not for light blue, but rather for "a very light brown" eye, which would properly be described as "blue" when observing "Jews and Slavs." In this last example, one sees in particularly starkly the ways in which racial taxonomy and color taxonomy could be deeply entwined: the very names of colors necessarily changed according to that which they were typifying. In any case, as Sullivan had to admit, color was rarely a stable aspect of human physicality. Sun, saltwater, and fashionable dyes might all alter subjects' hair color. Eye color, meanwhile, "varies from time to time in different states of health, temper, and age," as Sullivan noted somewhat ruefully.[15]

As for skin tone, the von Luschan scale provided a useful metric, though it, too, was subject to variations of human morphology and behavior. For one thing, human beings don't exhibit uniform skin hues across their bodies, and results could vary greatly depending on which portions of their subject's body anthropologists chose to measure. Standard procedure was to take skin color readings from the inner side of the upper arm in order to minimize the effects of sun and other environmental factors, though a good scientist had to admit that even among notionally stable "races," things like a subject's occupation, diet, health, and other factors would seriously limit the utility of skin color metrics.

More important still was the fact that many subjects were keenly aware of the fact that von Luschan's scale began with "white" skin as a baseline. They took umbrage at the tacit comparison between ideal, "white," Caucasian skin and their own skin hue. These spontaneous and startlingly ingenuous realizations of Fanon's epidermalization created difficulties for scientists, who depended on the goodwill of their "'victims'" (as one anthropologist rather tellingly put it) to complete their work.[16] The solution frequently recommended by field researchers was to use a color "top" of the variety used to demonstrate the optical combination of colors to children. The color top would be loaded with adjustable segments of colored papers—usually red, yellow, black, and white, which corresponded with what anthropologists generally understood to be "basic" racial categories. When the scientist spun the top next to the subject's skin—usually the forearm, resting on a table next to the top—the colors on the top would blend. By adjusting the size of each segment between spins, the anthropometrist could, in theory,

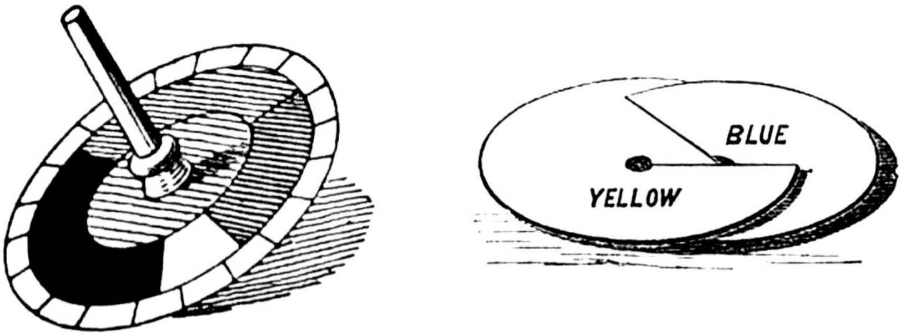

FIGURE 5.3
Milton Bradley's "color top." The sections of colored paper (exemplified here by "BLUE" and "YELLOW") could be adjusted and spun to produce new colors. The gradations around the side of the color top could be used to provide the numerical percentage of each color used.

match the color on the spinning top with the color of their subject's skin. Noting the percentage of each exposed segment of color would provide a numerical transcription of the subject's skin color, which could then be recorded and re-created at will using other color tops.

Anthropologists generally understood the color top method to be more precise than von Luschan's blocks, as well as being less fraught. It also had the advantage of being useful for estimating hair and eye color if the anthropometrist put different colors of paper (e.g., blues and greens) on the top. Working against its favor, however, was its reputation as being time consuming and difficult to master, which made it somewhat unwieldy for field use.

Sullivan and his peers therefore saw the creation of information about color and form as essential to the project of anthropometry. However, whereas anthropometric measurements were relatively easy to standardize—in theory, if not always in practice—color was more difficult to use as a classificatory device. In spite of this, Sullivan never let go of the idea that color was an integral part of understanding race. This was because understanding skin color—a folk metric of race—as scientific information assisted the task of rendering race a measurable, classifiable variety of socially-useful information. In this sense, the work of classifying colors was not dissimilar to the work classifying human beings—both tasks were vague and fraught, but ultimately and undeniably vital to the project of eugenic social reform.

PAINT AND PLASTER

By 1923, Sullivan was confident enough about his skill at turning humans into information that he felt able to write a manual. But when he disembarked on Oahu in 1920 the methods through which he should secure Hawaiian racial typologies were not at all clear. Director Osborn had given Sullivan strict marching orders. He was, firstly, to measure and photograph the "races" of Hawai'i, in order to establish a taxonomy of race mixing. Perhaps more importantly, he was to create full-color, lifelike plaster casts of Hawaiians—heads at minimum, full bodies if possible—for display at the Second International Congress of Eugenics, to be held at the AMNH in 1921. This last was something of a pet project of Osborn's as it would bring together some of his most powerful eugenicist peers from around the world. It was important that his museum and his projects look good, and lifelike, accurately-colored plaster casts were a key part of this mission. Nevertheless, the task was a daunting one, as Osborn warned Sullivan at the outset. With respect to capturing perfect racial types in plaster and paint, Osborn wrote, "the coloring of the skin and the expression of the eyes and face seem to present a great obstacle, which I hardly see how we shall be able to overcome."[17]

Casting was an important—and, in some ways, counterintuitive—solution to at least one part of this obstacle. In one stroke, museum exhibitors understood casts to capture both the essence of the singular person, and the essential characteristics of the racial "type." To take a cast of a person's face—much less their entire body—was to capture those features that made the person unlike any other. A cast recorded not only gross features like nose length, cheekbone height, or distance between eyes. A good cast captured every pore, every slight scar, every wrinkle in the corner of the eye, the slightest turn of the lips. Sometimes tiny hairs from the subject's face became embedded in the mold. In this way, the facial cast could be seen as something of an apotheosis of anthropometry: a way of generating data from every point of skin.

At the same time, exhibitors saw the individual as the taproot of the type. To select an individual for casting meant, necessarily, that the scientist was selecting a representative of a race, whose features would reliably stand for whatever they considered salient in racial typography. In this way, as one exhibitor wrote, "the value of a plaster cast, as a portrait of the dead or living face cannot for a moment he questioned. It must of a necessity be absolutely true to nature. It cannot flatter; it cannot caricature. It shows the subject as he was, not *only as others saw him in the actual flesh, but as he saw himself*."[18] The cast, that is, provided an essential, interior truth about the subject—one which the subject himself could not avoid. One thinks again of Fanon and the epidermalization of race as a superficial, but unavoidably internalized mode of simultaneously external and internal classification. Indeed, as the AMNH's own journal wrote in 1912, new methods of facial casting had made it so that the cast could "be gained without distress to the subject, therefore the expression remains true to life and all measurements are accurate." This natural mode of expression, the article concluded, was a "a most important factor for truth in the ethnological [i.e. racial] investigations."[19]

But it was not enough simply to capture the expressions of his subjects. Sullivan also had to pigment the white casts in a manner that "truly" represented the "types" that he was charged with obtaining. This was a vexing problem from two standpoints. In the first instance, it wasn't obvious what color a "true" Hawaiian ought to be. As professor Vaughn MacCaughey of the University of Hawaii noted in 1917, the "Hawaiian race" itself had been in decline since ancient times, and in this decline, one of the finest "physical types" in the Pacific was becoming rare.[20] In ancient time, "the color of the Hawaiian," MacCaughey remarked, was "an olive-brown or a rich brown, never black nor conspicuously reddish . . . The variation in hue was considerable, ranging from a light coffee brown to a dark reddish-brown." He noted that "occasionally there was a distinct olive tint," which—after contact with Europeans, became more prevalent.[21] It's worth noting here that MacCaughey was a botanist, and might not be fully qualified as an authority on anthropometry. Nevertheless, Sullivan seems to have accepted in rough outlines MacCaughey's assessment of Hawaiian skin pigmentation, echoing his verdict of the generally more reddish-brown character of Polynesian people in the early twentieth century.

More to the point, von Luschan's glass tiles turned out to be more difficult to use than Sullivan would later indicate. Sullivan took a set of the scales into the field, and attempted to get accurate readings that he could translate into face colors for casting. As a check on his own, subjective vision, he also asked other observers to weigh in on color readings, but to his evident horror found that different observers reported wildly different assessments of skin color. For instance, in a related expedition to study the anthropometry of Māori and Samoan people—understood as related groups of Polynesian people relevant to his Hawaiian work—Sullivan noted that reports of lighter skin colors ranged wildly, with values ranging from 6, to 11, 12, 13, 14, and 15 on von Luschan's scale. Darker colors likewise ranged from 17, to 18, 22, 23, and 24.[22] In other words, the scale, while useful at attaching numbers to colored tiles, did little for the project of classifying race according to indexes of skin color. Indeed, for the purpose of producing racial "information" it wasn't at all clear that color scales were any better than color names. The numbers attached to von Luschan's scales were literally insignificant—that is, they signified nothing more than the color to which they were applied, which, itself, was merely self-referential.

In the end, despite his later insistence on the use of color scales as metrics for describing skin hue, it seems that Sullivan abandoned the use of any particular colorimetric information technology for his project of racial typology, and instead relied on his own judgment and that of skilled artists. He contracted with a local sculptor named Gregory Usborn to assist with casting and exhibition and his expenses for the casting project include line items for supplies such as ochre paint and shellac, and for the payment of $15 for a "painter" to help with exhibition materials.[23] It seems that to secure scientifically accurate paints, he simply visited the local dry goods store, though it's not clear what precise pigments he used. With such material assistance, he was able to forecast a happy outcome. "I shall be able to bring back about 40 busts," he wrote back to the AMNH, "and one or two full figures. It will not be wholly a race mixture exhibit but will represent the Pacific stocks of Hawaii. There are so many different kinds that we will secure only a few of each kind."[24]

The resulting casts were, indeed, lifelike and masterful. Sullivan exceeded his projected forty cast heads, and returned to New York with fifty-four cast heads, and one full-body cast figure. Each expressive face appeared to capture the moment of its creation. One man seemed distressed, his brows furrowed, barely holding his composure against the weight of the plaster on his face. A woman smirked ever so slightly, as if amused the profound weirdness of being covered with plaster by an anthropologist. Not all of the faces were painted, but those that were showed signs of meticulous attention to verisimilitude. Cast number 52, for instance—an individual noted only as "M[ale] Hawaiian"—has glittering eyes that seem almost to dart around. A slight blush to his cheeks hints at warm blood coursing beneath his skin. The slight darkening of his nose shows the effects of the sun. Usborn painted the sclera of his eyes to be slightly yellowish—a quality which Sullivan and others had remarked upon as characteristic of Polynesian people. (The precise number that Sullivan assigned to this eye color was "76.5"—though it is not clear to what scale the number referred, and anyway Sullivan noted that the color was "not quantitatively measurable.")[25] But other than this nod to "type," one senses a collapse of the anthropological artifact—the information it conveys seems more about biography and history than typology or taxonomy.

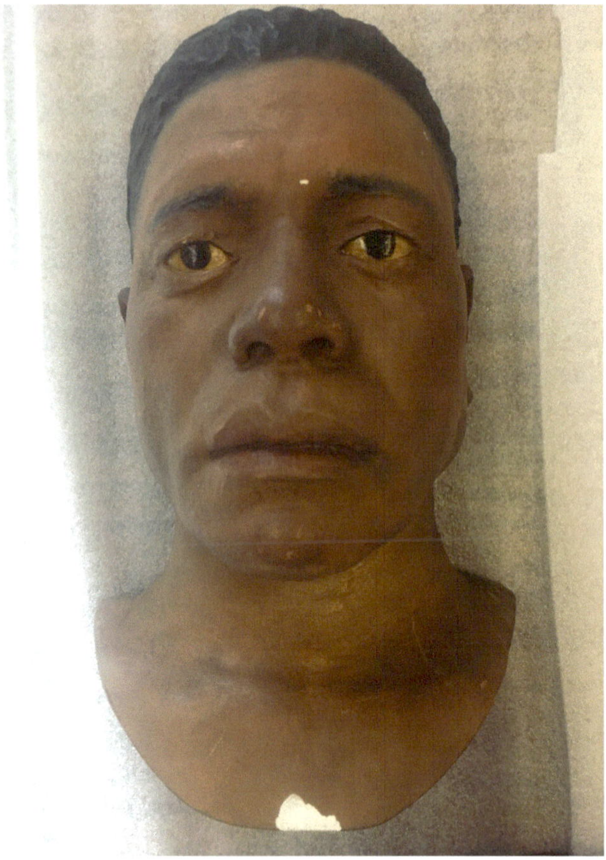

FIGURE 5.4
"Hawaiian Male," subject #54 in Louis Sullivan's collection of anthropological casts. This cast was one of many taken during 1920, as part of Sullivan's anthropological expedition to Hawaii. Courtesy of Anthropology Archives, American Museum of Natural History.

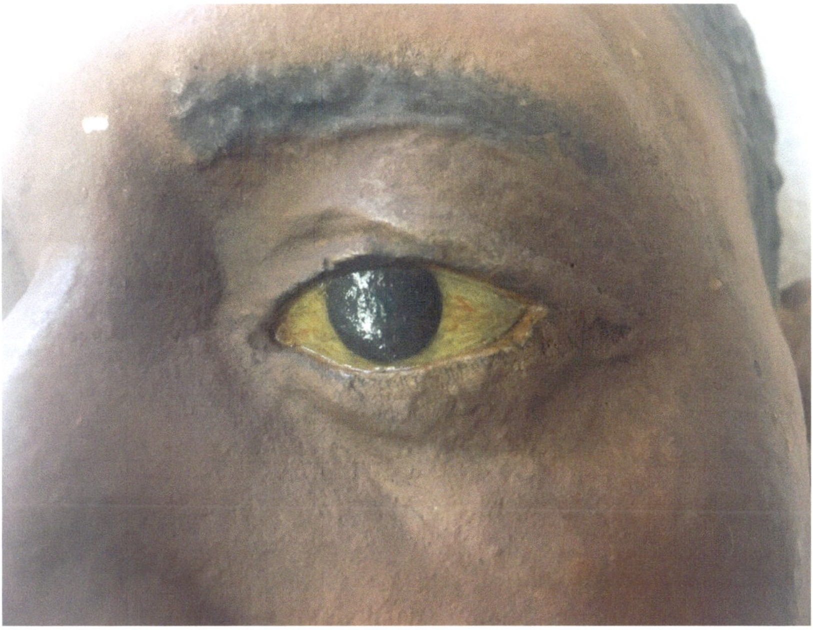

FIGURE 5.5
Close up of cast head #54, showing detailing of the eyes. Courtesy of Anthropology Archives, American Museum of Natural History.

CONCLUSION

In the hands of men like Sullivan and Osborn, the problem of race became one of informational praxis—which is to say, philosophy on the one hand, and practice on the other. In terms of philosophy, the superficial form of race was obvious: it could be captured in measurements, described in numerical terms, and forced into being through the tension of the mold and the cast. Indeed, casting was the ideal form of measurement, insofar as it provided data that came asymptotically close to actual being. In so doing, this data reified race as ontologically real in a way that was both scientifically comprehensible and socially applicable. At the same time, however, casting also flattened (so to speak) one of the most salient features of racial typography for American anthropologists—that is, skin color. The possibility of producing information about race was contingent on veridical coloration—and this could not be guaranteed by the casting.

In practical terms, Sullivan and his team attempted to navigate this philosophical conundrum through artistry, if not outright artifice. The information conveyed through coloration could be simulated by careful painting. Osborn's injunction that the "coloring of the skin" posed a nearly insurmountable problem did not stop Sullivan from attempting to give an informative account of the racial makeup of the Hawaiian people in paint and plaster. His success, of course, was ultimately equivocal.

True, Sullivan's research artifacts—ported by steamship and railroad from Hawai'i to New York—were a success at the Second International Eugenics Conference. Scientific racists the world over could admire his diligence, accuracy, and industriousness as represented in the vibrant and lifelike casts that the museum displayed. Had Sullivan not died four years later—of lung complications due to having contracted influenza during the pandemic of 1918—all agreed that he would have become a leading anthropometrist—a producer of information about the nature of race, humanity, and society.

At the same time, information can be fragile; fleeting. Even the ostensible, classificatory certainty provided by the pigmented cast could not insure lasting clarity. In the wake of Sullivan's death in 1925, Osborn's death in 1935, and the collapse of American eugenics in the face of Nazism, Sullivan's casts faded in relevance—their information lost, or rather transformed from factual certainty about racial taxonomy to an acknowledgement of the diversity of human types, and the errors of scientific racism. As recently as 2021 the Bishop Museum in Honolulu staged an exhibition entitled "(Re)Generations: Challenging Scientific Racism in Hawaii," which revisited the legacy of Sullivan's casting work. In this exhibition, Sullivan's casts—not colored, but in their unpigmented, white plaster forms—featured prominently. The point of the 2021 exhibition was precisely not to detail racial classification, but to demonstrate the ways in which the particularity of individual people and families could be subsumed by this powerful technology into an ultimately questionable representation of social reality.

At the time of this writing, a little more than a hundred years after Sullivan endeavored to catalog the racial types of Hawai'i in plaster and pigment, a great many of his successors—researchers in genetics and allied fields—are at pains to emphasize that ideas of "race" and "racial types" are figments of social imagination. The eugenics of the past has nothing to do with how researchers think about human variation today. Instead of measuring and casting, scientists who study human populations can rely on data provided by molecular biology. Instead of converting perceptual data into information about racialized color types, researchers can look at statistical data about invisible, genetic differences between groups of people. This "information," moreover, seems far removed from the work of eugenicists, with their measurements and tables and plaster and pigments. And, indeed, contemporary biomedicine resists easy comparisons with its antecedents of a century ago. Nevertheless, thinking about color as an information technology—different from, but of a species with, current genetic technologies—reminds us both of the mutability of information, and of its durability and salience.

NOTES

1. To suggest that information in this specialized sense existed before the middle of the twentieth century is, of course, to court anachronism. However, as Lida Zeitlin-Wu and Carolyn L. Kane, among others, have shown, the fantasy of color as an algorithmic, calculable, controllable entity predates the work of midcentury cyberneticians by more than fifty years. See Kane, *Chromatic Algorithms,* and Zeitlin-Wu, "Seeing by Numbers."

2. See, e.g., Gaskill, *Chromographia*; Kalba, *Color in the Age of Impressionism*; and Rossi, *The Republic of Color.*

3. Henry Fairfield Osborn to Clark Wissler, Oct. 22, 1921.

4. Thank you, Lida Zeitlin-Wu, for pointing out that "skin deep" is, indeed, the best way to qualify Sullivan's research. It is also worth noting that not all attempts to coordinate morphology with skin color were exclusively "skin deep." For example, in his 1828 *Elements of Physiology,* German physician J. F. Blumenbach attempted to match skin color with skull size and shape, which he found interesting in part for the way in which skull size suggested brain size. For that matter, it's worth mentioning that Sullivan also undertook serological studies of his subjects, though these featured far less prominently in his anthropometry.

5. See especially Simondon, *Individuation in Light of Notions of Form and Information* and *Communication et Information.* For more on Simondon, see, e.g., Barthélémy, "Individuation and Knowledge"; De Boever, *Gilbert Simondon*; Guchet, "Technology, Sociology, Humanism"; Hayward and Geoghegan, "Introduction: Catching up with Simondon"; and Woodward et al., *Gilbert Simondon.* It is worth mentioning, of course, that Simondon's attention to information vis-à-vis technology is rooted in much older, Aristotelian notions of hylomorphism.

6. On bricks, see Simondon, *Individuation,* 21–25.

7. Simondon, "L'Amplification dans les Processus d'Information," in *Communication et Information,* 159.

8. Simondon, "L'Amplification dans les Processus d'Information."

9. Greisemer, "Modeling in the Museum."

10. Fanon, *Black Skin, White Masks,* 89.

11. Indeed, one can't help but think about Fanon's expression of anger and dismay at the "epidermalization" of racial inferiority. Upon being identified solely as a "black" man, he writes, "[m]y body was given back to me sprawled out, distorted, *recolored*" (*Black Skin, White Masks,* 89; emphasis in original). Of course, Sullivan and Osborn would not have accepted their casts as "sprawled out," much less "distorted" representations of Native Hawaiians, but this is precisely Fanon's point: the "recoloring"—literally in the case of plaster casts, or more figuratively in the case of racializing humans themselves—is part and parcel with a taxonomic dissection—one that, by focusing entirely on the epidermis, inverts the entire human being.

12. Sullivan and Shapiro. *Essentials of Anthropometry,* 17.

13. Sullivan and Shapiro. *Essentials of Anthropometry,* 20.

14. Sullivan and Shapiro. *Essentials of Anthropometry,* 37.

15. Sullivan and Shapiro. *Essentials of Anthropometry,* 48–49.

16. Blackwood, "Racial Differences in Skin-Colour as Recorded by the Colour Top."

17. Osborn to Sullivan, April 15, 1920.

18. Hutton, *Portraits in Plaster,* xiii–xiv (italics added).

19. "Indian Heads Cast from Life."

20. Vaughan MacCaughey, "The Physique of the Ancient Hawaiians."

21. MacCaughey, "The Physique of the Ancient Hawaiians," 168.

22. Sullivan et al., *Marquesan Somatology with Comparative Notes on Samoa and Tonga,* 22.

23. Sullivan to Osborn, Sept 13, 1921.

24. Sullivan to Wissler, May 16, 1921.

25. Sullivan, "Summary of Somatological Characteristics of Samoans." Again, the question of whether people from different Pacific islands were distinct "races" or should all be considered "Polynesian" was up for debate. In this instance, he appears to have defaulted to generality rather than specificity.

BIBLIOGRAPHY

Barthélémy, Jean-Hugues. "Individuation and Knowledge: The 'Refutation of Idealism' in Simondon's Heritage in France." *SubStance* 41, no. 3 (2012): 60–75.

Blackwood, Beatrice. "Racial Differences in Skin-Colour as Recorded by the Colour Top." *Journal of the Royal Anthropological Institute of Great Britain and Ireland* 60 (1930): 137–168.

De Boever, Arne, ed. *Gilbert Simondon: Being and Technology*. Edinburgh: Edinburgh University Press, 2012.

Fanon, Franz. *Black Skin, White Masks*. Translated by Richard Philcox. Berkeley, CA: Grove Press, 2008 (orig. 1952).

Gaskill, Nicholas. *Chromographia: American Literature and the Modernization of Color*. Minneapolis: University of Minnesota Press, 2018.

Greisemer, James R. "Modeling in the Museum: On the Role of Remnant Models in the Work of Joseph Grinnell." *Biology and Philosophy* (1990): 3–36.

Guchet, Xavier. "Technology, Sociology, Humanism: Simondon and the Problem of the Human Sciences." *SubStance* 41, no. 3 (2012): 76–92.

Hayward, Mark, and Bernard Dionysius Geoghegan. "Introduction: Catching up with Simondon." *SubStance* 41, no. 3 (2012): 3–15.

Hutton, Laurence. *Portraits in Plaster: From the Collection of Laurence Hutton*. New York: Harper & Brothers, 1894.

"Indian Heads Cast from Life: An Interesting Plan of the American Museum of Natural History to Further Ethological Research." *Scientific American*, March 2, 1912.

Kalba, Laura Anne. *Color in the Age of Impressionism: Commerce, Technology, and Art*. University Park: Penn State University Press, 2017.

Kane, Carolyn L. *Chromatic Algorithms: Synthetic Color, Computer Art, and Aesthetics after Code*. Chicago: University of Chicago Press, 2014.

MacCaughey, Vaughan. "The Physique of the Ancient Hawaiians." *Scientific Monthly* 5, no. 2 (1917): 166–174.

Osborn, Henry Fairfield. Letter to Clark Wissler. October 22, 1921, Box: 10, Central Archives, AMNH, New York.

Osborn, Henry Fairfield. Letter to Louis Sullivan. April 15, 1920, Box: 55; Folder 13, Anthropology Archives, AMNH, New York.

Rossi, Michael. *The Republic of Color: Science, Perception, and the Making of Modern America*. Chicago: University of Chicago Press, 2019.

Simondon, Gilbert. *Communication et Information: Cours et Conférences*. Edited by Nathalie Simondon. Paris: Presses Universitaires de France, 2015.

Simondon, Gilbert. *Individuation in Light of Notions of Form and Information*. Translated by Taylor Adkins. Minneapolis: University of Minnesota Press, 2020.

Sullivan, Louis R. Letter to Henry Fairfield Osborn. September 13, 1921, Box: 55; Folder 13, Anthropology Archives, AMNH, New York.

Sullivan Louis R. Letter to Clark Wissler. May 16, 1921, Box: 55; Folder 13, Anthropology Archives, AMNH, New York.

Sullivan, Louis R. "Summary of Somatological Characteristics of Samoans." Box: 55; Folder 13, Anthropology Archives, AMNH, New York, c.1921–1922.

Sullivan, Louis R., and Harry L. Shapiro. *Essentials of Anthropometry: A Handbook for Explorers and Museum Collectors*. New York: American Museum of Natural History, 1928.

Sullivan, Louis R., Edward Smith, Craighill Handy, and Willowdean C. Handy. *Marquesan Somatology with Comparative Notes on Samoa and Tonga*. Honolulu: Bishop Museum Press, 1923.

Woodward, Ashley, Alex Murray, and Jon Roffe. *Gilbert Simondon: Being and Technology*. Edinburgh: Edinburgh University Press, 2012.

Zeitlin-Wu, Lida. "Seeing by Numbers: Color Systems and the Digitization of Perception." PhD dissertation, University of California, Berkeley, 2022.

6 ORNAMENTALISM: A FEMINIST THEORY FOR THE YELLOW WOMAN (2018)

ANNE ANLIN CHENG

For a long time now there have been two primary conceptual frameworks through which many of us conceptualize racial embodiment: Frantz Fanon's "epidermal racial schema"[1] and Hortense Spillers's "hieroglyphics of the flesh."[2] The former denaturalizes black skin as the product of a shattering white gaze; the latter has been particularly instructive in training our gaze on the black female body and the ineluctable matter of ungendered, jeopardized flesh. Yet, has the "epidermal racial schema" hardened for us into a thing of untroubled legibility? To what extent have the "hieroglyphics of the flesh" prevented us from seeing an alternative materialism of the body?

This chapter is driven by the haunting of a different kind of racialized female body whose "flesh" survives through abstract and synthetic rather than organic means and whose personhood is animated, rather than eviscerated, by aesthetic congealment. Culturally encrusted and ontologically implicated by representations, the yellow woman is persistently sexualized yet barred from sexuality, simultaneously made and unmade by the aesthetic project. *She* denotes a person but connotes a style, a naming that promises but supplants skin and flesh. Simultaneously consecrated and desecrated as an inherently aesthetic object, the yellow woman troubles the certitude of racial embodiment and jeopardizes the "fact" of yellowness, pushing us to reconsider a theory of person thingness that could accommodate the politics of a human ontology indebted to commodity, artifice, and objectness.

Although the yellow woman, like the black woman, has suffered long histories of brutal denigration and relentless prurience, her discursive construct is qualitatively different. Consider these two iconic nineteenth-century

FIGURE 6.1

Christopher Crupper Rumford, *Love and Beauty—Sartjee the Venus Hottentot* (1811). Hand-colored etching. Courtesy of Library of Congress, Prints and Photographs Division.

FIGURE 6.2
Afong Moy, *The Chinese Lady* (1835). Published by Risso and Browne (New York). Afong Moy was the first known female Chinese immigrant to the United States. The Miriam and Ira D. Wallach Division of Art, Prints and Photographs. Courtesy of New York Public Library Digital Collections.

images of racialized femininity: on one hand, Sarah "Saartjie" Baartman, the so-called Hottentot Venus, was reduced to bare flesh, what Spillers calls "the zero degree of social conceptualization";[3] on the other hand, Afong Moy, a young Chinese woman imported by the Carne brothers to tour major U.S. cities in the 1830–1850s as a living museum tableau and known simply as The Chinese Lady, offered a scopic pleasure that centered on her textual thickness: her material, synthetic affinities. Her appeal does not derive from her naked flesh but from her decorative (and projected ontological) sameness to the silk, damask, mahogany, and ceramics alongside which she sits. While primitivism rehearses the rhetoric of ineluctable flesh, Orientalism, by contrast, relies on a decorative grammar, a fantasmatic corporeal syntax that is artificial and layered. Where black femininity is "vestibular"/ bare flesh/weighted, Asiatic femininity is ornamental/surface/portable.[4]

The point here is not to posit a naturalized difference between Africanist and Asiatic femininities—indeed, my argument will insist on how modes of racialized representations mean to index racial difference but are in fact wholly promiscuous in application—but I do want to underscore here the specificity of a racial imaginary that has been at once pervasive and yet taken for granted. The tying of ornamental artifice to Asiatic femininity in Euro-American visual and literary cultures is ancient and enduring, reaching as far back as Plato, through the writings of Marco Polo in the thirteenth century, the novels of Joris-Karl Huysman and Oscar Wilde, the visual expressions of art nouveau, French symbolism, American rococo, all the way up to wide-ranging iterations in the twentieth and twenty-first centuries. Asiatic femininity is, above all, a style. As such, it claims specificity but lends itself to transferability. It designates a racial category but can be applied to different racial subjects. It can be enlisted by those wielding power and, more disturbingly, by those deprived of it.

It is astounding that so little has been done to consider the production of this particular form of personhood. This vast and tenacious history of Oriental female objectification is refracted through the lenses of commodity and sexual fetishism. Yet, the truth is, we barely know how to process the political, racial, and ontic complications of confronting a human figure that emerges *as* and *through* ornament. Neither mere flesh nor mere thing, she/it applies tremendous pressures on politically treasured notions such as agency, feminist "enfleshment," and human ontology.

FIGURE 6.3
Gustav Klimt, *Woman with Fan* (1917–1918).

When it comes to this exquisitely defiled subject, caught in the haunting convergence between aesthetic value and material abuse, it is not enough to say that she serves as a tool for power or its sublimation, or that she offers nothing but the congealment of commodification. Both statements are undoubtedly true, but neither is sufficient to address the profound, intrinsic, and often unspeakable ways in which the subjugator and the subjugated find themselves or meet one another in and through aesthetic form. What follows then is not just a story about how we use people as things, or how things dictate our uses of them, but a drama involving a deeper, stranger, more intricate, and more ineffable fusion between thingness and personhood.

FIGURE 6.4
George F. Cannons, *Photograph of Anna May Wong* (1933). Image courtesy of Luigi Tiriticco / "Chronorama 10" and the Creative Commons.

Almost two hundred years after New Yorkers flocked to the American Museum in New York to witness the living tableau of The Chinese Lady, throngs from all over the world lined up in Manhattan to see another spectacular display of Asiatic femininity qua ornament: this time, the 2015 exhibition at the Metropolitan Museum of Art entitled *China: Through the Looking Glass*. The exhibition featured more than 140 examples of Euro-American haute couture and avant-garde ready-to-wear from fashion luminaries such as Paul Poiret, Yves Saint Laurent, John Galliano, Alexander McQueen, and more. Interspersed among these resplendent sartorial creations are scattered decontextualized Chinese and Japanese artifacts from the museum's Asian Art Collection. An eighteenth-century yellow silk satin imperial robe shadows a sequined yellow silk satin evening gown by Tom Ford for Yves Saint Laurent; a Qing Dynasty snuff bottle and a twelfth-century kimono echo a Cartier perfume flask; an early fifteenth-century cobalt blue dragon jar flanks a Roberto Cavalli creation in blue and white.

This sumptuous collection rehearses for the twenty-first-century audience the basic tenets of nineteenth-century Orientalism: that opulence and sensuality are the signature components of Asiatic character; that Asia is always ancient, excessive, feminine, available, and decadent; that material consumption promises cultural possession; that there is no room in the Orientalist imagination for national, ethnic, or historical specificities. Most of all, the show reminds us that *China* (conflated throughout the show with Asia at large) equals *ornament*. The Metropolitan had accomplished quite an ornamental feat of its own: a byzantine build-out replete with intricate corridors and concentric spaces; large decorative arches leading to secret, enfolded rooms; a tall, transparent, plastinated bamboo forest—all nestled deep inside the museum's bowels, filling up both the Anna Wintour Costume Center in the basement and the Chinese galleries on the second floor and claiming a large portion of the repurposed Egyptian Gallery. Through this labyrinth, one would, for instance, turn and suddenly find oneself in a hushed "garden," standing on a meditation bridge overlooking a large glistening pond encased by winding corridors. No natural light disturbs this dreamlike, highly man-made environment. The lacquer pond glimmers but does not flow. In this beautiful graveyard, Eastern nature appears wholly continuous with Eastern artifice.

From the get-go we realize the limits of the critical framework of Saidean *Orientalism*, as the exhibition itself has already coopted the term as its own internal critique and alibi. A large sign greets the visitors as they enter through a set of imposing two-story-tall red lacquer doors:

EMPIRE OF SIGNS

> For the designers in this exhibition, China represents a land of free-floating symbols, a land where postmodernity finds its natural expression. Like Marco Polo or Gulliver, they are itinerant travelers to another country, reflecting on its artistic and cultural traditions as an exoticized extension of their own . . . When quoting

FIGURE 6.5
The Metropolitan Museum of New York. Exhibition hall of *China: Through the Looking Glass* (2015). Photograph by Anne Anlin Cheng.

> Chinese artifacts or costumes, these designers are not reproducing literal copies or accurate facsimiles. Rather, they reinterpret them through seemingly paradoxical postmodern constructions.[5]

The show recuses itself from the burden of authenticity even as its mise-en-scène fetishizes the artifactually and culturally real.[6] Invoking and recursively enacting Roland Barthes's famously cool dismissal of the notion of (Japanese) cultural authenticity—"to me the Orient is a matter of indifference"—this exhibit disaggregates aesthetic pleasure from politics and reclaims postmodernism as cure to Orientalism, performing exactly what Hal Foster has referred to as capitalism's revenge on postmodernism.[7]

There is much criticism one can offer on the grounds of racial appropriation, inauthenticity, commercialism, and neoliberal bad faith.[8] Yet we should be less surprised by the tenacity of this racial imaginary and more concerned with the limits of our response to it. For a long time now the concepts of Asian authenticity and of Orientalist commodification have remained our only safeguards against the vice of racist consumption. But neither the longing for the former nor the allegation of the latter can address the complex relations between appropriation and susceptibility, or between embodiment and style. Moral outrage in the face of consumption or fetishization, however warranted, cannot address or relieve the truly striking, idiosyncratic, and passionate exchange between thingness and personhood into which a display like this draws us. The exhibition's visual and sensorial extravagance—its dizzying invitation for visitors to lose themselves in the fastidiousness of extreme aestheticism; the completeness of severe, scrupulous details; the sensorium of textiles and materials; the seductive dynamism of vivacious inanimateness and synthetic pleasures—all work to facilitate that slide between things and persons, an erotic and erratic plunge that both preconditions the making of Chinese/Asiatic femininity and renders inadequate binary critiques when it comes to this kind of racial objection. Thus while Saidean Orientalism and the Foucauldian critique that it embodies help us identify symptoms and locate political culpability, neither can address the profound, queasily seductive entanglement between organic corporeality and aesthetic abstraction imputed to yellow womanhood. . . .

So how do we begin to think about racialized bodies that remain insistently synthetic and artificial? How do we take seriously the life of a subject who lives as an object, and how do we do so without either resigning that figure to the annals of commodity fetishism or assigning it to the sinecure of reassuring corporeality?

The challenge here is to negotiate between, without abandoning, the very human stakes of deanimated persons and the very material history of animated things. Of the many "enigmatic objects" (a term used by the exhibition) displayed in this extensive exhibition, the most mesmerizing and confounding one is surely the specter of the yellow woman, synecdochized through faceless and at times headless mannequins and metonymized through luxuriantly sensuous fabrics.[9] What is this charismatic sensorial presence that does not require a biological body or nature? Quite the opposite, Asiatic femininity is radiantly reproduced through inorganic and insensate mediums. Consider, for example, the evocation of Asiatic femininity via ceramics.

In a series of rooms devoted to the theme of blue willow, an imitation Chinese china pattern made popular by Thomas Minton in the 1790s, we find a grouping of blue and cream silk gowns designed by contemporary designers such as Cavalli and McQueen. We can dismiss this association between Asiatic femininity and Chinese ceramic as yet another Orientalist cliché. But if we do so, we would miss a much more intricate and intriguing proposition: the affinity among racialization, imagined personhood, and synthetic invention.

I am interested in how the primacy of the ornament, as artifact and as gesture, in this show acts as a crucial, transitional node through which the human is simultaneously invoked and displaced and how racialization functions in this transfer. I want to think through, rather than shy away from, that intractable intimacy between being a person and being a thing. Let us then build a different historiography of raced bodies: one constructed through fabrics, ornaments, and "skins" that never enjoyed the fantasy of organicity; one populated by nonsubjects who endure *as* ornamental appendages. Let us substitute *ornament* for *flesh* as the germinal matter for the making of racialized gender. Let us, in short, formulate a feminist theory of and for the yellow woman.

To begin to do so, I offer *ornamentalism* as a conceptual lens through which to attend to the afterlife of a racialized and aestheticized object that remains very much an object, even as the human stakes remain chillingly high.[10] At the most basic level, ornamentalism, with its almost homophonic echo of Orientalism, names for me the critically conjoined presences of the oriental, the feminine, and the decorative. But more than naming a symptom, it identifies a process whereby personhood is conceived and suggested (legally, materially, and imaginatively) through ornamental gestures: gestures that speak through the minute, the sartorial, the prosthetic, and the decorative. Ornamentalism is thus an admittedly rather inelegant word that describes a very elegant (that is, seamless) alchemy between the borrowing properties of thingness and personhood. . . .

The ornament and the ornamental gesture in *China: Through the Looking Glass*, far from being incidental or merely decorative, enacts critical labor through which powerful ideas of personhood, race, and objectness transfer. Let us take for example the intimacy between Asiatic femininity and cool ceramics. Recent scholarship in the area of material culture has revealed the complex history of Chinese porcelain: its importance in early global imperial trade; its role in spurring European technological invention and decorative design; its impact on growing economic, social, and cultural values in Denmark, Germany, France, and England and its American colonies.[11] To this richly documented history, I would add the wrought/ fraught intimacy between this "'White Gold'" and the making of yellow flesh.[12] For more than economic or social values, Chinese porcelain personifies a set of affective and somatic values forged out of the kiln of what Gordon Change aptly called the centuries-old "fateful ties" between China and the

West.[13] Connoting old-world exoticism and modern material, civilization and decadence, durability and fragility, heat and coolness, imperviousness and susceptibility, Chinese ceramic was thought to embody characteristics that are mapped onto Asiatic persons and bodies.[14] And the fates of Chinese bodies and Chinese porcelain run parallel. When this much valued and coveted material came to represent the precarity of a system of Western wealth based on importing novel Eastern goods, this prized object started to lose its radiance, along with other things Asiatic. As Euro-American acquisitiveness began to run in excess of what it could offer China in return, the early romance with china/China began to deteriorate in a breakdown that left putative and lasting traces in American law and economic policy (from U.S. foreign policy and trade agreements in the eighteenth century to discriminating immigration laws in the nineteenth and twentieth centuries); in the American popular imagination (think of the fate of Chinese porcelain itself, its fall from denoting precious goods to connoting tacky crockery); and, finally, for our discussion, on the bodies of Asiatic women. In 1996, the *New York Times* has this to say about the Chinese women's gymnastics team at the Summer Olympics: "The Chinese remain the world's most erratic top gymnasts, and today, like many a Ming vase, their routines looked lovely but had cracks in several places."[15]

Thus more than exemplifying an incidental decorative motif or revealing the limits of the imperial imagination, the citation—indeed, the embodiment objecthood—of Chinese porcelain in this exhibition, regardless of curatorial intentions, revivifies this long, expansive history of human imbrication with racialized and manufactured materials, fueling the fraught amalgamation between inorganic commodity and Asiatic female flesh. We might say that the ubiquitous presence of fine Chinese porcelain throughout the show generates a specific epidermal schema of its own.

This creation by McQueen for his autumn/winter 2011–2012 collection offers a play in simulation and contrast, juxtaposing the inorganic and the organic, the insensate and the sensorial, the hard and the soft. Most of all, it invokes a particular vision of the racialized female body, one that sustains these contradictions. The eruption of the fluffy, layered extravagance that is the skirt, at first glance, looks like supple feathers but turns out to be shredded silk organza. The exposed underskirt, suggestive of layers of artificial skins turned inside out, at once beautiful and violent, is then offset by the startlingly weight of the torso, which on closer inspection reveals

(a)

FIGURE 6.6
McQueen and Burton, evening dress. From *China: Through the Looking Glass*. Photographs by Anne Anlin Cheng.

(b)

FIGURE 6.6 (Continued)

the bodice to be made not out of the proverbial bone but of hundreds of reconstituted shards of blue and white porcelain: porcelain as flesh and flesh as porcelain.

Dare we say it? Ornament becomes—*is*—flesh for Asian American female personhood. Commodification and fetishization, the dominant critical paradigms we have for understanding representations of racialized femininity, simply do not ask the harder question of what *being* is at the interface of ontology and objectness. Here Chinese femininity is not only more and less than human but also man-made; not only assembled but also reassembled. This reassemblage, by virtue of its materiality, memorializes the practice of ornamentalism and the techniques of race making. This image of the flexible yet brittle body reminds us that this aesthetic discourse is fastened to a fractured history of craft, labor, and bodies in transit. If Eric Hayot has traced for us the persistent Euro-American conceptualization of the Chinese male body as infinitely and stoically capable of sustaining pain and suffering, an association that led to the image of the coolie as at once animal and superhuman, an ideal laborer, then we are tracing here a Chinese female version of this discursive formation—a smooth beauty that bears the lines of its own wreckage, a delicacy that is also impermeable and insensate.[16] The dream of the yellow woman is thus really a dream about the inorganic. The yellow woman is an, if not *the*, original cyborg. From the thingliness of Anna May Wong (whom Walter Benjamin once referred to as a "moon" and a "bowl") to Nany Kwan's trademark Porcelain Skin, for the Asian American woman, porcelain has always been flesh *and* not flesh.[17]

Going back to our blue willow room, we can now see a broader fascination with the inhuman human structuring the grammar of this stage. The McQueen gown is placed next to a Cavalli dress, and they in turn stand next to and echo an early fifteenth-century Chinese porcelain jar painted with a cobalt blue dragon—urn as beauty as death as dress as corporeal gesture. The sensorial and somatic realization of Asiatic femininity fulfills itself, paradoxically, through the forms of these empty-yet-orthopedic vessels. There is no ordinary flesh here, by which I do not mean the obvious (that there are no real bodies on display) but that this show is palpably not that interested in the human, much less the woman. These garments do not need the human; indeed, the human would disrupt their composure. In fact, we are looking at a peculiar form of anthropomorphism or prosopopoeia whereby the human is being used to recall objectness rather

FIGURE 6.7
Roberto Cavalli. Evening dress, blue and white silk satin (2005–2006), as featured in *China: Through the Looking Glass*, 2015, Metropolitan Museum of Art, New York.

than vice versa. The objects on display, from frocks to vases, invite neither wearability nor usage. Instead of objects that function as appendages to the human (as one would expect fashion and furnishing to do), what we find here instead are objects that reference other objects.

In this room, the human is the ornamental gesture. And the ornamental in turn acquires its ancillary human aura by being "Chinese." In other words, beauty here comes from the primacy of the object; the human or the anthropomorphic is the incidental alibi for, or an afterthought of, relishing this pure objectness. Crucially, at the same time, what renders this pure objectness legible as such is precisely the invocation of racial difference. While Orientalism is about turning persons into things that can be possessed and dominated, ornamentalism is about a fantasy of turning things into persons through the conduit of racial meaning in order, paradoxically, to allow us to abandon our humanness. . . .

To point to this enchantment of the inhuman is not to rehearse the problem of objectification or to downplay the issue of race but to point to a provocative dilemma about how the object preconditions, rather than being the product of, the human figure—a modern crisis that Asiatic femininity personifies. This is why ornamentalism is not only an object of feminist critique but can be also a vector of feminism. The eradication of the human, even as raced and gendered corporeality are being imputed, compels a reconceptualization of feminized, racialized flesh . . . What happens to our notions of the subject when carnality is cultivated not out of flesh but of its fusion with inorganic matter? What happens when we accept that style (mediated through yet detached from a racial referent) may be not simply the excess or the opposite of ontology but a precondition for embodiment, an insight that challenges the very foundation of the category of the human? What is at stake here is not just the objectification of people but how that objectification opens up a constitutive estrangement within the articulation of proper personhood and life.

More than denoting an aesthetic practice and a technology of power then, ornamentalism—the forging of the sense of personness through artificial and prosthetic extensions—provides an allegory for the crisis of personhood that the modern ideal of an integrated, organic, individual person was meant to alleviate. If we think of the Anglo conceptualization of modern personhood as indebted to an Enlightenment notion of natural and integral bodies (posited through John Locke, Thomas Hobbes, René Descartes, Montesquieu, William Blackstone)—the idea, for example, that the person is a living, organic, and organized human body "such as the God of nature formed us"—then we are tracing here another kind of body that confronts us: one that not only poses a challenge to this ideal but also insists on the primacy of aggregated objectness in the experience of the human.[18]

As if in answer to this bracing realization, we find something else in the blue willow assembly. Here, aesthetic congealment has grown into full-bodied edifice. The to-be-used Chinese female body seems to have petrified into domestic and collectible things (teapots, cups, plates) whose value now resides in their aggressive uselessness. At the same time, this congealed and fractured domesticity, offering repurposed purposelessness, transcends its own mundanity to lay claim to art. Made by contemporary artist Li Xiaofeng (born 1965), this piece is clearly not human, but it is also not entirely a thing. The weight (of material, history, domesticity, femininity) implies petrification, but the form suggests flight. Is this dress or armor? Is it winged victory or the madwoman in the attic? Is it a tribute to monumentality or a concession to the mundane? What has died here—the human or the ornament? And, finally, has the human outmoded itself or has the object outrun the human demand?

We cannot read this piece from the Chinese artist as a rebuke of the Western commercial designers with whom he shares the space because the work is itself a meditation on troubled authenticity.[19] The sculpture *Beijing Memory No. 5* (2009) by Li was part of a series known for utilizing ceramics excavated from authenticated archaeological sites throughout China. The museum catalogue attributes the ceramic fragments to the Qing Dynasty (1644–1912), commonly known as the last great Chinese dynastic empire, suggesting that the memory being recomposed (or shattered) here is attached to the memory of a lost imperial China and making this piece something of an exercise in what Rey Chow calls the "ethnic detail."[20] Chow notes with her usual acuity that the Chinese detail is always already an *ethnic* detail—even for the Chinese. For Chow, this is so because of the pastness assigned to the Chinese by Anglo modernity; consequently, the Chinese detail functions for Western modernity as a kind of *Nachträglichkeit*. The historicity of Li's version of the Chinese detail, however, has been subjected to much manipulation. On a closer look, among the rubble that is the body of this ceramic woman, we discern a scattering of some intact Chinese ideograms. Many are out of context, and several are positioned upside down (an ironic statement of value? a jab at the old joke about Chinese illegibility for Western viewers?), with words like "precious," "tea," "superior." Others offer seemingly precise self-authentication with stamped reign dates, such as "Xuande Reign of the Great Ming Dynasty (1426–1435)," purporting to indicate a royal workshop.

The shards thus speak to us and date themselves even further back in time, to the Ming Dynasty, prior to the Qing. But we cannot rest easy in their promise of more precious pastness because these signatures of provenance, both inside and outside of the artwork, only serve to underscore the homelessness of these fragments. Given the fact that fairly early on the practice of such temporal inscription on objects degraded into anything within the spectral range of affectation to forgery, this kind of designation is likely to be more misleading than not. And since an earlier date would signal a

FIGURE 6.8

Li Xiaofeng, *Beijing Memory No. 5* (2009), Qing period shards. From *China: Through the Looking Glass*. Photograph by Anne Anlin Cheng.

FIGURE 6.9
Detail of Li Xiaofeng, *Beijing Memory No. 5*. Photograph by Anne Anlin Cheng.

grander affectation and the Xuande era was exactly when such inscriptions first became common, it is probable that these fragments may be anything but Xuande.

Even more intriguing than this little drama/trauma of authenticity, more-over, is the possibility that this twenty-first-century Chinese porcelain body made for an international audience may be seeking to petrify itself as a gambit for its continued relevance. Just as *Beijing Memory No. 5* invokes a vanished China only to unsettle the very thing it claims has disappeared, the sculpture also summons via those calcified, fragmented ceramics the memory of yellow female flesh only to replace it with a more insistent, inorganic presence. This feels less like nostalgia for yellow female flesh than like a consecration of its fossilized (raced and gendered) afterlife. This is flesh congealed into porcelain *and* porcelain invoking the possibility of flesh. *No. 5*'s tensile balance between formal preoccupation and suffocating mass asks us to consider the simultaneously dematerializing *and* materializing processes through which raced female bodies come to matter. It insists that commodity is *also* art, body, thingness, memory, and its evaporation. It instantiates the salvage, the reassemblage, and the subjunctiveness behind Asiatic, female corporeality. If the yellow woman has always been simultaneously embodied and erased through ornamental objectness, then this piece asks what life is or could be after such devastation.

Sigmund Freud tells us that one of the most unsettling effects for human ontology is to be confronted by a machine that comes to life.[21] Here, with *No. 5*, what is uncanny is that the machine refuses to come to life *and*, in its lifelessness, imagines what life might have been. And it is in this very *might have been* that we experience the prospective and prosthetic quality of our ontology. In other words, *No. 5* does not give us a memory of something; it *is* memory: the layered encrustations of absent but imagined lives. (It gives us not the human hand but the empty handle.) . . . More than memorializing bodies that may otherwise not be remembered, Li's porcelain woman explores what it would mean to instantiate through excess materiality the dematerialized nonbody. It marks a kind of third nature, one that surrounds and approximates the human and one that manages to survive despite or through commodification. The perihumanity of Asiatic femininity (that is, something at once inside and outside of civilizational first principles) is why she is often a figure enlisted to represent contemporary apprehensions about more-than-human entanglements.

The aesthetic language—the entanglement between the animate and the inanimate—with which the yellow woman has been infused draws from and sustains a dynamic but disturbing principle of artificial life that, rather than being peripheral to, intensifies and haunts modernism itself. That is, the Asiatic ornamental object person is often seen as opposite to modernity, but it actually contains a forgotten genealogy about the coming together of life and nonlife, labor and style, that conditions the modern human conceit. It suggests a different genealogy of modern personhood: one that is not traceable to an ideal of a biological and organized body bequeathed from a long line of Enlightenment thinkers, but one that is peculiarly inorganic, aggregated, and non-European. This synthetic being, relegated to the margins of modernity and discounted precisely as a nonperson, holds the key to what I see as the inorganic animating the heart of the modern organic subject. She/it brings into view an alternative form of life, not at the site of the free and individualist modern subject, but, contrarily, at the encrusted edges and crevices of defiled, ornamented bodies. . . .

We have at last arrived at an understanding of ornamentalism, not as a theory of thingness but as theory about the profound imbrication of things and persons. It tells a different story about a different kind of flesh in the history of race making. We can now return to the juxtaposition between Asiatic and Africanist femininity with which this essay began and reconsider the paths of their divergence. I started this essay by distinguishing the wounded, flesh-laden black body from the immaculate, synthetic, ornamental yellow body. Our journey through the alternative logic of a racialized embodiment that is also not necessarily enfleshment, however, suggests that in the end there may be a potential for ornamentalism to speak also to Africanist female enfleshment. The body of labor (sexual, reproductive, economic) exemplified by the black female body—ungendered and excluded from the realms of kinship, state, and aesthetic value—is nonetheless *not* wholly alien to the practice and afterlife of ornamentalism. Or to put it differently, ornamentalism may help elucidate for the black woman a set of different though related issues about the fraught con- vergence of material violence and aesthetic congealment. . . .

From the divergence between black flesh and yellow ornament, we have arrived at this convergence: flesh that passed through objecthood needs "ornament" to get back to itself. When it comes to racialized flesh that has been mortified, ornamentalism points us not only to a history of disciplinary usage (where you mistake someone for something for invidious gains) but also to what it might take to reconceptualize personhood for unmade persons. If feminist scholars have been committed to the flesh in order to undo the taxonomy of gender, then ornamentalism points us toward a consideration of object life that not only undoes but suspends the taxonomy of the human. In our eagerness not to abandon the flesh, we have not been willing to attend to life's a priori enmeshment with nonlife. Instead of considering what it means for a person to have been turned into a thing with an implicit nostalgia for that lost subject, I am suggesting that we must also consider the reversed process whereby things have been made into persons, thereby revealing the fundamental logic of abstracted decoration that constitutes the category of personhood in the first place.

Ornamentalism identifies both an epistemology and its fugitive meanings—both instrumentality and unexpected opportunities. It is tied to the practice and aesthetics of Orientalism, but it also offers a critical framework beyond the assumed racial categories and periodization implied by terms such as Orientalism, primitivism, and modernism. By opening up a broader and historically deeper set of inquiries about how the aesthetic entails the political and how the political entails the aesthetic, the coercion and the enchantment of ornamentalism allow the superfluous and the not-living that are integral parts of the human to come into view. It is precisely at the interface between ontology and objectness, animated by the ornament, that we are most compelled to confront the horizons and the limits of the politics of personhood.

NOTES

Chapter 6, "Ornamentalism: A Feminist Theory for the Yellow Woman" by Anne Anlin Cheng originally published in *Critical Inquiry*, Volume 44, number 3, © 2018. Used with permission of The University of Chicago Press; permission conveyed through Copyright Clearance Center, Inc.

The full version of this article was published in 2018 in *Critical Inquiry* (volume 44, number 3).

1. Fanon, *Black Skin, White Masks*, 92.

2. Spillers, "Mama's Baby, Papa's Maybe: An American Grammar Book," 61; hereafter abbreviated "MB."

3. The focus on Baartman's flesh extends well beyond her death; as we know, her reproductive organs, dissected by Georges Cuvier, were on display at the Musée de

L'Homme in Paris until as recently as 2002. Spillers famously made the distinction between "body" and "flesh," the latter being pure matter that exists outside of any kinship or state apparatus even as it is enlisted to serve both: "In that sense, before the 'body' there is the 'flesh,' that zero degree of social conceptualization" ("MB," 61).

4. "MB," 73. Spillers also famously conceptualizes black female flesh in spatial terms; as the black female is barred from crossing the symbolic threshold into personification, she is both stuck in and a mechanical transit for the threshold dividing the human and the not human, rendering her "vestibular" to culture.

5. From the exhibition *China: Through the Looking Glass*, Metropolitan Museum of Modern Art, May 7–September 7, 2015

6. As the exhibition overview on the museum website states: "This exhibition explores . . . how China has fueled the fashionable imagination for centuries. In this collaboration between The Costume Institute and the Department of Asian Art, high fashion is juxtaposed with Chinese costumes, paintings, porcelains, and other art, including films, to reveal enchanting reflections of Chinese imagery."

7. Barthes, *Empire of Signs*, 3. See also Foster, "Postmodernism in Parallax."

8. There was much public protest. *The New York Times* published a scathing review, offering even a history lesson about the so-called China trade and its legacy for American imperialism; see Cotter, "In 'China: Through the Looking Glass,' Eastern Culture Meets Western Fashion." The Asian American Arts Centre wrote a letter of protest to the Metropolitan Museum; see Lee, "China: Through the Looking Glass—An Open Letter."

9. "Exhibition Overview," www.metmuseum.org/exhibitions/listings/2015/china-through-the-looking-glass.

10. The term *ornamentalism* has surfaced among art historians and those engaged in aesthetic philosophy to refer to the deployment of ornament for decorative purposes, especially when done in excess, but I wish to recall it precisely because of what has gone completely unheard in its previous iterations: its suggestive and almost-homophonic entanglement with Said's deployment of Orientalism. . . . While scholars have noted how gendered the discourse of the ornament in the decorative arts has been, much less attention has been focused on how deeply racialized (indeed, how Orientalized) this history has been as well. It is my sense that retrieving the Oriental logic of ornamentation, and vice versa, will alter some of the most fundamental terms behind how we think about race and personhood, as well as impact larger concepts such as modernism and Orientalism. See Schor, *Reading in Detail* and Wigley, *White Walls, Designer Dresses*.

11. See Anderson, *Mahogany;* Cavanaugh and Yonan, eds., *The Cultural Aesthetics of Eighteenth-Century Porcelain;* Dobie, *Trading Places*; Yang, *Performing China*.

12. "With the exceptions of tapestry and silver, no artistic medium was more coveted or valuable [than porcelain]." Its "cultural associations" with "exotic lands"

made porcelain "a semi-magical material, a 'White Gold'" (Cavanaugh and Yonman, "Introduction," in *The Cultural Aesthetics of Eighteenth-Century Porcelain,* 4, 3, 4).

13. See Chang, *Fateful Ties.*

14. A noted characteristic of true fine *kaolin* Chinese porcelain is that it can sustain tremendous heat while still being receptive to and sustaining brilliant, clear colors.

15. Clarey, "Atlanta: Day 3," l.

16. See Hayot, *The Hypothetical Mandarin.*

17. Quoted in Lim, "Speaking German Like Nobody's Business."

18. Blackstone, *Commentaries on the Laws of England, 1765–1779.* For a study of ornamentalism in law—that is, the formative role of Asiatic female ornamentation in nineteenth-century juridical ideas about legal personhood—see Cheng, "Law and Ornament."

19. Li is himself an equally complicated figure who works in China, Hong Kong, and the US and whose work traverses the realms of art, commerce, and kitsch. In 2010, for example, he collaborated with Lacoste to create a collection of polo shirts.

20. In speaking about the ornamental details in the films of Zhang Zimou, Chow observes: "Like Barthe's loquacious lion, the ethnic details in these films are not there simply to 'mean' themselves; rather, they are there for a second order of articulation. They are there to signify 'I am an ethnic detail; I am feudal China'" (*Primitive Passion,* 145).

21. See Freud, "The Uncanny" (1919). It is beyond the scope of this chapter, but there is much more that can be said about ornamentalism and the racial logic of the machine, especially in relation to contemporary debates about technology, the human, and the ethics of artificial intelligence.

BIBLIOGRAPHY

Anderson, Jennifer L. *Mahogany: The Costs of Luxury in Early America.* Cambridge, MA: Harvard University Press, 2012.

Barthes, Roland. *Empire of Signs.* Translated by Richard Howard. New York: Hill and Wang, 1983.

Blackstone, William. *Commentaries on the Laws of England, 1765–1779,* 15th ed. London: T. Caddell and W. Davies, 1765.

Cavanaugh, Alden, and Michael E. Yonan, eds. *The Cultural Aesthetics of Eighteenth Century Porcelain.* Surrey, UK: Routledge, 2010.

Chang, Gordon H. *Fateful Ties: A History of America's Preoccupation with China.* Cambridge, MA: Harvard University Press, 2015.

Cheng, Anne Anlin. "Law and Ornament." In *New Directions in Law and Literature,* edited by Elizabeth S. Anker and Bernadette Meyler, 229–251. New York: Oxford University Press, 2017.

Chow, Rey. *Primitive Passion: Visuality, Sexuality, Ethnography, and Contemporary Chinese Cinema.* New York: Columbia University Press, 1995.

Clarey, Christopher. "Atlanta: Day 3—Gymnastics; Miller Gives United States High Hopes for a Gold." *The New York Times,* July 22, 1996. www.nytimes.com/1996/07/22/sports/atlanta-day-3-gymnastics-miller-gives-united-states-high-hopes-for-a-gold.html.

Cotter, Holland. "In 'China: Through the Looking Glass,' Eastern Culture Meets Western Fashion." *The New York Times,* May 7, 2015. www.nytimes.com/2015/05/08/arts/design/review-in-china-through-the-looking-glass-eastern-culture-meets-western-fashion.html?_r=0.

Dobie, Madeleine. *Trading Places: Colonization and Slavery in Eighteenth-Century French Culture.* Ithaca, NY: Cornell University Press, 2010.

Fanon, Frantz. *Black Skin, White Masks.* Translated by Richard Philcox. New York: Grove Press, 2008.

Foster, Hal. "Postmodernism in Parallax." *October* 63 (Winter 1993): 3–20.

Freud, Sigmund. "The Uncanny" (1919). In *The Standard Edition of the Complete Psychological Works of Sigmund Freud,* vol. 17, translated and edited by James Strachey, 219–252. London: The Hogart Press, 1955.

Hayot, Eric. *The Hypothetical Mandarin: Sympathy, Modernity, and Chinese Pain.* New York: Oxford University Press, 2009.

Lee, Robert. "China: Through the Looking Glass—An Open Letter." Asian American Arts Centre. artspiral.blogspot.com/2015/07/china-through-looking-glass-open-letter_20.html.

Lim, Shirley Jennifer. "'Speaking German like Nobody's Business': Anna May Wong, Walter Benjamin, and the Possibilities of Asian America Cosmopolitanism." *Journal of Transnational America Studies* 4, no. 1 (2012): 1–17. https://escholarship.org/uc/item/16v4g0b1.

Metropolitan Museum of Art. "Exhibition Overview: China: Through the Looking Glass." http://www.metmuseum.org/exhibitions/listings/2015/chinathrough-the-looking-glass.

Said, Edward W. *Orientalism.* New York: Vintage Books, 1979.

Schor, Naomi. *Reading in Detail: Aesthetics and the Feminine.* New York: Routledge, 2007.

Spillers, Hortense J. "Mama's Baby, Papa's Maybe: An American Grammar Book." In *The Black Feminist Reader,* edited by Joy James and T. Denean Sharpley-Whiting, 57–86. Malden, MA: Blackwell Publishers, 2000.

Wigley, Mark. *White Walls, Designer Dresses: The Fashioning of Modern Architecture.* Cambridge, MA: MIT Press, 2001.

Yang, Chi-Ming. *Performing China: Virtue, Commerce, and Orientalism in Eighteenth Century England, 1660–1760.* Baltimore: Johns Hopkins University Press, 2011.

II IMAGES

INTRODUCTION TO PART II: LIGHTING FOR WHITENESS IN MOVING IMAGE MEDIA

With the advent of the twentieth century, we find ourselves in the heart of industrialization. The immersion into a broader culture of radically new forms of mechanization—from cinema to cars, clocks, trains, and telephones—left many uncertain about the adoption of new behaviors and perceptual experiences both at home and at work. In the creative arts, these fears and anxieties were expressed on the canvas as wild brushstrokes in eye-popping, vivid colors and on the silver screen by way of rapid editing and magical montages. Yet in industry, there was also a marked drive to codify color across media into a globally exchangeable, reproducible standard, and this obsession with standardization soon began to encroach into visual culture. While color in this optical paradigm in many ways perpetuated and reproduced—albeit in mass-media form—the same Enlightenment framework that saw it as supplementary to human experience and separated it from truth and Form (i.e., the distinction between line, or *disegno,* versus *colore* in Renaissance painting), color was nevertheless a key aspect of emergent visual technologies throughout the twentieth century.[1]

With Isaac Newton's *Opticks* (1704), seeing and knowing the world based on visible color became a cornerstone of optics, and later psychophysics, in the nineteenth century. With predecessors including, but not limited to, Goethe's *Zür Farbenlehre* (*Theory of Colors*, 1810), James Clerk Maxwell's protocolor photographs (1856), and Eugène Chevreul's *The Principles of Harmony and Contrast of Colours* ([1828], 1839), ordering systems for color use, knowledge, and spaces became de rigueur in applied science, art, industry, and the humanities and social sciences. As these systems were adopted as standards

FIGURE PII.1
Milton Bradley. Color charts from *Elementary Color*.

FIGURE PII.2
Several editions of the *Munsell Book of Color*. The atlas is arranged into removable pages of color swatches of varying value and chroma for forty distinct hues. Image courtesy of Mark Fairchild and CC BY-SA 3.0.

across a wide range of industries, they were embraced as neutral design tools, despite their technical compromises and latent but problematic cultural assumptions.

Two key examples of this are art educator A. H. Munsell (1858–1918) and game designer and businessman Milton Bradley (1836–1911; discussed in part I), both of whom celebrated color's variegated nature and affective power, but nonetheless tended to order their colors according to deep-seated racial ideologies rooted in colonial narratives of "civilized" progress over "primitivism." Munsell's three-dimensional "tree" model, for instance, was used as an educational tool to "civilize" children's proclivity for "savage" and "primitive" (i.e., bright and bold) colors. In line with early twentieth-century color education in the progressive-era US, Munsell's goal was cultivating the "color sense" in (white) children to mold them into idealized modern subjects, or simply curbing their "taste" by moving them toward five "middle colors" that contained high percentages of white.[2] Similarly, Bradley, as previously noted, used optically derived color systems to classify textiles and pigments, also primarily for classroom use. Although neither Bradley nor Munsell explicitly classified or ranked human beings based on skin color, there was an implicit, racialized connection between color perception and so-called civility in their work, which, unsurprisingly, was later adopted for explicitly racist purposes in the making of modern, reproducible color media (Munsell's swatches, for instance, were used to calibrate supposedly "normal" skin tones in the development of early color photography and film—see the discussion of Shirley Cards that follows here and in chapter 9).[3]

Part II of *Color Protocols* grapples with some of these issues by transitioning from the largely text- and linguistic-based color grammars of part I to the optic and visual codification of color in the Industrial Age. The same discretizing and normalizing logic of color returns—typically by way of abstraction and numerical indexing, albeit in the form of visual calibration systems based on skin tone—to the infrastructure of moving image technologies in film, television, and video.

While it might seem as if race would be *more* apparent in visual media due to the optical nature of these imaging technologies, dominant histories of screen media, somewhat counterintuitively, rarely address the ways in which color-based media naturalize the culturally constructed equation between skin color and race.[4] A number of the chapters in part II illustrate this by analyzing how media ranging from television, video, and computational technology have historically privileged paler skin tones as part of a longer history of "lighting for whiteness." Richard Dyer, Lorna Roth, and Genevieve Yue, for instance, all insist that the light skin bias has led to a fundamental inability to accurately represent bodies of color that deviate from this norm. As a result, dominant histories of color media are falsely taken as apolitical and color-blind (i.e., "raceless") and the mythos of modern progress continues to march through today's high-tech landscape.[5]

THE ESSAYS

Picking up chronologically where part I left off, part II begins with Aileen Robinson's "'Black Art:' Light, Magic, and the Racial Techniques of Innovation," an essay locating the roots of lighting for whiteness in the costuming and lighting techniques of nineteenth-century theatrical performance and magic shows. Of all the performing arts, she argues, stage magic's technique of "Black Art" acutely reflects the deceptions of Western modernity as it wrestled with concepts of self, nation, and otherness. On stage, imperialist ideologies, coupled with the rising concerns of scientific secularism, manifest through color-driven racial stereotypes and Orientalized displays of magic. As a result, Robinson concludes, perceptions of racial difference and colonial modalities actively made *what* was seen on stage inextricable from *how* they were seen off stage.

Robinson's piece is followed by "The Light of the World," excerpted from British cultural critic Richard Dyer's canonical *White: Essays on Race and Culture* (1997; reprinted in 2017). Dyer's work was one of the first to critically interrogate how photography and cinema, as light-based media, fix and naturalize an aesthetic of whiteness. By putting lightness-as-whiteness at the center of the conversation, he reveals how both the color white and whiteness as a racial category go unmarked and unseen, precisely because they are taken as a natural default—a blank canvas that is seen to bear no color in and of itself.

Dyer's case studies focus on classical Hollywood lighting: by examining the formal and generic conventions of film stock calibration, cinematic lighting, and makeup, he reveals how both black-and-white and color film (and later, broadcast television) were purposely designed to highlight the visibility of lighter skin tones, referred to by industry practitioners with the

FIGURE PII.3
A still image from *Rising Sun* (Philip Kaufman, 1993). Richard Dyer cites this film as an example of the white face as the norm ("get Connery right and the rest will fall into place"; Dyer, 1997, p. 101).

seemingly innocuous term "color balance." This light-skinned bias simultaneously carries thematic and narrative repercussions: not only are actors of color not lit as well, constituting a technical "problem," but they are also typically less developed as characters, with less screen time.

Traditionally, it is the white leading lady's face, Dyer observes, that provides the litmus test for "standard flesh tones." Even when a heterosexual white couple is in the same frame, the woman will almost always be more illuminated than the man, constructing a form of compulsory white feminine heterosexuality where white women "glow." Dyer further posits that the technical affordances of light-based technology simultaneously reinforce several racial-cultural myths: the Judeo-Christian symbolism of whiteness as pure and good, angelic white femininity, and a singular, free individual whose autonomy and uniqueness are valued above those deemed less important.[6] Dyer's work, along with the precursors that he alludes to in this excerpt, lays the groundwork for a new genre of critical race–based analysis of screen media rooted in the materiality of film technology, further developed by such scholars as Lorna Roth, Genevieve Yue, and Ali Feser, whose work also appears in part II.

Roth's "Looking at Shirley, the Ultimate Norm: Colour Balance, Image Technologies, and Cognitive Equity" (2009) also addresses the light-skin bias socially and technologically embedded in color film stock emulsions and in digital camera design. She traces the origins of the once-obscure reference cards known as the "Shirley Card" and "China Girl" in the photographic and film and television industries, respectively. The central problem with these devices, Roth argues, is their deployment of white women's flesh tones as the keystone for adjusting all other colors in the image ("This colour girl is always white").[7] Because color engineering standards tend toward efficiency and compression, color information on the red or yellow end of the spectrum is often reduced or eliminated, since it is less crucial for accurately representing the norm of Caucasian flesh tones. This racial bias toward whiteness in image technology has clear social and technical implications, making the rendering of non-Caucasian skin tones problematic, or visually weak at best.

The Shirley Card/China Girl norm was implemented at the beginning of the twentieth century (a bit later for color television), and circa 2009, it—along with variations that include "Lenna," a test card used in digital image processing that was initially cropped from a *Playboy* centerfold[8]—has remained the dominant model of color standardization until recently. Things began to shift in the twenty-first century, when, as Roth articulates, compensatory measures were introduced; and yet, she concludes, these changes were implemented in response to consumer demand rather than directly addressing issues of equity or racial bias. Since its 2009 publication in the *Canadian Journal of Communication*, the essay has had wide-ranging effects from CRT scholarship to image design in the media industries.

Like Roth's work, Genevieve Yue's 2015 article "The China Girl on the Margins of Film" (originally published in *October*) excavates the origins of the "China Girl" reference card image that was used to balance color and exposure in the film laboratory. In this excerpt, she argues that the China Girl is a hybrid subject-object—one that brings together filmic materiality, race, and gender and offers new ways of thinking through the afterlives of film postdigitization. She points out that, despite the orientalist roots of the name "China girl" (which likely referenced both porcelainlike white skin and racialized stereotypes of Asiatic female submissiveness), these text card subjects were nearly always white. The notion that "woman's flesh alone . . . [provided] an index of ideal skin tone" that served to "[produce a] fiction of a single color against which all others [could] and must be measured."[9]

Ultimately, Yue argues, this creation of "naturalness" in the form of a white female standard subject "signals the artificial and arbitrary premise of racial categorization."[10] Her archival approach diverges from the majority of feminist film analysis, which has focused primarily on issues of representation as in Laura Mulvey's discussion of Hollywood narrative cinema (1975). Instead, she opts to draw out the gendered and racialized nature of film's material dimensions that are often overlooked, from the lab-based assembly process of film to its tangible industrial output in the form of celluloid or acetate.[11] Whereas China Girls are only ever seen in the theater due to error, experimental filmmakers and contemporary artists have also embraced and repurposed them as archival material. When viewing works by artists such as Morgan Fisher and Cécile Fontaine, Yue concludes that "we realize there is a different system of signification at work behind the images we typically see: For every *leading* lady, there is an attendant *leader* lady."[12]

Like Roth and Yue, cultural anthropologist Ali Feser considers the material substrate of chromatic media, and Kodachrome specifically, to unearth racialized social and technical practices otherwise lost to history. Her essay, "Making Fantasy Material: Kodachrome, Eugenics, and the

Chemoaesthetics of Racial Capitalism," offers a pathbreaking integration of photographic color technologies, ethnographic studies, and the history of eugenics and white supremacy in the postindustrial US. Drawing on fieldwork and archival research conducted in Rochester, New York (the birthplace of the Kodak Corporation), Feser narrates the little-known story of founder George Eastman's longtime support for the American Eugenics Society (AES) while simultaneously revealing how Kodak branded itself as the white, middle-class American family ideal through its hiring practices and corporate structure.

Eugenics and color photography, Feser argues, contain multiple historical and philosophical parallels: both were fundamentally future-looking enterprises that constructed a particular version of whiteness by "[manipulating] the molecules of the emulsion to best image the objects of their desire." The technical color palette and idealized advertising photos for Kodachrome promoted a vision of the "good life" that justified the ills and violence of racial capitalism. "A fantasy of the white good life materialized as a lived social reality in Rochester," Feser writes, "and . . . this lived reality of whiteness was transposed into the technologies that underlie the image and consumer culture of the American Century."[13]

Finally, Tina Post's essay, "Advantage Black: Hyperblackness in Performative Self-Portraiture," moves beyond mass media circuits to the gallery space, interrogating the ways in which a selection of contemporary artists from the African diaspora renegotiate Blackness in their work as both race and sensory hue. Beginning with a material and philosophical exploration of Surrey NanoSystem's Vantablack—a pigment that has been called "the blackest black in existence" due to its ability to absorb over 99.6 percent of incoming light—Post observes how Surrey NanoSystems consciously sidesteps any references to racial Blackness in its description of the technology, while simultaneously conjuring the "darkness of uncivilized wilds." From there, she theorizes how hyperblackness, in its subtle negotiation between race and color, might be viewed as a "minor gesture" that opens up a space of potential that she locates in the work of Zanele Muholi, Lina Iris Viktor, and Mohau Modisakeng. The hyperpigmentation of their subjects (which, as self-portraits, are the artists themselves) simultaneously reference the enduring racist legacy of minstrelsy and blackface while reimagining it through performance and material experimentation. As Post convincingly puts it, "To the extent that hyperblackness works as a minor gesture, it

FIGURE PII.4
Still image from *Insecure* (HBO, 2016–2021).

requires the linkage between race and hue *and* our profound cultural memory of this linkage's horrific historical . . . consequences in order to perform its repetition with a difference."[14]

As a whole, the authors in part II analyze color-based differentiation in racial and gendered physical attributes that have become embedded within biased imaging systems, falsely presenting themselves as neutral. And yet, as Yue and Post show, experimental artists and filmmakers have been able to critique these false contentions by appropriating the same technologies in new ways. We anticipate that future scholarship will pay increasing attention to these nuances, as evidenced, for example, in cinematographer Ava Berkovsky of HBO's *Insecure* (2016–2021). Berkovsky found creative ways to light the show's majority Black cast properly, including special cosmetics, using a whiteboard with light-emitting diode (LED) filters to reflect light onto the actors' faces rather than shining it directly, placing a polarizer over the camera lens, and applying customized digital color settings known as Look-up Tables (LUTs) in postproduction.[15] Still, *Insecure* remains an exception rather than a norm, and today's digital cameras and filter settings (which will be discussed in part III) exhibit the same difficulty detecting and accurately representing darker skin tones that they did nearly a decade ago. There is much work to be done, and as optical media appear to remain hegemonic in contemporary media cultures, they must nonetheless be positioned alongside previous and emerging media technologies, from the written word and text (part I) to the algorithmic and computational (part III). In short, color is a technology of vision, where vision is always already invisibly coded through history, politics, and the social.

NOTES

1. See cornerstones of visual studies such as Berger, *Ways of Seeing* (1990), Jay, *Downcast Eyes* (1993), Crary, *Techniques of the Observer* (1992), Mitchell, *Iconology* (1986), and Mirzoeff, *The Right to Look* (2011).

2. Gaskill, *Chromographia*, 21.

3. See Keevak, *Becoming Yellow*. American eugenicist Charles Davenport also appropriated these same color systems that were initially intended for childhood pedagogy.

4. Exceptions include Murray, *Bright Signals*; Dootson, *The Rainbow's Gravity*, and Street and Yumibe, *Global Film Culture*.

5. Goldberg, *The Racial State*; Cheng quoted in Cheng et al., *Race and Modern Architecture*, 150.

6. Dyer, *White*, 84, 102.

7. Jan Kasoff, as cited in Roth, "Looking at Shirley," 112.

8. Called "Lenna" or "Lena" for short, the test card is a photo of Swedish model Lena Forsén posing for *Playboy* in 1972 with her nude body cropped out. This hypersexualized image and its ubiquity in the computing world is discussed in Mulvin's *Proxies*, as well as the documentary *Losing Lena* (2019), directed by Kyra Bartley.

9. Yue, "The China Girl," 104.

10. Yue, "The China Girl," 104.

11. In Yue's 2020 monograph *Girl Head: Feminism and Film Materiality*, she situates the China Girl within a larger discussion of how the female body is embedded within the technical process prior to the actual screening of the film, from the archive to the laboratory to the editing room.

12. Yue, "The China Girl," 108.

13. See page #.

14. See page # (emphasis in original).

15. Harding, "Keeping 'Insecure'"; Mulcahey, "*Insecure* DP Ava Berkovsky."

BIBLIOGRAPHY

Bartley, Kyra, dir. *Losing Lena*. Sydney: FINCH, 2019.

Berger, John. *Ways of Seeing*. New York: Penguin Books, 1990.

Cheng, Irene, Charles L. Davis II, and Mabel O. Wilson, eds. *Race and Modern Architecture: A Critical History from the Enlightenment to the Present*. Pittsburgh: University of Pittsburgh Press, 2020.

Crary, Jonathan. *Techniques of the Observer: On Vision and Modernity in the 19th Century*. Cambridge, MA: MIT Press, 1992.

Dootson, Kirsty Sinclair. *The Rainbow's Gravity: Colour, Materiality and British Modernity*. London: Paul Mellon Centre for Studies in British Art, 2023.

Dyer, Richard. *White: Essays on Race and Culture*. London: Routledge, 1997.

Gaskill, Nicholas. *Chromographia: American Literature and the Modernization of Color*. Minneapolis: University of Minnesota Press, 2018.

Goldberg, David Theo. *The Racial State*. Malden, MA: Wiley-Blackwell, 2001.

Harding, Xavier. "Keeping 'Insecure' Lit: HBO Cinematographer Ava Berkovsky on Properly Lighting Black Faces." *Mic*, September 6, 2017. https://www.mic.com/articles/184244/keeping-insecure-lit-hbo-cinematographer-ava-berkofsky-on-properly-lighting-black-faces.

Jay, Martin. *Downcast Eyes: The Denigration of Vision in Twentieth-Century French Thought*. Berkeley: University of California Press, 1993.

Keevak, Michael. *Becoming Yellow: A Short History of Racial Thinking*. Princeton, NJ: Princeton University Press, 2011.

Mirzoeff, Nicholas. *The Right to Look: A Counterhistory of Visuality*. Durham, NC: Duke University Press, 2011.

Mitchell, W. J. T. *Iconology: Image, Text, Ideology*. Chicago: University of Chicago Press, 1986.

Mulcahey, Matt. "*Insecure* DP Ava Berkovsky on Avoiding '*Seinfeld* Shots,' Faking Coachella and Lighting Mega-Churches." *Filmmaker Magazine*, October 10, 2018. https://filmmakermagazine.com/106136-insecure-dp-ava-berkofsky-on-avoiding-seinfeld-shots-faking-coachella-and-lighting-mega-churches/#.Y0lo_ezMLOQ.

Mulvey, Laura. "Visual Pleasure and Narrative Cinema." *Screen* 16, no. 4 (1975): 6–18.

Mulvin, Dylan. *Proxies: The Cultural Work of Standing in*. Cambridge, MA: MIT Press, 2022.

Murray, Susan. *Bright Signals: A History of Color Television*. Durham, NC: Duke University Press, 2018.

Roth, Lorna. "Looking at Shirley, the Ultimate Norm: Colour Balance, Image Technologies, and Cognitive Equity." *Canadian Journal of Communication* 34, no. 1 (2009): 111–136.

Street, Sarah, and Joshua Yumibe, eds. *Global Film Color: The Monopack Revolution at Midcentury*. Rutgers, NJ: Rutgers University Press, 2024.

Yue, Genevieve. "The China Girl on the Margins of Film." *October* 153 (Summer 2015): 96–116.

Yue, Genevieve. *Girl Head: Feminism and Film Materiality*. New York: Fordham University Press, 2020.

7 "BLACK ART": LIGHT, MAGIC, AND THE RACIAL TECHNIQUES OF INNOVATION

AILEEN ROBINSON

> I am invisible, understand, simply because people refuse to see me. Like the bodiless heads you see sometimes in circus sideshows, it is as though I have been surrounded by mirrors of hard, distorting glass. When they approach me, they see only my surroundings, themselves, or figments of their imagination—indeed, everything and anything except me.
>
> —Ralph Ellison, *Invisible Man*, Prologue, 3

OPENING TRICK

In December 1907, Harry Houdini published Austrian magician Ottokar Fischer's editorial, "The Inventor of Black Art: Ben-Ali-Bey (Max Auzinger)," as a contribution to his section of *The Conjuror's Monthly Magazine,* "History Makers in the World of Magic." Fischer outlines the theatrical accident that led to "the most profound sensation of any of the modern mysteries," coined Black Art. Black Art emerged around 1886 as a trick of rapid appearance/disappearance, where a white-clad magician and a stage assistant, disguised in black, coordinated quick reveals of white objects in a darkened stage: "In this illusion the entire stage from the first groove to the rear is hung with black velvet, the floor covered with black felt, and the top is covered with black velvet, thus forming a room lined entirely in black."[1] Operating on a simple yet carefully timed theatrical choreography, objects would suddenly appear and disappear; beautiful women would be conjured, decapitated, and restored; and spirits could arise. The act worked

FIGURE 7.1
The Enquirer Job Printing Company, 1898. Magic Poster. McCord Stewart Museum.

through the simple reveal of a contrasting object—a white wand, a white ghostly hand, or a white woman.

According to Fischer, Auzinger developed the illusion after working at the National Theatre in Berlin as stage manager and actor. During a scene in a dungeon, draped in black velvet, a black slave, most likely performed in blackface, enters to save a captured maiden, Lucy. As Fischer recounts: "The stage manager, Herr Auzinger, during this part of the act was waiting for the applause to come from the audience, but he waited in vain, and on close scrutiny, instead of the negro he just managed to see a row of shining teeth, belonging to the negro, and if ever a German had a brain storm, that man was Auzinger." Fischer crafts the tale as an "aha!" moment of ingenuity and aligns innovation with the imagined black body, imperceptible and flattened, erasing distinction and accuracy, evoking, in its wake, a racist joke.[2] Black Art emerged only a few years after artist Paul Bilhaud premiered his 1882 *Negroes Fighting in a Tunnel at Night* (*Combat de nègres dans un tunnel*), a black monochrome square printed on white paper, which writer Alphonse Allais extended in his portfolio of color-coded racist monochromes.[3] In his work on technologized darkness, Noam Elcott connects the two: "The Incoherent artists Paul Bilhaud and Emile Cohl reduced the "negro in the dark" to a sight gag. But where Bilhaud and Cohl saw a joke, Auzinger saw an opportunity."[4]

As sight gag, joke, or opportunity, Auzinger's Black Art appears as another site of racial collapse where racist rhetoric "taints" technological innovation. The racism displayed is thus separate from the technology and its effects. However, what Jim Steinmeyer calls a "suspiciously neat" story enacts the very logic that developed through technological experimentation, performance practice, and cultural hierarchies.[5] Here, the black figure, which indexes the black body and person, becomes a generative source that corresponds with the racial cultural logic that ascribes blackness to the background. Auzinger trades on a genealogy of what I call *technical racialized vision,* a mode of perception combining technical, artistic, and performative practices. Black Art, as a final product in a long history of color-coded performance practices, is both a trick of perception and an example of racialized thinking. Magic performance, for all of its wonder, grew up with race, in discourse, performance, and technology.

In this chapter, I examine how Black Art emerged within a repertoire of tricks and techniques, all called *black art,* that exemplify this complex association between race, illusion, and technology. While Black art today commonly refers to Black aesthetic production, emerging from 1960's civil rights and Black power, the nineteenth-century concept, *black art,* refers to a collection of concepts, techniques, and shows that became staple elements of Victorian magic performance: 1) the concept of "dark" or evil magic; 2) the technique of using darkly pigmented materials for absorption; and 3) the specific performance that used paints, drapery, and lights to craft illusions of appearance, disappearance, disembodiment, and more.[6] (Throughout this chapter, I use 'the black art' to indicate 1, "black art" to indicate 2, and Black Art to indicate the third. I use italics when referencing the whole category.)

While magic historiography deals mostly with the second and third conceptual categories, Elcott examines Black Art as a theatrical permutation of the black screen that "conjoined physical bodies and virtual images and created model subjects in an era of artificial darkness."[7] In an addendum article, he addresses an omission in his book on the implications of race and artificial darkness. Retaining his claim that artificial darkness developed independently from race discourses, he nevertheless highlights the racist rhetoric that surrounded the history of technologized darkness through modern artistic practice.[8] I argue here that the legacy of *black art*, as performance and technology, reveals the ways discourse and technique are inextricable from its racial logics.

Black Art is an important example of the place magic performance holds in the history of racialized media and technological development. Late-Victorian stage magic, which celebrated a golden age from the 1880's, reflected the modern ethos of the imperial European world with its racialized "color comedy," technological and scientific innovations, and participation in colonial expansion. Magicians, predominately white, could celebrate a rising scientific secularism while trading on a newly configured "irrationality" of magic reserved for an imagined "East." White European men styled themselves as Arab, Indian, Chinese, and Japanese (and sometimes all four), manifesting the idea that "the Orient then seems to be . . . a theatrical stage affixed to Europe."[9] Auzinger thus performs Black Art as Ben Ali Bey. As race hierarchies resulted in stereotyped "colors," the language of black, red, yellow, and brown became refracted through magic's prism, resulting in costumes, make-up, and magic tricks that circulated colonial thoughts and behaviors.

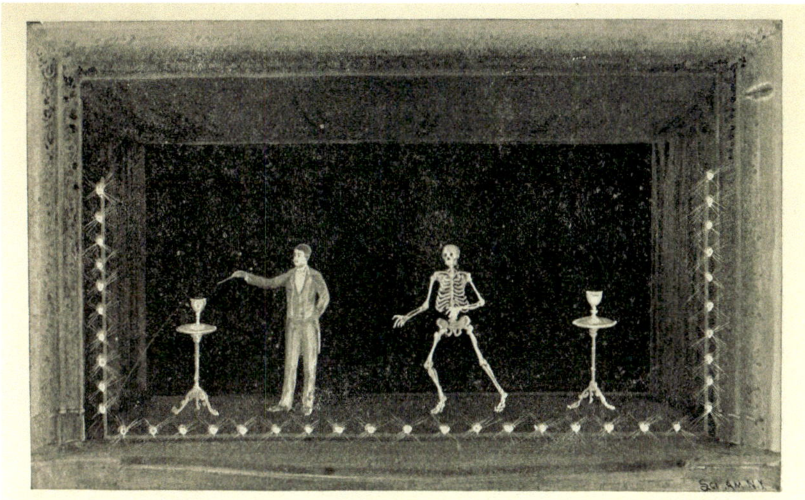

THE STAGE SETTING FOR BLACK ART.

FIGURE 7.2
Albert A. Hopkins. 1901. Black Art Setup.

Rather than simply stunning tricks and humorous asides, magic performance was thus an essential medium for conveying and normalizing racial hierarchies. Capitalizing on the complex relationships between sight and perception, magic tricks created narratives that privileged objects rendered white while also reifying negative associations for objects rendered black. The alignment between technology and objectivity further concretized these binaries. The seemingly objective functionality of lighting technology can suggest a lack of cultural relevancy: these lights are simply bright, and light, and white. Blackness, as the opposite of whiteness, becomes a simple contrasting fact that can be used to create illusory effects. However, magic illusions and the development of modern media were entwined with perceptions on racial difference or imperial practices of hierarchy that made *what* we look at inextricable from *how* we see.

In some respects, Black Art explores visual principles within the theatrical stage; in others, it reveals how the deep-seated development of stage illusions are imbricated with beliefs about skin tone, race, and perception. That is, stage illusions, in their current features, were built upon strategies of light and dark that presumed a fair toned actor, privileged whiteness as the condition of the object seen, witnessed, and acknowledged, and made "natural" the racist rhetoric and assignation of Black skin to (and with) the background. Even though Black Art was an optical trick, the perceptual alignments with racial color symbology were not an illusion at all. The relationships of color, figure, and person were recorded in magic and in language, as lexicographer Noah Webster attests: "Black, n.1. That which is destitute of light or whiteness, the darkest color, or rather a destitution of all color. 2. A negro; a person whose skin is *black.*"[10]

In the following sections, I briefly outline the paradigm of good/evil: black/white that became standard fare for magician's script (patter) and tricks. I then examine the use of "black art" techniques within mid-Victorian magic and science performances, using Pepper's Ghost as an exemplum. I finally turn to the "invention" and development of the Black Art shows in the 1880s, examining the full spectrum of *black art* at the turn of the century. I show how the repertoire of rhetorical, performative, and technological binaries (black/white; dark/light) fashioned a technical racialized vision.

SETTING THE SCENE: BLACK MAGIC, DARK MAGIC; WHITE MAGIC, LIGHT MAGIC

Magic, either as skill-based performance or religious theological concept, has played with the divisions between dark/light, black/white, shadow/brightness throughout its slow Western trajectory. Theologians used the theological analogy, Jesus = light/white, resulting in representations of Jesus as light-skinned and blue-eyed, regardless of his historical identity as a Palestinian Jew. The devil, on the other hand, was often depicted in connection with darkness and was easily depicted as in shadow or black.[11] Within Western Christian theologies, magic was often demonized as a contractual association with the devil, best codified in the treatises identifying and classifying witchcraft, Heinrich Kramer's *Malleus Maleficarum* (1449) and Reginald Scots *Discoverie of Witchcraft* (1584).[12] Kramer and Scots distinguished between artifice and real magic, separating jugglers from magicians. However, the alignment between trickery and true manifestation was unstable, producing concern over illusions too much like the "real thing." Early modern magicians negotiated a careful line between enticing people with the promise of "real magic" and causing panic over the threat of sorcery.[13]

Trading on skill acquisition and technological knowledge, modern magicians positioned themselves as stagecraft technicians and talented performers while the dyad of magic became contemporaneously attached to bodies and non-Christian religions. Black magic as malevolent divination became directly aligned with European skin-tone racisms, allowing colonizers the opportunity to use good/bad binaries for expansion, conversion, and resource extraction.[14] The language of black magic ("mumbo-jumbo," "voodoo," "negromancy") enacted its own colonial deception in its totalizing stance: "The African creed is demonology,—the specters of the dead,—witchcraft, or black magic."[15] As Charles Mills argues, "Given the racialization that accompanies modernity, it is then unsurprising that metaphor, color symbolism, and Euro-identity all fuse: whiteness becomes the identity of both enlightenment and of the human bearers of enlightenment."[16]

By the end of the eighteenth century, conjurers fit easily into the itinerant entertainment culture of Europe and its colonies. As performance style, theatrical technique, and conman's repertoire, conjurers benefited from embracing extremes and playing on affective resonances for their audiences. They could elect to trade on suspicion as an advertising ploy for their acts, styling themselves, like Gustavus Katterfelto, as masters of the occult. Katterfelto's 1780's performances were also marketed as philosophical amusements, centering objects such as the solar microscope at their core.[17] Performance magic's growth was due to this interplay between performance and explorations around technology, knowledge, vision, and identity. Magic experiments shared stage space with theatrical illusion-making and scientific investigation as magicians sought to exploit discoveries about vision and perception. Magic performances—both in their narratives and in their illusions—benefited from walking the edge of fear and delight, sight and blindness, misperception and seeing. As a result, black art developed throughout the nineteenth century not only in the performing arts, but as a descriptor for everything from technological wonder and white European ingenuity to the invocation of the Devil.

The relationship between dark/black: light/white was further codified in the performing and representational arts. Spectacular theatre practice, including melodrama, pantomime, and extravaganza, traded on new lighting and illusory technologies to create dramatic effects. Footlights and special effects were crucial not only for visibility for the audience, but also for conjuring the dramatic extremes that spectacular theatre required.[18] Within gothic melodrama, for example, the codes of good and evil, heroine and villain, were easily established through costume signifiers (white/black) and lighting effects.[19] Stage narratives reiterated the binaries of dark/light, dangerous/ innocent, while lighting punctuated the scene in its illusory use. This symbolic deployment of light (as goodness, innocence, and the angelic) developed and sustained the alignment of God's light with the sun, artificial light, and thus white skin through intersecting cultural practices.[20] And so visual performance practices that utilized light technologies, refused performance and creative rights to diverse bodies, and embraced metaphoric corollaries developed illusory techniques that reflected this system.

From this history, the nineteenth-century magician emerged, developing as a secular wonder-worker with expertise in legerdemain, stage technology, and illusory experiments. No longer a fairground performer, the modern magician shared knowledge, displaying technological marvels, wondrous natural phenomena, and polite skills of deception. The ascription of true magic to the imagined East and colonized Africa allowed these predominantly white magicians to perform at the margin of whiteness. The post-1830's magician could thus, with relative security, align themself with mysterious supernatural forces as narrative fodder while maintaining the focus on skill, technique, and stagecraft. Collapsing distinctions between geography, religion, and people, white magicians crafted stage performances that included beheadings, resurrections, ghosts, and seances.

Within these performances, white magicians instrumentalized the negative valences of black magic while escaping its patina. Popular magicians Alexander and Adelaide Herrmann explicitly championed their *black art* in acts such as "Black Art." Herrmann enticed crowds with a promise of "dreadful" acts: "I cut off my wife's head. I—in short, I raise the devil. Come and see me. I am sure you will enjoy yourself."[21] The casual juxtaposition of the demonic with stage magic was echoed in mid-Victorian magic manuals such as *The black art! or, Magic made easy* (1869) and *The black art fully exposed and laid bare* (1875). Herrmann also employed several Black Americans, including Isaac Willis, Alfonso Moore, Milton Hudson Everett, as assistants from 1870 to the 1920's. Nicknamed "Boomsky," the performers worked as apprentices for the large-scale Herrmann magic show and performed comic minstrel bits within the performance, reifying the white magician, black servant binary of Anglo-American racisms.[22]

The plasticity of magic, to build on Goto-Jones' analysis, allowed for opposites to be held together: secular rationality/ divine intervention, technological innovation/ supernatural influence.[23] This flexibility and versatility also

FIGURE 7.3
Forbes Co., Boston & New York, 1889. Magic Poster. McCord Stewart Museum.

allowed white magicians the ability to craft performances that, a century earlier, would have been suspect. They could exploit the supernatural for sensational effect; the associations of black magic, black art, and black bodies remained fused.[24] That is, *black art*, as technology, technique, and show, was aligned alongside arguments regarding primitivism, blackness, and supernatural belief. This relationship made possible the most effective disappearance trick: that of the Black inventors, poets, thinkers from the stage at all. It was within this larger magical apparatus that "black art" became a widespread technique for a new kind of white manifestation.[25]

CONJURING GHOSTS: BLACK ART IN TECHNIQUE

By the 1880's, the show that became known as Black Art had become a popular sensation with professional magic. Built on stage techniques and larger ideological concepts, Black Art introduced a different *mise-en-scène* for stage magic that inverted the brightly lit stage and simple table arrangement that had been the magician's staple. Instead, more akin to Robertson's 1780 *Phantasmagoria*, the set-up required a black room and a set of contrasting light-colored objects and people. The effects were impressive: "Figures could appear and disappear; inanimate objects could come to life; gravity was defied."[26] And popular magicians, including well-known London-based performers Bautier de Kolta, William John Nevil Maskelyne, Alexander Herrmann, and Harry Kellar capitalized on its possibilities. The Egyptian Hall in London featured many performances of "Black Art," at times renamed as "Modern Black Magic;" William Robinson, who became famous as Chung Ling Soo, incorporated a wildly popular Black Art act into his show.[27]

A novel sensation, the act reconfigured the *mise-en-scène*: as mentioned above, the brightly-lit stage was exchanged for a black background, Elcott's black screen.[28] The magic in Black Art was in the contrast of object to background, the flattening of the backdrop to a uniform line. Depth appeared and disappeared alongside objects, the stage assistant's hand and form completely invisible to the audience. A 1907 *Scientific American* editorial, "Black Art Redivivus," proclaims that "the number and style of tricks performed in the mysterious black chamber are almost unlimited."[29] The trick, while expensive in its set-up, was also versatile, allowing magicians easy room to develop their own illusions. By the early twentieth century, Black Art had

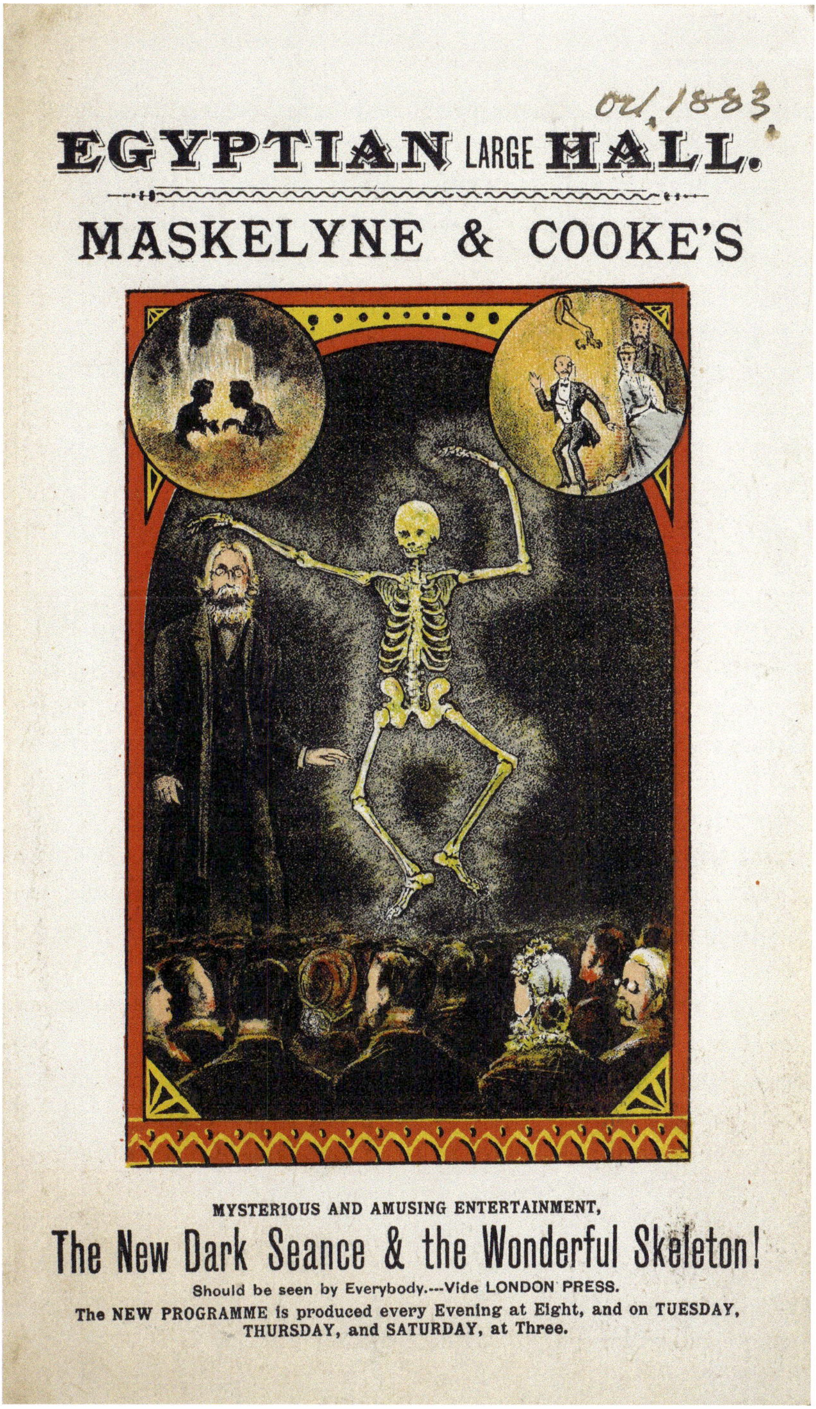

FIGURE 7.4

Stafford and Co., Egyptian Large Hall Card, October 1833. © The British Library Board.

become commonplace, with magic purveyors marketing "Black Art Kits" and magazines printing editorial "How-Tos."[30]

As magic historians acknowledge, Black Art reconstituted earlier practices of concealment and disguise,[31] where black drapery and brightly lit objects had been combined to create a variety of different illusions—the most famous being Pepper's Ghost. Bringing together vision science, popular theatrics, and magic performance, this illusion created the conditions for standardizing theatrical black art. Crafted at the Royal Polytechnic Institution in 1863, Pepper's Ghost "actually reproduced the living but impalpable, 'modern' everyday experience of bodies displaced onto public glass . . ."[32] Polytechnic director John Henry Pepper and civil engineer Henry Dircks installed a 45-degree angled piece of plate glass, that, would create a translucent reflection when a brightly illuminated object was situated underneath in the right place. The effect astonished audiences with its verisimilitude, as they watched a ghostly figure appear, disappear, and with later modifications, walk across the stage, be stabbed, lose their heads, and more: "Another phantom was a supposed sculptor, at work upon a bust; the bust intensely white and brilliant . . . But such was the solidity of appearance, and the play of light and shade, that it was with the utmost difficulty the mind was induced to shake off the delusion . . ."[33]

The "solidity of appearance" that the ghost possessed was due to the relationship between object, light, and reflection. The manipulation of the ghostly figures, and their ability to move around, was due to the use of carts and carefully manipulated drapes to conceal, and reveal, parts of various bodies. The ghost could appear to be floating when positioned on a rolling cart; a head appeared severed when the body was obscured with light-absorbing fabric. The guidelines for Pepper's Ghost and its permutations centered around the black/white binary: black as absorption and disguise, white as reflection and reveal. Albert Hopkins details such an illusion called "The Three-headed Woman" in *Magic: Stage Illusions and Other Scientific Diversions* (1901). The trick, a permutation of Pepper's Ghost, featured the figure of a three-headed woman, moving their eyes, singing, and saluting the audience. As shown in figure 7.6, three girls were closely positioned together under the stage, the woman in the middle wearing a light-colored silk dress. Two other girls on either side would be covered up to their necks with a "fabric of a dead black color . . . The heads, hair, and arms of the "body" are covered with powder so as to present completely white surfaces."[34] The

FIGURE 7.5
Albert A. Hopkins. 1901. Three-Headed Lady Ghost illusion.

"dead black," or non-reflective cloth, absorbed the light to reflect it more fully; as Hopkins concludes, "the image is thus made upon the spectator."

The innovative limits of these "black art" tricks replicated, in material form, color binaries comprising a magic epistemology. The "black art" methods focused on enhancing reflection, capitalizing on shadow and light to both complicate and erase depth and create form within reflection. The technique reached a boundary line because of the conditions of its material—white objects, white ghosts, white skin. The technique thus held an experimental limit that defines a core aspect of the technical racialized vision. The trick followed a formula of binaries: an illuminated object reflects

FIGURE 7.6
Albert A. Hopkins. 1901. Pepper's Ghost Illusion.

and is itself a *white* object. The principle, however, is not dependent on white objects, but on illumination. The trick—created within the popular science gallery, which championed an "accessible" and amusing scientific practice—reproduced cultural binaries as technical fact. That is, the reflective white object, observed so many times on the street and in the home, took center stage. And the black object receded into the background.

FINAL ACT: "BLACK ART" AS BLACK BOX

The "black art" of Pepper's Ghost remained a background technique until the 1880's when Black Art inverted the show, further reifying the foreground/background of magic's white/black color logic. Black Art revealed the earlier trick by bringing the black landscape of Pepper's Ghost onto the stage. The stage, transformed into a black box, became the background

FIGURE 7.7
Hardin Burlingame, Alexander Herrmann Black Art Illusion. 1897.

upon which white objects could be manipulated and manifested. This produced startling effects, where "black art" conjured ghosts, Black Art corporealized objects and people who "[seemed] to suddenly start into being."[35] Magicians would blend such rapid appearances with violent and impossible effects. In "Black Art Decapitation," Herrmann, under the instructions of his stage assistant playing Mephistopheles, cuts off Adelaide Herrmann's head and "taking it with one hand under the chin and the other holding her hair, he carries it across the stage . . . she walking along with him . . . only her head showing."[36] Black magic becomes Black Art which displays a white technological masculine mastery.[37]

Steinmeyer and Elcott record how Max Auzinger (Ben Ali Bey), dressed in a white wizard costume, performed an act that included an animated skeleton, a conjured silver platter with real coffee, and a translucent bubble act called "Indian Soap Bubbles." This final trick surprised audiences with "translucent spheres that hovered in the darkness and arranged themselves into perfect pyramids. A cloud briefly descended upon the Oriental "high priest," who, as soon as the haze lifted, faced the audience cross-armed and headless; he bowed to the audience as his head sat on a nearby table."[38] While the stunning *mise-en-scène* highlights technical ingenuity, the show also concretizes the relationship between the imagined magical "Arab," the black background, and the always visible white European performer.

And what of the stage assistant, that figure in the black box of Black Art? The show works only with the aid of a disguised figure, working silently and unseen in the "impenetrable gloom" of the darkened stage. The assistant is dressed in black with gloves and a black hood. While white skin is rarely mentioned directly, the instructions for the assistant imply a light-toned stage assistant. Even here, the black box dissolves Black labor into rhetoric and metaphor.

For Herrmann and Auzinger, the Black Art show, with the focus on a dark backstage and light object, triangulated the white magician, his manipulated objects (including white women), and blackness. The magician, visible *as* white, controls what and how the black box displays its objects; the hidden labor of the stage assistant echoes the desired invisibility of the Black laboring body. By manifesting "decidedly white" objects in front of the "dead black" box, Black Art crafted a living tableau reifying white's associations with light, visibility, and technical neutrality. The show rehearses the deception that the white object is simply the visible object.

Black Art persisted through the twentieth century, extending, as Elcott traces, through different aesthetic mediums. As a theatrical technique, "black art" became ubiquitous, the strategy of disguise and dampening onstage. Black Art also continued to have a theatrical life as T. J. Newlin's theatrical stagecraft series exemplifies. "The Black Art Entertainment" (*Popular Electricity Magazine*, 1912) focuses on optics, or scotomy, "a dark spot on the visual field," and the electrical and technical requirements for Black Art ("for it is electricity and the laws of light or optics that practically make the illusion feasible"). However, Newlin turns to the repertoires of *black art* in his explanation of the performances. He refers to the stage assistant as the "imp of darkness," rehearsing the connection between "imp," "demon," and "black," while also assuming all light-skinned performers.[39] The title of the accompanying image, "The Imp and the Black Bag," refocuses our attention on this slight figure in the back of the image, the ghostly outline of which conjures the invisible man, the chalk outline.

CLOSING ACT

Technical racialized vision developed through performative strategies of stagecraft and contrast illusions. The alignment between skin, sight, and invisibility, rendered as artistic and scientific innovation through special

FIGURE 7.8
T. J. Newlin, Black Art illusion. 1912, *Popular Electricity Magazine.*

effects, surfaces again and again in *black art,* bubbling up through long-standing associations between belief, religion, and skin-tone. The contrast illusion of Black Art is, thus, not incidental but coheres in magic's affinity with racial theorizing and color-coded racisms. The laboratory of Victorian magic reified perceptual practices of racial stereotype and hierarchy even while testing the boundaries of "objective" sight. The mechanics of Black Art worked as part of a system of perception that included the human body amid its environs—the white body emerging from the dark room; the black body dissolving into the background. The magic trick thus emulated the perceptive and representational practice of a racial logic while reiterating the common—and expected—dynamic of racial color in society.

In beginning this chapter with the eponymous opening to *Invisible Man,* I rehearse another repertoire of perception and the visible. Ellison here conjures, in his striking style, the estranging experience of being present but unseen. The condition of invisibility he describes is cast in the language of a magic trick—mirrors, reflections, and ghosts. The narrator, concealed in a one-way mirrored box, watches an audience preen themselves, unaware of his presence. Unlike H. G. Wells' Griffin, whose invisibility is a condition of a "biochemical accident to [his] epidermis," it is the "peculiar disposition of the [watchers] eyes," the "*inner* eyes," that renders the Black subject of *Invisible Man* imperceptible.[40] This is a matter of perception, a trained social vision that, like sleight-of-hand, makes the figure disappear.[41]

This is the impact of the technical racialized vision that limits the innovative possibility of the appearing Black body. Ellison found descriptive purchase in the language of the fairground, the early home of the magician. Here magic offers a method for understanding and rendering the condition of invisibility. Ellison's work is one of many, in literature, theater, film, photography, and more, that assert the presence of Black individuals outside of a limiting visual field. The renaming, and rethinking, of Black magic as Black thriving, of Black Art as Black creative expression, enacts another evocative stance: *now see us*.

NOTES

1. Hopkins, *Magic*.

2. Sharpe, *In the Wake*.

3. Bilhaud, *Combat de nègres dans un tunnel*.

4. Milhaud and Cohl, referenced in the first few pages of the introduction, crafted cinematic animation that often included racist jokes with monochrome screens. "Chinaman" transporting corn on the Yellow River. . . . ; "Negroes making shoe polish at night." Elcott, *Artificial Darkness*, 58.

5. Steinmeyer, *The Glorious Deception*, 75.

6. Contacting the dead (necromancy), conjuring demons, and harming others were all a part of "dark" or "black" magic. See also Chireau, *Black Magic*; Hazzard-Donald, *Mojo Workin.'*

7. He defines the black screen, which he traces through photography, cinema, and theater, as "a strictly circumscribed dark chamber opposite a darkened hall; controlled lighting; white clothing and props; maximum contrast; illusion. Black Art was another manifestation of the black screen." Elcott, *Artificial Darkness*,103. See also Burlingame, *Herrmann the Magician*; Hopkins and Evans, *Magic*.

8. Elcott, "A Brief History of Artificial Darkness and Race."

9. Said, *Orientalism*, 63.

10. Webster, *An American Dictionary of the English Language* (1844).

11. Verkerk, "Black Servant, Black Demon."

12. Scot, *The Discouerie of Witchcraft* (1584); Burton and Grandy, *Magic, Mystery, and Science.*

13. Butterworth, *Magic on the Early English Stage.*

14. McMillan, "Black Magic."

15. Spring, *The Negro at Home*, 101. See also During, *Modern Enchantments*, 10.

16. Mills, "The Illumination of Blackness."

17. Paton-Williams, *Katterfelto: Prince of Puff.*

18. Bowers, *Lengthening the Day*; Butterworth, *Theatre of Fire.*

19. Malchow, *Gothic Images of Race in Nineteenth-Century England.*

20. Dyer, *White.*

21. Burlingame, *Herrmann the Magician*, 86. Here, colonial violence and desire, also a feature of Victorian magic performance come together. See also Goto-Jones, *Conjuring Asia.*

22. It is also in the presentation of African American performers and their pictorial representation on later magic posters that we see the blackamoor tradition manifest. Steinmeyer does not mention this tradition of early modern decorative objects, but the terminology he chooses is significant.

23. Haskins and Benson. *Conjure Times: Black Magicians in America.*

24. For explicit connections to religion, see Murphy, "Black Religion and 'Black Magic'."

25. Herrmann, *Adelaide Herrmann Queen of Magic*. See also During, *Modern Enchantments.*

26. During, *Modern Enchantments*, 148.

27. "The Sphinx at the Egyptian Hall" (1865); Kellar, *A Magician's Tour Up and Down and Round about the Earth*; Herrmann, *Adelaide Herrmann Queen of Magic.*

28. Elcott, *Artificial Darkness.*

29. "Black Art Redivivus" (1907).

30. "Barter and Exchange" (1908)

31. Goto-Jones, *Conjuring Asia;* During, *Modern Enchantments*; Burlingame, *Herrmann the Magician*; Herrmann, *Herrmann's Black Art.*

32. Armstrong, *Victorian Glassworlds*, 30.

33. "Ghosts Everywhere" (1864).

34. Hopkins and Evans, *Magic.*

35. "Done by Magicians" (1889).

36. Hardin J. Burlingame, *Leaves from Conjurers' Scrap Books*, 162.

37. Magic performance included a great deal of violent tricks and set-ups, especially around women. For a detailed study of women and magic, see: Karen Beckman, *Vanishing Women*.

38. Elcott, *Artificial Darkness*, 103.

39. Newlin, "The Black Art Entertainment," 678.

40. Ellison, *Invisible Man*, 3. See also Wells, *The Invisible Man*.

41. Daston and Galison, *Objectivity*.

BIBLIOGRAPHY

Armstrong, Isobel. *Victorian Glassworlds: Glass Culture and the Imagination 1830–1880*. Oxford: Oxford University Press, 2008.

Auzinger, Max. "Our Mail Bag." *The Conjurors Monthly Magazine*, May 15, 1908.

Beckman, Karen. *Vanishing Women: Magic, Film, and Feminism*. Durham, NC: Duke University Press, 2003.

Bilhaud, Paul. *Combat de nègres dans un tunnel*. 1882. Reproduced as *Combat de nègres dans une cave, pendant la nuit* in Alphonse Allais, Album Primo-Avrilesque. Paris: Ollendorf, 1897.

Bowers, Brian. *Lengthening the Day: A History of Lighting Technology*. Oxford: Oxford University Press, 1998.

Brown, J. H. *Spectropia, or, Surprising Spectral Illusions Showing Ghosts Everywhere and of Any Colour*. London: Griffith and Farran, 1864.

Burlingame, H. J. *Herrmann the Magician: His Life, His Secrets*. Lee's Library of Occult Science. Chicago: Laird & Lee, 1897.

Burlingame, H. J. *Leaves from Conjurers' Scrap Books, Or, Modern Magicians and Their Works*. Chicago: Donohue, Henneberry & Company, 1891.

Burton, Danny Ethus, and David A. Grandy. *Magic, Mystery, and Science: The Occult in Western Civilization*. Bloomington: Indiana University Press, 2003.

Butterworth, Philip. *Magic on the Early English Stage*. Cambridge: Cambridge University Press, 2005.

Butterworth, Philip. *Theatre of Fire: Special Effects in Early English and Scottish Theatre*. London: Society for Theatre Research, 1998.

Chireau, Yvonne P. *Black Magic: Religion and the African American Conjuring Tradition*. Berkeley: University of California Press, 2006.

The Conjurors Monthly Magazine. "Barter and Exchange," March 15, 1908.

Daston, Lorraine, and Peter Galison. *Objectivity*. Cambridge, MA: Zone Books, 2007.

"Done by Magicians." *Milwaukee Daily Journal*, May 18, 1889. *Nineteenth-Century U.S Newspapers.*

During, Simon. *Modern Enchantments*. Cambridge, MA, and London: Harvard University Press, 2002.

Dyer, Richard. *White: Twentieth Anniversary Edition*. 2nd ed. New York: Routledge, 2017.

Elcott, Noam M. *Artificial Darkness: An Obscure History of Modern Art and Media*. Chicago: University of Chicago Press, 2016.

Elcott, Noam M. "A Brief History of Artificial Darkness and Race." In *Rethinking Darkness: Cultures, Histories, Practices*, 61–76. Edited by Nick Dunn and Tim Edensor. New York: Routledge, 2020.

Ellison, Ralph. *Invisible Man*. New York: Vintage International, 1980.

Fischer, Ottokar. "History Makers in the World of Magic: The Inventor of Black Art Ben-Ali-Bey (Max Auzinger)." *The Conjurors Monthly Magazine*, December 15, 1907.

Goto-Jones, Chris. *Conjuring Asia: Magic, Orientalism, and the Making of the Modern World*. Cambridge: Cambridge University Press, 2016.

Goto-Jones, Chris. "Magic, Modernity, and Orientalism: Conjuring Representations of Asia." *Modern Asian Studies* 48, no. 6 (November 2014): 1451–1476.

Haskins, James, and Kathleen Benson. *Conjure Times: Black Magicians in America*. New York: Walker & Co, 2001.

Hazzard-Donald, Katrina. *Mojo Workin': The Old African American Hoodoo System*. 1st edition. Urbana: University of Illinois Press, 2012.

Herrmann, Adelaide. *Adelaide Herrmann Queen of Magic: Memoirs, Published Writings and Collected Ephemera*. Edited by Margaret B. Steele. Place of publication not identified: Bramble Books, 2011.

Herrmann, Alexander. *Herrmann's Book of Magic: Black Art Fully Exposed*. Chicago: Frederick J. Drake & Company, 1903.

Hopkins, Albert Allis, and Henry Ridgely Evans. *Magic: Stage Illusions and Scientific Diversions, Including Trick Photography*. New York: Munn & Co., 1901.

Kellar, Harry. *A Magician's Tour Up and Down and Round about the Earth: Being the Life and Adventures of the American Nostradamus*. Chicago: R. R. Donnelley & Sons, 1886.

Levack, Brian, ed. *The Oxford Handbook of Witchcraft in Early Modern Europe and Colonial America*. Oxford: Oxford University Press, 2013.

Malchow, Howard L. *Gothic Images of Race in Nineteenth-Century England*. Palo Alto, CA: Stanford University Press, 1996.

Maskelyne, J. N. "Our Mail Bag." *The Conjurors Monthly Magazine*, February 15, 1908.

McMillan, Timothy J. "Black Magic: Witchcraft, Race, and Resistance in Colonial New England." *Journal of Black Studies* 25, no. 1 (1994): 99–117.

Mills, Charles W. "The Illumination of Blackness." *The Black Scholar* 43, no. 4, 32–37.

Murphy, Joseph M. "Black Religion and 'Black Magic': Prejudice and Projection in Images of African-Derived Religions." *Religion* 20, no. 4 (October 1, 1990): 323–337.

Newlin, T. J. "The Black Art Entertainment." In *Popular Electricity and the World's Advance*, 1912.

Paton-Williams, David. *Katterfelto: Prince of Puff*. Leicestershire, UK: Troubador Publishing Ltd, 2008.

Said, Edward W. *Orientalism*. New York: Vintage Books, 1979.

Scientific American. "Black Art Redivivus," August 31, 1907.

Scot, Reginald, 1538? *The Discouerie of Witchcraft Vvherein the Lewde Dealing of Witches and Witchmongers Is Notablie Detected, the Knauerie of Coniurors, the Impietie of Inchantors, the Follie of Soothsaiers, the Impudent Falshood of Cousenors, the Infidelitie of Atheists, the Pestilent Practises of Pythonists, the Curiositie of Figurecasters, the Vanitie of Dreamers, the Beggerlie Art of Alcumystrie, the Abhomination of Idolatrie, the Horrible Art of Poisoning, the Vertue and Power of Naturall Magike, and All the Conueiances of Legierdemaine and Iuggling Are Deciphered: And Many Other Things Opened, Which Have Long Lien Hidden, Howbeit Verie Necessarie to Be Knowne. Heerevnto Is Added a Treatise Vpon the Nature and Substance of Spirits and Diuels, &c: All Latelie Written by Reginald Scot Esquire.* Early English Books, 1475–1640 / 1116:07. Imprinted at London: by [Henry Denham for] William Brome, 1584., 1584.

Sharpe, Christina. *In the Wake: On Blackness and Being*. Durham, NC: Duke University Press, 2016.

"The Sphinx at the Egyptian Hall." *London Play Places* 10, no. 27 (November 18, 1865).

Spring, Lindley. *The Negro at Home: An Inquiry After His Capacity for Self-Government and the Government of Whites for Controlling, Leading, Directing, or Cooperating in, the Civilization of the Age*. New York: Author, 1868.

Steinmeyer, Jim. *The Glorious Deception: The Double Life of William Robinson, aka Chung Ling Soo, the Marvelous Chinese Conjuror*. Cambridge, MA: Da Capo Press, 2006.

Verkerk, Dorothy Hoogland. "Black Servant, Black Demon" *Journal of Medieval and Early Modern Studies*, 31, no. 1 (Winter 2001).

Webster, Noah. *An American Dictionary of the English Language: Exhibiting the Origin, Orthography, Pronunciation, and Definition of Words*. Harper, 1844.

Wells, H. G. *The Invisible Man*. Washington, DC: National Geographic Books, 2005.

8 THE LIGHT OF THE WORLD (1997)

RICHARD DYER

A television company is about to shoot a panel discussion before a studio audience. The producer, from the control room, is discussing with the floor manager in the studio how the audience looks in his monitor. The producer says something about the number of black people at the front of the audience. "You're worried there are not too many whites obviously there?", asks the floor manager. No, says the producer, it's nothing like that, a mere technical matter, a question of lighting—"it just looks a bit down."

This exchange occurred in the preparation of a programme about the street fighting that took place in Handsworth, Birmingham, in September 1985, fighting that was largely understood to be about race and which was the most vivid and controversial of many such incidents throughout Britain that year. The exchange was recorded by the Black Audio Film Collective and included in their film *Handsworth Songs* (1987),[1] which explores the cultural construction of "race riots". That construction is embedded in part in the professional common sense of media production, two items of which are registered in this exchange.

One item, which is less germane to the subject of this chapter, is that of "balance". The floor manager cannot at first understand what the producer is getting at. Is it perhaps the racial composition of the audience in numerical and representative terms? The topic of the programme has been constructed as race riots, and to have "balance" one has to thinks in terms of sides and ensure that equal numbers on each side are represented. This lies behind the floor manager's remark about there perhaps not being "enough white people obviously there", and the producer understands what he is getting at. However, it is not what concerns him or us here, whereas the second notion at play in the exchange goes straight to the heart of the matter.

For the producer it is a purely aesthetic matter. The image looks "down": dull, dingy, lacking sparkle. There is no reason to presume he is saying this because he finds black people dislike able or uninteresting. He is, in the terms of professional common sense, right: shoot the scene in the usual way with the usual technology with that audience and it will look "down". The corollary is that if you do it the usual way with a white audience, it will look "up", bright, sparkling. What I want to explore in this chapter is how this comes to be and what it signifies.

What is at issue is an aesthetic technology. The producer is making a statement about the formal quality, the look, of a set-up. The technology he is using and the habitual ways of using it both produce a look that assumes, privileges and constructs an image of white people.

All technologies are at once technical in the most limited sense (to do with their material properties and functioning) and also always social (economic, cultural, ideological). Cultural historians sometimes ride roughshod over the former, unwilling to accept the stubborn resistance of matter, the sheer time and effort expended in the trial and error processes of technological discovery, the internal dynamics of technical knowledge. Yet the technically minded can also underestimate, or even entirely discount, the role of the social in technology. Why a technology is even explored, why that exploration is funded, what is actually done with the result (out of all the possible things that could be done with it), these are not determined by purely technical considerations. Given tools and media do set limitations to what can be done with them, but these are very broad; in the immediacy and instantaneity of using technologies we don't stop to consider them culturally, we just use them as we know how—but the history, the social inscription, is there all the same.

Several writers have traced the interplay of factors involved in the development of the photographic media (e.g. Altman 1984, Coleman 1985, Neale 1985, Williams 1974) and this chapter is part of that endeavour. I am trying to add two things, in addition to the specificity of a focus on light. The first is a sense of the racial character of technologies, supplementing the emphasis on class and gender in previous work. Thus just as perspective as an artistic technique has been argued to be implicated in an individualistic world view that privileges both men and the bourgeoisie, so I want to argue that photography and cinema, as media of light, at the very least lend themselves to privileging white people. Second, I also want to insist on the aesthetic, on the technological construction of beauty and pleasure, as well as on the representation of the world. Much historical work on media technology is concerned with how media construct images of the world. This is generally too sophisticatedly conceptualised to be concerned with anything so vulgar as whether a medium represents the world accurately (though in practice, and properly, this lingers as an issue) but is concerned with how an ideology—a way of seeing the world that serves particular social interests—is implicated in the mode of representation. I have no quarrel with this as such, but I do want to recognise that cultural media are only sometimes concerned with reality and are at least as much concerned with ideals and indulgence, that are themselves socially constructed. It is important to understand this too and, indeed, to understand how representation is actually implicated in inspirations and pleasures.

The aesthetic technology at issue in this chapter is light and lighting. This is fundamental to all photographic media; why this is so and what it involves are discussed in the next section ("Light and the Photographic Media"). Mainstream cinema (above all, Hollywood) developed a particular style of lighting that may be called "movie lighting", and this will be described in the same section. Both the technology of lighting and the specific mode of movie lighting have racial implications. The third section of the chapter ("Lighting for Whiteness") discusses these in historical and current practice, looking at the way the aesthetic technology of light has a tendency to assume, privilege and construct an idea of the white person. The fourth section ("A Culture of Light") traces the development of the technology and its implication in white ways of seeing whiteness. The argument there is that the photographic media are centrepieces in a whole culture of light that is founded on two particular notions, namely that reality can be represented as being on a ground of white, and that light comes from above; these notions have the effect not only of advantaging white people in representation and of discriminating between and within them, but also of suggesting a special affinity between them and the light. In the final section ("The Glow of White Women"), I take a particularly striking example of the use of light in white representation, namely the glow of the woman within the construction of white heterosexuality.

LIGHT AND THE PHOTOGRAPHIC MEDIA

Photography and film are media of light. Writing about them frequently asserts this. Consider these titles, chosen from what comes to hand as I write: "Painting with Light" (Milner 1930, about film lighting), *Painting in Light* (Alton 1949, about cinematography), *Printed Light* (Ward and Stevenson 1986, about early photography), *Narration in Light* (Wilson 1986, about point of view in film and not dealing specifically with light at all); or these statements:

> Lighting is the very essence of the motion picture. Figuratively it is the palette of the art director and occupies the same relation as pigment does to painting. It is the medium.
>
> (Ihnen and Atwater 1925: 27)

> As writers work with words, and directors with players, [the cameraman] works with a more elusive medium: light.
>
> (Hoadley 1939: 52)

> [P]hotography is building with light; light is the medium, a medium of infinite plasticity.
>
> (De Maré 1970: 55)

> Basically, the visual part of the illusion we call the motion picture is nothing but the accurate control of light.
>
> (Handley 1967: 120)

> Films are light.
>
> (Federico Fellini, quoted in Malkiewicz 1986: 1)

Such titles and statements are not metaphors: photography and film really are technologies of light. A photographic image is the product of the effect of light on a chemically prepared surface (the stock); a single frame of a projected film is one such surface with light shining through it on to a screen. No light, no photos or films.

The apparatus of light does not only concern stock and projection. Schematically, the following are involved.

- The light thrown at the subject of the image, in this sense the *lighting*. This may simply be the available

  light in the location, be it natural or artificial, which may be assisted (e.g. by reflectors and other kinds of bouncing) or controlled (e.g. by filtering sunlight though muslin or glass). It may equally involve a wide array of kinds of lighting apparatus, produced by a variety of means, using various lenses, reflectors and so on, all with different qualities of light.

- The properties of the *subject*. Objects (including people) reflect and absorb light differently, as well as being of different colours.
- The stock. Different kinds of stock are sensitive to different kinds and intensities of light.
- *Exposure*. The length of time the photographic stock is exposed to the light and the size of the aperture through which light passes both affect how light is registered on the stock and therefore how it looks when developed and projected.
- The *development* of the stock, during which many adjustments and transformations in light qualities can be achieved.
- *Projection*. Different kinds of projectors produce different qualities of light, and of course film seen on video is different again.

All these elements are in play in the history and practice of the photographic media, including film. A change in stock to register certain colour and textural qualities entails changes in, say, lighting and make-up, which in turn call for modifications in developing procedures and eventually the stock itself. The introduction of sound and colour were particularly intense periods of such alteration, but the technology has never stayed still.[2] Nor was this a purely internal, technical history, since it is entwined with issues of economics, fashion, social pressure and aesthetics.

The history of light technology, even if we just confine ourselves now to film, is highly intricate. Yet through it one can see operating a fairly consistent sense of what light in film should be, which acted as both benchmark and goal for most innovations and variations. This is embodied in a style of lighting developed by the 1920s which became and remains so widespread that it can be referred to as, for instance, "cinema lighting" (Coutard 1966: 9) or the "film look" (Malkiewicz 1986: 100). I shall call it "movie lighting" and will draw examples principally from Hollywood films, since by the 1920s they had come to dominate and set standards for most other film production.

Although this lighting style involves all the elements described above as contributing to the medium's technology of light, it is most easily described in relation to the light thrown at the subject, "lighting" in this sense. Its guiding principal is controlled visibility. At a very basic level, movie lighting wants to ensure that what is important in a shot is clearly visible to the audience. This may seem laboriously elementary but, on the one hand, simply shooting in available light may not achieve this (as all casual photographers know) and, on the other, what constitutes importance in a shot is not a given, but a matter of the film-makers' choices. It may certainly not be story clarity in any obvious sense. In a mystery film, it may be important that the audience cannot see something. In a star vehicle, seeing the star to best advantage may outweigh other narrative considerations. In a drama in which no character is supposed to take precedence over the others in the audience's sympathies, equal visibility for all concerned may be required, something that is quite hard to achieve (given the way natural, and normal domestic, light fall). It is in relation to such problems of film expression that the elaborately controlled method of movie lighting evolved.

Movie lighting nearly always considers people to be the most important element in a shot. In practice, this means lighting people so that they are clearly separated from their surroundings, and do not appear to merge with the scenery. Most often, lighting is set up in two stages, once for the overall setting, once for the people in it, with the latter referred to as the "figure lighting". It is common in movies for there to be a perceptible difference between the figure lighting in a long or medium shot and that in a closeup. This is so even in films as early as *Hearts of the World* (USA 1918) and *Way Down East* (USA 1920) (both directed by D. W. Griffith), where the full panoply of light technology was not yet in place. In the former, there is a scene where the heroine (Lillian Gish) is menaced by a German officer (Eric von Stroheim). The lighting in the shots of them together is an even, overall illumination. However, in the close-ups of her, cringing from his advances against a wall, she is lit slightly from below, with no light on the wall—the light catches her wide eyes and the whiteness of her face is emphasised by the contrast with the darkness of the wall. Such a change in light for expressive purposes between mid and close shots is unnaturalistic but wholly within the convention of movie lighting. In *Way Down East*, when the hero and heroine (Robert Harron and Lillian Gish) first meet, the long and medium shots of them are taken in unassisted sunlight, with only the placing of the performers ensuring that at least each face is visible. However, the close-ups of Harron and Gish use a variety of techniques (soft focus, gauzes, placement against plain white backgrounds, side lighting) to make them special; editing in combination with lighting makes them stand out, as befits both their role as the lovers and their star status. Both these examples also construct the characteristic glow of white women, contrasted especially in the first case with a dark masculine desire that, under the pressure of war propaganda, would also have been felt as racially other.

Figure lighting is the main concern of this chapter, since what is at issue is the image of people. Movie figure lighting ensures the proper visibility of each performer (according to the needs of narration, star status and so on) by a use of several lights, classically in a three-point system consisting of a primary light (the key), giving general illumination of the figure, a second, softer light (the fill), eliminating some of the shadows created by the key and other set lighting, and backlighting, which serves to keep the figure separate from the background as well as creating, when wanted, the rim and halo effects of heroic and glamour lighting.

In practice, movie lighting is a good deal more complex and flexible than this. Three lights for figure lighting is minimal and the picture is always complicated by the infinite range of placements possible for all three positions and by their relationship to other lighting in the scene and to all the other elements that constitute light in film. Nor was the pattern of lighting fixed in the 1920s to remain utterly unchanged ever since. Hollywood, the paradigm of movie style, was influenced by, for instance, both 1920s German ("expressionist") lighting styles, with their much greater emphasis on chiaroscuro effects cutting across the figure (as in, especially, film noir and the horror film), and the bright, bounced, overall white light of the French new wave, less concerned with picking out the figures in a hierarchy of importance.[3] Yet such influences modified movie lighting rather than displacing it, and I have been struck while writing this chapter how extraordinarily pervasive the style remains, in ordinary television as much as in contemporary art cinema and still overwhelmingly in Hollywood movies. . . .

LIGHTING FOR WHITENESS

The photographic media and, *a fortiori*, movie lighting assume, privilege and construct whiteness. The apparatus was developed with white people in mind and habitual use and instruction continue in the same vein, so much so that photographing non-white people is typically construed as a problem. All technologies work within material parameters that cannot be wished away. Human skin does have different colours which reflect light differently. Methods of calculating this differ, but the degree of difference registered is roughly the same: Millerson (1972: 31), discussing colour television, gives light skin 43 per cent light reflectance and dark skin 29 per cent; Malkiewicz (1986: 53) states that "a Caucasian face has about 35 percent reflectance but a black face reflects less than 16 percent". This creates problems if shooting very light and very dark people in the same frame. Writing in *Scientific American* in 1921, Frederick Mills, "electrical illuminating engineer at the Lasky Studios," noted that

> when there are two persons in [a] scene, possibly a star and a leading player, if one has a dark make-up and the other a light, much care must be exercised in so regulating the light that it neither "burns up" the light make-up nor is of insufficient strength to light up the dark make-up. (1921: 148)

The problem is memorably attested in a racial context in school photos where either the black pupils' faces look like blobs or the white pupils have theirs bleached out.

The technology at one's disposal also sets limits. The chemistry of different stocks registers shades and colours differently. Cameras offer varying degrees of flexibility with regard to exposure (effecting their ability to take a wide lightness/darkness range). Different kinds of lighting have different colours and degrees of warmth, with concomitant effects on different skins. However, what is at one's disposal is not all that could exist. Stocks, cameras and lighting were developed taking the white face as the touchstone. The resultant apparatus came to be seen as fixed and inevitable, existing independently of the fact that it was humanly constructed. It may be—certainly was—true that photo and film apparatuses have seemed to work better with light-skinned peoples, but that is because they were made that way, not because they could be no other way.

All this is complicated still further by the habitual practices and uses of the apparatus. Certain exposures and lighting set-ups, as well as make-ups and developing processes, have become established as normal. They are constituted as the way to use the medium. Anything else becomes a departure from the norm, or even a problem. In practice, such normality is white.

The question of the relationship between the variously coloured human subject and the apparatus of photography is not simply one of accuracy. This is certainly how it is most commonly discussed, in accounts of innovation or advice to photographers and film-makers. There are indeed parameters to be recognised. If someone took a photo of me and made it look as if I had olive skin and black hair, I should be grateful but have to acknowledge that it was inaccurate. However, we also find acceptable considerable departures from how we "really" look in what we regard as accurate photos, and this must be all the more so with photography of people whom we don't know, such as celebrities, stars and models. In the history of photography and film, getting the right image meant getting the one which conformed to prevalent ideas of humanity. This included ideas of whiteness, of what colour—what range of hue—white people wanted white people to be . . .

Innovation in the photographic media has generally taken the human face as its touchstone, and the white face as the norm of that. The very early experimenters did not take the face as subject at all, but once they and their followers turned to portraits, and especially once photographic portraiture replaced painted portraits in popularity (from the 1840s on), the issue of the "right" technology (apparatus, consumables, practice) focused on the face and, given the clientele, the white face. Experiment with, for instance, the chemistry of photographic stock, aperture size, length of development and artificial light all proceeded on the assumption that what had to be got right was the look of the white face. This is where the big money lay, in the everyday practices of professional portraiture and amateur snapshots. By the time of film (some sixty years after the first photographs), technologies and practices were already well established. Film borrowed these, gradually and selectively, carrying forward the assumptions that had gone into them. In turn, film history involves many refinements, variations and innovations, always keeping the white face central as a touchstone and occasionally revealing this quite explicitly, when it is not implicit within such terms as "beauty," "glamour," and "truthfulness." Let me provide some instances of this.

The interactions of film stock, lighting and make-up illustrate the assumption of the white face at various points in film history. Film stock repeatedly failed to get the whiteness of the white face. The earliest stock, orthochromatic, was insensitive to red and yellow, rendering both colours dark. Charles Handley, looking back in 1954, noted that with orthochromatic stock, "even a reasonably light-red object would photograph black" (1967: 121). White skin is reasonably light-red. Fashion in make-up also had to be guarded against, as noted in one of the standard manuals of the era, Carl Louis Gregory's *Condensed Course in Motion Picture Photography* (1920):

> Be very sparing in the use of lip rouge. Remember that red photographs black and that a heavy application of rouge shows an unnaturally black mouth on the screen.
>
> (316)

Yellow also posed problems. One derived from theatrical practices of make-up, against which Gregory inveighs in a passage of remarkable racial resonance:

> Another myth that numerous actors entertain is the yellow greasepaint theory. Nobody can explain why a performer should make-up in chinese yellow. . . . The objections to yellow are that it is non-actinic and if the actor happens to step out of the rays of the arcs for a moment or if he is shaded from the distinct force of the light by another actor, his face photographs BLACK instantly.
>
> (ibid.: 317; emphasis in original)

The solution to these problems was a "dreadful white make-up" (actress Geraldine Farrar, interviewed in Brownlow 1968: 418) worn under carbon arc lights so hot that they made the make-up run, involving endless retouching. This was unpleasant for performers and exacerbated by fine dust and ultraviolet light from the arcs, making the eyes swollen and pink (so-called "Klieg eyes" after the Kliegl company which was the main supplier of arc lights at the time (Salt 1983: 136)). These eyes filmed big and dark, in other words, not very "white", and involved the performers in endless "trooping down to the infirmary" (Brownlow 1968: 418), constantly interrupting shooting for their well-being and to avoid the (racially) wrong look. [89–92] . . .

Movie lighting in effect discriminates on the basis of race. As the rest of this chapter will argue, such discrimination has much to do with the conceptualisation of whiteness. There is also a rather different level at which movie lighting's discrimination may be said to operate. What is at issue here is not how white is shown and seen, so much as the assumptions at work in the way that movie lighting disposes people in space.

Movie lighting relates people to each other and to setting according to notions of the human that have historically excluded non-white people. Movie lighting focuses on the individual. Each person has lighting tailored to his or her personality (character, star image, actorly attributes). Each important person, that is. At a minimum, in a culture in which whites are the important people, in which those who have, rather than are, servants, occupy centre stage, one would expect movie lighting to discriminate against non-white people in terms of visibility, individualisation and centrality. I want however to push the argument a bit further. Movie lighting valorises the notion of the unique and special character of the individual, of the individuality of the individual. It is at the least arguable that white society has found it hard to see non-white people as individuals; the very notion of the individual, of the freely developing, autonomous human person, is only applicable to those who are seen to be free and autonomous, who are not slaves or subject peoples. Movie lighting discriminates against non-white people because it is used in a cinema and a culture that finds it hard to recognise them as appropriate subjects for such lighting, that is, as individuals.

Further, movie lighting hierarchises. It indicates who is important and who is not. It is not just that in white racist society, those who are not white will be lit to be at the bottom of the hierarchy, but that the very process of hierarchisation is an exercise of power. Other and non-white societies have hierarchies, of course; it is not innate to white nature. However, hierarchy, the aspirational structure, is one of the forms that power has taken in the era of white Western society.

Movie lighting also separates the individual, not only from all other individuals, but from her/his environment. The sense of separation from the environment, of the world as the object of a disembodied human gaze and control, runs deep in white culture. The prime reason for introducing backlighting in film was to ensure that the figures were distinguished from their ground, to make them stand out from each other and their setting. This was regarded as an obvious necessity, so clearly part of how to see life that it was an unquestionable imperative. Yet it expresses a view of humanity pioneered by white culture; it lies behind its highly successful technology and the terrible price the environment now pays for this. People who are not white can and are lit to be individualised, arranged hierarchically and kept separate from their environment. But this is only to indicate the triumph of white culture and its readiness to allow some people in, some non-white people to be in this sense white. Yet not only is there still a high degree of control over who gets let in, but, as I want to argue in the rest of this chapter, the technology and culture of light is so constructed as to be both fundamental to the construction of the human image and yet felt to be uniquely appropriate to those who are white.

A CULTURE OF LIGHT

The aesthetic technology of photography, as it has been invented, refined and elaborated, and the dominant uses of that technology, as they have become fixed and naturalised, assume and privilege the white subject. They also construct that subject, that is, draw on and contribute to a perception of what it means to be white. They do this as part of a much more general culture of light. This has produced both an astonishing set of technologies of light and certain fundamental philosophical, scientific and aesthetic perceptions of the nature of light. White people are central to it, to the extent that they come to seem to have a special relationship to light.

The culture of light is part of the still wider characterisation of modern Western culture as one which privileges seeing above all other senses. In his *History of Bourgeois Perception,* Donald Lowe argues that different epochs are characterised by the primacy given to different senses, embodied in the apparatuses developed to express and extend them. In oral cultures, hearing and touching have primacy over seeing, and even with the introduction of writing, this remains true. It is printing (making the seen word more widespread and available) and perspective (emphasising verisimilitude in imagery and a notion of knowledge through looking rather than from revelation or authority) that begin to move the (bourgeois, Western, white) world towards an emphasis on seeing as the epistemic sense par excellence. This is evident in the pervasiveness of sight metaphors about knowledge: seeing is believing, I see what you mean, as far as I can see, the evidence of one's own eyes, taking the long view, a short-sighted decision, she saw right through it, oh I see!

The equation of seeing with knowledge is also one with power. Michel Foucault points to the exercise of power in modern societies through surveillance (literally "overseeing"), the organisation and control of the social order by means of enhanced and highly elaborated mechanisms of sight. This is realised in the panopticon, in the first instance a prototypical eighteenth-century prison architecture with wardens situated in a high centre and cells ranged all around, thus affording the former a constant view of the latter and instilling in prisoners a constant sense of being surveilled, which becomes internalised as self-surveillance (Foucault 1975). This has been taken as a paradigm of a leading tendency of modern societies, their elaboration of bureaucracies, databanks and surveillance mechanisms, habits of self scrutiny (how do I look? why am I doing this? who will they think I am?) and, not least, both the use of enhanced technologies of light as a form of "subjection by illumination" (Foucault 1980: 154) and the development of photography and film as modes for recording, classifying and controlling the populace and observing and presenting oneself.

At the technological level, seeing has been enhanced and transformed such that human beings can see better, can record what they see more accurately and can reproduce that record more or less ad infinitum. Instruments enabled humans to see ever further (the telescope) and in ever more detail (the microscope). The camera obscura simultaneously reproduced and framed a sight, giving an image that could readily be copied. Photography directly and chemically fixed the image in a camera obscura; film, an exponential multiplication of photographs, caught the image of movement. The photographic media permit the endless reproduction of these fixed and caught impressions.

Such a stress on sight poses a problem, however, in relation to that which cannot be seen. In a humanist culture, the cardinal instance of this is human personality. No amount of looking at someone gives authoritative access to their inner being. Yet just such scrutiny in search of personality has characterised the past two centuries. From the growth of surgery, not least as demonstrated in, literally, theatres of anatomy, through the sciences of phrenology and the rest, Western society has had a positive mania for trying to see what's inside the human being, body and soul (Clair 1994). The photographic media, too, so clearly at the cutting edge in capturing appearances, have also sought to see and show past them. The earliest social success in photography was its use in portraiture, starting around 1840 and rapidly expanding throughout international bourgeois society. Portraiture, whether in Matthew Brady's hugely successful Broadway studios (a must for any mid-nineteenth-century visitor to New York) or Julia Margaret Cameron's intense, private studies of friends and the famous, was founded on the belief that the apparatus could reveal the inner being (or could be made to). . . .

NOTES

1. I'm grateful to Ann Gray for bringing this example to my attention.

2. Overviews of this history can be found in Handley (1967), Salt (1983), and Revault d'Ailones (1991).

3. In his study of lighting in film, Fabrice Revault d'Allones distinguishes movie or "classical" lighting (which can be more highly wrought to the point of being "baroque") from what he terms "modernist" lighting. The latter does not seek to control meaning or to hierarchise elements in the image; nor does light itself carry meaning (of, for example, darkness being sinister). He takes the example of the film *Thérèse* (France 1986), about the life of a young nun. . . . [Allones] finds this kind of lighting in the work of individual directors at various points in the history of cinema, but sees it as characteristic of only a brief period of cinema, around the French new wave in the 1960s. In its indifference to (racial) character, such lighting will reproduce the privileging of the white face discussed in this chapter.

BIBLIOGRAPHY

Altman, Rick (1984) "Towards a Theory of the History of Representational Technologies," *Iris*, 2(2): 111–124.

Alton, John. *Painting with Light.* New York: Macmillan, 1949.

Brownlow, Kevin. *The Parade's Gone By.* London: Seeker & Warburg, 1968.

Clair, Jean, ed. *L'Ame au Corps: Arts et Sciences 1793–1993*, Paris: Gallimard, 1994.

Coleman, A. D. "Lentil Soup." *Et cetera* Spring(1985): 19–31.

Coutard, Raoul. "Light of Day." *Sight and Sound* 35(1)(1966): 9–11.

De Maré, Eric. *Photography*, Harmondsworth, UK: Penguin, 1970.

Originally published in 1957.

Foucault, Michel. *Surveiller et punir. Naissance de la prison.* Paris: Gallimard, 1975.

Foucault, Michel. "The Eye of Power." In *Power/Knowledge: Selected Interviews and Other Writings 1972–1977*, edited by Colin Gordon, 146–165, New York: Harvester, 1980.

Gregory, Carl Louis, ed. *A Condensed Course in Motion Picture Photography.* New York: New York Institute of Photography, 1920.

Handley, C. W. "History of Motion-Picture Studio Lighting." In Fielding, 1967: 120–124.

First published in the *Journal of the Society of Motion Picture and Television Engineers*, October 1954.

Handley, C. W. "Lighting for Technicolor Motion Pictures." *Journal of the Society of Motion Picture Engineers*, 25(5)(1935): 423–431.

Hoadley, Ray (1939) *How They Make a Motion Picture.* New York: Thomas Y. Crowell.

Hoadley, Ray. *How They Make a Motion Picture.* New York: Thomas Y. Crowell, 1939.

Inhen, Wiard B., and D. W. Atwater. "The Artistic Utilization of Light in the Photography of Motion Pictures." *Transactions of the Society of Motion Picture Engineers*, 21(1925): 21–37.

Malkiewicz, Kris. *Film Lighting*, New York: Prentice-Hall, 1986.

Millerson, Gerald. *The Technique of Lighting for Television and Motion Pictures*, London: Focal Press, 1972.

Mills, Frederick S. "Film Lighting as a Fine Art: Explaining Why the Fireplace Glows and Why Films Stars Wear Halos." *Scientific American*, 124(1921): 148,157–158.

Milner, Victor. "Painting with Light." In Hal Hall (ed.), *Cinematographic Annual*, 91–108. Los Angeles: American Society of Cinematographers, 1930.

Neale, Steve. *Cinema and Technology: Image, Sound, Colour.* London: Macmillan/British Film Institute, 1985.

Revault d'Ailones, Fabrice. *La Lumiere au cinema*, Paris: Seuil/Cahiers du cmema, 1991.

Salt, Barry. *Film Style and Technology: History and Analysis.* London: Starword, 1983.

Ward, John, and Sara Stevenson. *Printed Light.* Edinburgh: HMSO, 1986.

Williams, Raymond. *Television: Technology and Cultural Form.* London: Fontana, 1974.

Wilson, George. *Narration in Light: Studies in Cinematic Point of View.* Baltimore: Johns Hopkins University Press, 1986.

9 LOOKING AT SHIRLEY, THE ULTIMATE NORM: COLOUR BALANCE, IMAGE TECHNOLOGIES, AND COGNITIVE EQUITY (2009)

LORNA ROTH

> According to Jan Kasoff, an NBC colour-television cameraman on the program *Saturday Night Live*, . . . "a good VCR person will have a colour girl stand in front of the cameras and stay there while the technicians focus on her flesh tones to do their fine adjustments to balance the cameras. This colour girl is always white."
> —Jan Kasoff, former cameraman, NBC[1]

"Skin-colour balance" in still photography printing refers historically to a process in which a norm reference card showing a "Caucasian"[2] woman wearing a colourful, high-contrast dress is used as a basis for measuring and calibrating the skin tones on the photograph being printed. The light skin tones of these women—named "Shirley" by male industry users after the name of the first colour test-strip-card model—have been the recognized skin ideal standard for most North American analogue photo labs since the early part of the twentieth century and they continue to function as the dominant norm. Figures 9.1 and 9.2 are typical of widely-circulated Shirley cards used in most North American analogue photo labs until very recently.

The colour-TV industry in North America also had its version of "Shirley" in the form of a white porcelain "China Girl" in use until the 1950s, at which time it was replaced by BBC cardboard flesh-tone cards especially designed for compatibility with National Television System Committee

FIGURE 9.1
Polaroid Shirley card. Image courtesy of Polaroid.

(NTSC) and Phase Alternating Line (PAL) broadcast technologies (http://www.videointerchange.com/color_correction1.htm). The official name for this flesh-tone reference card is BBC Chart 61P. It is used in conjunction with a chip chart in all NTSC broadcasting networks. For PAL-based technologies, BBC Test Card F is used in over 30 nations around the globe. It pictures the daughter of its designer, George Hersee of the BBC.[3] This girl-image does not differ much in the lightness of complexion or ethnic

FIGURE 9.2
Traditional Kodak Shirley card, widely circulated since the 1950s. Image courtesy of Kodak.

appearance[4] from the other Shirley cards in circulation at photo labs at the time of its release.

Although I have researched multiple adjustments to flesh-tone-based colour products and technologies in the last decade, the core of my argument for this chapter stems from the experience of chemical- and digital-based photographers and their subjects' critiques. Central to this chapter is a series of Shirley images that can be seen as emblematic of the state of race-relations/aesthetics in the industries of visual representation and the

FIGURE 9.3
BBC flesh-tone reference chart of "Cardboard Kate," taken at Fuji Film & TV Research Lab headquarters in Mt. Fuji, Japan, in 2004. Image courtesy of Kodak; photo credit: Lorna Roth.

gradual shifts in technology targeted at improving the sensitivity of film emulsions and dynamic ranges (the difference between the lightest and the darkest of colours) in the digital media sphere. I am interested in two levels of historical analysis in the domain of colour balancing imagery. The first is that of the technical challenges presented by the limitations within imaging technologies and the ways in which an ensemble of practices emerged to address these deficiencies with reference to human skin tone reproduction quality. These deficiencies include the difficulty of imaging high contrasts in skin tones within the same screen shot—for example, a very dark-skinned person sitting next to a very pale-skinned person—and the lack of establishment and design of appropriate lighting and make-up for peoples of darker skin colours.

The lighting and make-up challenges have been tackled with a series of compensatory practices, which have addressed the issues with some level of success. The difficulty of colour balancing two extremely different skin tones in the same screen or on paper has been less successful. As you shall see from the material I have gathered for this chapter, the process of recognizing this challenge in 1959 and of informing a major film producer, Kodak, of the difficulty of achieving high-quality photo prints with multiracial content precipitated a lengthy socio-technical journey, which is not yet complete. In the digital camera environment, Philips (now Thomson Multimedia Broadcast Solutions) in Breda, the Netherlands, took up this challenge in the 1990s with its television studio cameras and has probably made the most progress to date on extending the dynamic range of its cameras. Philips' contribution to the solution of the contrast problem will be elaborated upon in the latter half of this chapter.[5]

The second level of analysis is cultural and racial in origin. Between 1959 and the present, there have been innumerable versions of Shirley as she has crossed the decades, continents, and skin colour lines. In this chapter, images of North American and Japanese Shirleys are seen as transforming to reflect the prevailing norms of skin colour beauty in the period in which the labs were using them. I shall argue here that the anthropological and sociological content of these images has always been emblematic of the period in which they were circulating. They reflect and reveal an order of domination and have had a social and psychological impact (Thierry Le Brun, Independent Cinematographer and Sociologist, Montreal, personal communication, January 7, 2009).

The relationship between the social and the technical in this story—how each randomly drove the other to redefine its object of interest—has been serendipitous. Although as a scholar, I would have preferred to have seen a patterned, ongoing attempt to improve skin rendition technologies by the various research labs, what history reveals instead is a random and messy search, based on ad hoc feedback from surprising publics. Peoples of colour, whose embodied imagery would have benefited from a more sensitive chemical emulsion in the case of still photography and a more dynamic range in the case of digital technology, were not the constituency group leading the visual engineers and scientists to further explore the dynamic range of their company's film products.

On a broader level, my reflections respond to the following key questions: What kinds of knowledge about human skin colour in the form of racialized imagery are manufacturers constructing and defending through the marketing of their "presumed innocent" products? What is the corporate stake in investing in or changing a colour aesthetic of Whiteness? What have been the precipitating factors that inspired companies to make recent skin colour modifications? Where are skin colour adjustments in chemical and digital photography taking place? How have colour balance procedures evolved since the 1970s in the techniques and social practices of photo cultures in North America and Japan? What are the larger socio-political and economic implications of colour balancing products and procedures?

ABOUT SHIRLEY

Shirley's skin became an industry standard in North American photo labs that were dominated by male employees in the 1940s and 1950s. It is no surprise then that the person who was imaged on a Shirley card would have had a look and a skin colour to conform to a popular masculinist notion of beauty, likely defined from a Western/European perspective. Alone or in groups, men wearing coloured shirts with similar skin-tone ranges and hair colours to those of female Shirleys could have worked as effectively as a reference standard—especially if they had had beards or moustaches. These actually might have provided technicians with the practice to deal more effectively with the contrast issue. Yet I have discovered few adult male reference images in my search through labs in the U.S., Canada, Britain, and Japan. The most gender-inclusive images I have found are those of either kids or brides and grooms used by Kodak in Japan and numerous companies in North America.[6] Consequently, at the very same time that technical and aesthetic decisions were being made by photo labs with regard to what constituted a "beautiful" skin colour norm, there emerged a masculinist collection of sexy female imagery to tinker with, pin up on lab walls, and use in the colour balancing process. This gendered bias is likely to continue until male visual designers begin to unstick themselves from the very powerful patterns of social relations into which they have been conditioned or until more women enter the labs and take on positions of power.

THE DISAPPEARANCE OF THE "TECHNOLOGICAL UNCONSCIOUS" (VACCARI, 1981)

In the early histories of photo and television technologies, it was possible to consider their design and the ensemble of production techniques and methods surrounding them as ideologically neutral. Until about the mid-1960s, it was probably assumed by most users that visual media were designed to "naturally" reproduce all skin tones equally well. As experience with the use of these photo technologies expanded to international markets, non-"Caucasian" communities identified shortcomings and became more critical and questioning of their visual quality. Problems for the African-American community, for example, have included reproduction of facial images without details, lighting challenges, and ashen-looking facial skin colours contrasted strikingly with the whites of eyes and teeth.

From a more technical perspective, evidence has been accumulating that the reason for these deficiencies is that film chemistry, photo lab procedures, video screen colour balancing practices, and digital cameras in general were originally developed with a global assumption of "Whiteness" embedded within their architectures and expected ensemble of practices. What had become a "White"-biased international standard for the ideal flesh tone had been used as a barometer against which the flesh tones of Blacks, Asians, First Peoples, and other "peoples of colour" had been read negatively as an aggravation—a deviation from this invisible norm. This, along with cross-cultural skin-colour-preference tests conducted by film manufacturers such as Kodak and Fuji, had confirmed an international preference for light complexions within the global consumer photo markets. The virtual public silence in Kodak's and Fuji's institutional discourses and professional literature on alternatives to traditional ways of colour balancing analogue prints is the most concrete evidence we have of this institutional oversight and resistance to change.[7]

It was from within the broader social context of professional visual technology users that new practices emerged. These included special lighting methods for Black skin, as well as trial-and-error colour balance techniques to compensate for the challenges of shooting and printing contrasting skin colours from within the same screen frame. Photographers of African-American and Asian subjects, who had developed methods for dealing with these "problems" independently, began to share their knowledge with each other and with the public some time around the late 1950s, at around the same time Kodak was experiencing some criticism of its photo emulsions regarding this very issue.

In what follows, I shall focus on the (inter)national driving forces that provoked Kodak, in particular, to rethink and redesign the range of its chemical representation of brown tones and to later multiracialize its

Shirley card as a gesture of inclusiveness to its broadening consumer photo markets.

THE COLOUR ADJUSTMENT PROCESS

THE CASE OF STILL PHOTOGRAPHY

> Colour photography is not bound to be "faithful" to the natural world. Choices are made in the development and production of photographic materials.
> —Winston, 1996

In the last decade, it has become clear to those who seek out this information that the chemistry for stock colour film for still cameras was designed originally with a positive bias toward "Caucasian" skin tones because of its high level of reflectivity (Personal interviews with multiple chemists and film designers at Kodak, Rochester, NY, 1995; Winston, 1985, 1996). This is not surprising, given that the dominant market in the early days of photography was perceived to be that of "Caucasians" by Kodak, the main film manufacturer in North America. This did not have to be the case. Had NASA, the U.S. intelligence service, or meteorological scientists already completed their research on photography of low-light areas at the time of the popular development of still photography, the evolution of film chemistry might have unfolded quite differently, as Brian Harris, lighting technician at the Black Entertainment Television (BET) network, pointed out to me (Brian Harris, Lighting Engineer, Black Entertainment Television, Washington, DC, personal communication, July 3, 1997).

Film emulsions could have been designed initially with more sensitivity to the continuum of yellow, brown, and reddish skin tones, but the design process would have had to be motivated by a recognition of the need for an extended dynamic range. At the time film emulsions were developing, the target consumer market would have been "Caucasians" in a segregated political scene; their skin tones would have been less likely to be the basis for thinking about dynamic range, because most subjects in a photograph would either have been all light-skinned or all darker-skinned. Thus, this was not an element of social consideration for film chemists. It was also believed at the time that physics was physics, chemistry was chemistry, and science was based on reasoned decisions without consideration of cultural or racial subtleties. It is now becoming acknowledged more widely within the industry that refinements to the chemistry of film emulsions have never been issues of physics or chemistry exclusively, but have been the result of cultural choices as well.

Five key issues were responsible for raising the level of awareness of Kodak research scientists and image technology developers with regard to skin colour reproduction in North America and elsewhere. The first had to do with the identification of problems in the 50s, when their film was being used for school graduation and class photos.

According to Jim Rice, the challenge became apparent only when students with contrasting skin colours were to be photographed in the same image frame (Jim Rice, former Technical Sales Representative and Marketing Manager, Kodak, and current McGee Endowed Professor, Rochester Institute of Technology, Rochester, NY, personal communication, August 18, 1995). When each student was photographed alone, differences in skin tones were easily accommodated through compensatory lighting and a range of technical adjustments learned through experience, but when a group portrait was set up and children of all races and ethnicities were photographed together, these techniques could not resolve the problem of the film bias in favour of "Caucasian" skin. Consequently, the picture results showed details on the White children's faces, but erased the contours and particularities of the faces of children with darker skin, except for the whites of their eyes and teeth. Parents complained about this situation and demanded a wider continuum of darker skin tones.

Figure 9.4 is another example of this phenomenon. This photo of two children in Senegal in the early seventies demonstrates that even years after complaints of uneven qualities apparent in images of dark and light skins, the issue remained uncorrected.

Kodak's drive to increase the dynamic range of its film products was motivated by two other (seemingly irrelevant) issues. These had to do with the photography of brown objects. Here is Kathy Connor's description of the experience of Earl Kage, former head of the Color Photo Studio at Kodak Park in the 60s and 70s, and former Manager of Kodak Research studios:

> "Well, he said that it was interesting, that in the mid-sixties and seventies there was a coincidental problem that the company was facing. Two of their biggest professional accounts were, he didn't name the company, but somebody said that they made chocolate candies. Apparently, in reproducing chocolate candies, Kodak was receiving complaints that they weren't getting the right brown tones on the chocolates. Also, furniture manufacturers were complaining that stains and

FIGURE 9.4
Two children in Senegal (Natalie Le Brun [right] and Guilado Sarr (or her sister) [left]). Circa 1973. Kodak film (125ASA) used with a Canon camera. An example of the challenges posed by photographing highly contrasted skin colors in the same frame. Image courtesy of Olivier Le Brun, Paris, France.

> wood grains in their advertisement photos were not true to life, and that they weren't appropriate, so the chemists did some work on that. Earl also said to a certain extent, that research to improve those professional markets and addressing their questions helped them to do a little bit better with ethnic skin colours. I was amazed." (Kathy Connor, Executive, Kodak, Rochester, NY, personal communication, August 16, 1995)

In his own words, Earl Kage remarked to me:

> "In the 4×5, 5×7, or even 8×10 colour transparency area that manufacturers of furniture were using to display their wares and to advertise their furniture for catalogues, they were having a good deal of difficulty in demonstrating the subtle differences of certain woods. Now, whether it was maple vs. oak vs. a couple of dark woods, this couldn't be distinguished in the photographs. This was also about the same time that we got some interesting observations from chocolate manufacturers who, in displaying Whitman's chocolate or whatever the names were in any case, the subtle variations between the dark and bittersweet and milk chocolates weren't as discernible and so some modifications were tried and consequently my little department became quite fat with chocolate, because what was in the front of the camera was consumed at the end of the shoot." (Earl Kage, former Manager, Kodak Research Studios, Rochester, NY, personal communication, August 21, 1995)

It is indeed interesting that the improvement of dark skin colour reproduction came about quite incidentally in this context, and Kage later admitted his own surprise at this submerged historical point. "Yes," he noted, "it is fascinating that this has never been said before, because it was never Black flesh that was addressed as a serious problem that I knew of at the time" (Earl Kage, former Manager, Kodak Research Studios, Rochester, NY, personal communication, August 21, 1995). Other than parents complaining about graduation photos, Kage did not recall pressures from the Black community to improve the image quality of Kodak's product. This is surprising in many ways. One would have thought that during the height of the civil rights movement in the 1960s and 70s, attention might have been turned to Kodak to demand better recognition of the communities' skin specificities. There were some economic conflicts between Kodak and its labourers in the 60s, many of whom were African-Americans, but the quality of the photo product was not contested in an organized manner by the Black communities, as far as I could discover. It is more likely that at the time, it was assumed by the public that such things were based on science and could not be changed, and so battles were fought on issues of economics, poverty, and other civil rights matters that were of higher priority to the African-American and African-Canadian communities.

The fourth and fifth factors motivating Kodak were the obvious desires to make some impact on the Japanese film stock market, which was strongly linked to Fuji films, and to extend its market to the global community. By the mid-90s, the Japanese had used data from their own colour preference tests to redesign the look of their Shirleys for their still photography labs. The new reference card look for analogue and digital photography, as well as video, was imagery of Japanese women with light yellow skin, although for their current television work, they also still use the BBC flesh tone card. Here, the Japanese Shirley for TV is described in detail by Toru Hasegawa:

> She is a Japanese lady, black hair, wearing a kind of light blue sweater, white shirt and the background is a kind of light gray. This picture is standard for NHK, Japanese TV stations. Every TV station uses the same photo to adjust their cameras. That is why NHK and some of the commercial TV stations appear almost the same. (Toru [Tom] Hasegawa, Video Engineer, NHK [Japan Broadcasting Corporation], New York, NY, personal communication, May 27, 1996)

FIGURE 9.5
Fuji's mannequin Shirley taken at Fuji Film & TV Research Lab headquarters in Mt. Fuji, Japan in 2004. Image courtesy of Fujifilm; photo credit: Lorna Roth.

FIGURE 9.6
Kodak Japan's bride and groom card. Image courtesy of Kodak; photo credit: Lorna Roth.

Even though most of NHK's and Fuji's current Shirley images have been facially ethnicized with the skin tone still remaining on the lighter end of the spectrum, the act of changing the cards showed their will to culturalize/indigenize their visual standards to match the dominant aesthetic of the widest population they serve: the Japanese and Asian markets. Figures 9.5 and 9.6 show images of Shirley (single and in bride and groom format) which are intended to reflect the ethnic majority and the preferred skin tones in two of the dominant Japanese industries of visual representation, that of Fuji Film and Kodak, Japan. These have been widely circulating since the mid-nineties in Japan.

Ongoing international skin colour preference tests conducted by Kodak have also generated much data useful in informing Kodak film chemists and "designers" of the skin colour biases preferred in different parts of the world. This information has shaped Kodak's geography of emulsions, which conform to these preferences, rather than to considerations of exact reproduction. The industry term used for this business choice is "optimum reproduction" (Winston, 1996). For this accommodation, film inventory is batched by region and distributed in accordance with these researched preferences. Although there are no explicit signs on Kodak's film boxes of where each target market is located geographically, the film is coded numerically to indicate countries or regions. So, for example, in 1996, several Kodak sources indicated the regional codes to be as follows: #1—U.S. & Canada; #2—Latin America, Central America, South America, and Mexico; #3—Asia Pacific, China, and Japan (often treated as a separate entity); #4—Europe, Middle East, Africa, India.

ONGOING DEVELOPMENTS

Skin-tone rendition was further clarified with Kodak's VeriColor portrait film series, which has continued to expand its range of brown-ness and Black tones. VeriColor III, a professional portrait film developed in the early 80s by Richard Wien and his team at Kodak Park, was particularly notable for its flexible accommodation of a range of skin colours. Gold Max, a very popular consumer film, was also a leap forward in this regard from most previous films on the market: it was referred to at Kodak initially as being able "to photograph the details of a dark horse in low light" (Richard Wien, Executive, Kodak, Rochester, NY, personal communication, August 18, 1995). With my interest in the filming techniques applicable to darker skins, I take this to be a coded message, informing the public that this is "the right film for photographing 'peoples of colour.'"

Finally, there have been some cultural changes over time to the Shirley norm reference cards to make them more inclusive. From the single "Caucasian" woman surrounded by the necessary colour balancing information codes, Kodak's Shirley has evolved into an image of three women with different skin colours, dressed fashionably in brightly contrasted clothing.

FIGURE 9.7
Kodak's multiracial Shirley card—North America, circa 1995. Image courtesy of Kodak and CC.

As seen in figure 9.7, the women are visibly of "Caucasian," Asian, and African descent, though each of them has a fairly light complexion. That being said, the use of this multiracial norm reference card by some of the major photo-chemical labs is concrete recognition that they are invested financially and intellectually in addressing a multiracial clientele and are no longer willing to use trial-and-error techniques on a case-by-case basis to find the appropriate colour balancing methods for processing images of non-"Caucasians." Some resistance to changing over to the multiracial reference is evident in the fact that although it was designed in 1995, it still sells at a rather high premium, and it took a few years to become available through major photo lab suppliers. Consequently, many laboratories have still not switched cards, and it has not yet penetrated the global market. Ironically, at about the same time as the multinational film stock companies began to recognize the diversity of skin tones within their Shirley imagery, it became a lot easier to design one's own norm reference images digitally, so it is no longer as invaluable a product as it once was.

Whether the decision to stick with the traditional image in North American analogue photo labs is based on financial or sociocultural considerations would require more corporate lab research and analysis. This might be important sociocultural research, as it would provide us with a concrete barometer reading as to how racially marked issues are being treated in the current daily practices of North American photo laboratories.

HOW DOES SHIRLEY FARE IN THE DIGITAL MEDIA SPHERE?

In the digital media sphere of the Internet, there are many Shirley cards circulating. One such digital card, which reinforces the common, almost universal preference for light skin as an international photo standard, is DuPont's Digital WaterProof image of a Black female with a very light complexion who parallels the existing "Caucasian" Shirley prototypes. She is seen beside narrow stripes and squares of digital colours, as well as the grey scale. Another colour balance image (online and in paper form), which initially came with Adobe Illustrator's Photoshop software program and is shown in figure 9.8, is that of a Latin American–looking woman, conjuring up memories of Carmen Miranda[8] with a fanned headdress containing a basket of exotic fruits, playing what looks like a mandolin. She is standing in front of a grey background, beside a set of colour bars. Adobe has exoticized its Shirley's fashion look. However, though she is Latino in appearance, the skin colour of this woman is no darker than that of a typical "Caucasian" female. Thus, despite the ethnicization of the Adobe "Shirley's" facial features, the lightness of complexion attests to the reappearance and re-privileging of the "look" of Whiteness as a beauty norm in this "internationally ideal" photo.

FIGURE 9.8
Adobe Illustrator's Shirley card. Reprinted with permission from Adobe.

It is important here to recognize the variation of skin tones within each racial and ethnic community and the intraracial or intraethnic hierarchy of skin-tone preferences from lightness to darkness. Aside from its impact on film manufacturers, the almost universal preference for light and pale skin has motivated discriminatory behaviours. As well, it has been a factor in the production of multiple consumer items, such as skin bleach, lighter make-ups, and sun-blocks that contain skin bleach as an essential ingredient in most Asian countries.

In the last three years, two more complex cards have begun to circulate in the digital sector. These expand upon the Kodak multiracial card and offer both concrete recognition of some of the critical issues with which photographers of peoples of colour have struggled and a solution to some of the obstacles they have met in the past.

The first is a colour card in which there is a multiplicity of images, including face shots of four young children with varying skin colours (two children are "Caucasian," one is of African descent, the other is Asian). The current source of this card is http://www.inkjetart.com/custom. As evident from figure 9.9, the grouping of various images of brightly coloured objects and colour palettes surrounding the images of the children of different races offers a technical improvement over the existing single- or three-person norm reference cards, as it represents a wider range of colour subtleties to which the technician can refer.

Figure 9.10, part of the Getty Images Collection, is an excellent example of an international reference tool that would have been useful to the Kodak photographer doing graduation photos in the 1950s. It contains several charts and images of people of "Caucasian," Asian, and African descent, as well as a kangaroo and a variety of bright-coloured household items. What makes this reference image distinctive is both its sociocultural inclusivity and its vast skin colour range: it has by far the best dynamic range technically of the colour balance photos in circulation that I have researched. Furthermore, it is freely distributed over the Web and only requires an informal user agreement with its copyright owner.

Has the expansion of racial imagery and discourses into the cyber-world confounded questions of race relations as it transposes them from the political to the virtually personal, from imagery coded by others to that which can be coded more democratically, by all of us? The trend of custom-designing individual Shirleys based on the specific needs of the

FIGURE 9.9
Four-child color card. This image is a rather progressive recognition of multiracialism as a global social current. Image courtesy of Dry Creek Photo.

photographer and subjects will no doubt expand as computer software for digital image manipulation becomes more affordable and popular. This will make a significant difference to independent photographers. That said, I would argue strongly that as more skin-tone-range choices open up and as colour standards for digital monitors and printing evolve, there will likely be little difference in the actual content of skin-tone colour balance cards in the analogue and digital printing sectors. In both cases, despite

FIGURE 9.10
Image commonly used for calibration verification. Located within the Getty Collection of photos on the Internet. Retrieved in 2009, it is no longer available on the Internet. Image courtesy of Dry Creek Photo.

the recognition of ethnic subjects, lighter skins will likely prevail on most of the reference cards for social and normative reasons. Consequently, it is in these socio-cultural domains that we need to focus our attention to understand more effectively the relationship between social cognition, the technologies we use, and the practices that emerge from this linkage. . . .

CONCLUDING REFLECTIONS: BEYOND SHIRLEY

Within the digital marketplace, Philips/Thomson and Ikegami were the first in the industry to risk large sums of investment capital in the recognition of a future multiracial market that will very likely expand over time as industrial globalization captures and maintains the norm in business practices. Other camera and photo companies have followed suit and are now far more conscious of the need for dynamic range in their cameras. The proliferation of multiracial Shirley cards by Fuji, Kodak, and other suppliers of photographic apparatus parallel this recognition in their industries' social field. The dual-skin-contour camera feature, which can colour balance two skin tones within the same image, comes closer to a technology that would enable cognitive equity than any other that I have seen to date. This is more than an incremental step in opening up representational practices to a form of inclusiveness that is designed into the technology itself; it is a leap forward, initiated quietly from outside of the mainstream geographies of the visual industries. Typically, innovation has come from the margins.

Beyond the image industries, the multiracialization of other flesh tone objects, such as the Crayola crayons' multicultural collection, the ethnicization of the common store mannequin and children's dolls, the skin colour ranges now available in hearing aids and other prosthetics, and the darkening of tonal ranges available in the make-up industries, each contribute in their own small way to the social and cultural possibilities of achieving cognitive equity sometime in the future. In addition to diversity studies and anti-racist education, public institutional changes based on equality rights legislation, cultural and racial inclusiveness within the industries of visual representation and journalism, as well as the very recent (2008) sea change in American presidential politics, will these technical shifts encourage the development of an unspoken but embedded anti-racist common sense and consciousness? There are no guarantees. However, in contemplating the words of anthropologist Margaret Visser, "The extent to which we take everyday objects for granted is the precise extent to which they govern and inform our lives" (Visser, 1986), I would like to suggest that the more open we become to new possibilities for racial inclusiveness in commonplace objects and technologies, the closer we will likely get to building a next generation whose social and cultural cognitive processes will be multiracial in scope and practice. Acknowledgment and explicit discussion about the ethnocultural and racial choices we embed within our technologies, products, and practices will, I hope, serve the purpose of raising our consciousness about how important it is to transform the way we think about, engage with, and act upon the "historical fixtures of our existence" (Walter Benjamin, as cited in Kearney, 1994, p. 153).

NOTES

1. New York, personal communication with Roth, November 20, 1994.

2. In the industries of visual representation, "White" people continue to be called "Caucasian." I apologize in advance to those who might be offended by the name, but I am using it to be consistent with the terminology still in active use in the industry. As an acknowledgment of the problematic aspect of this word, it (as well as the term non-"Caucasian") is used within quotation marks throughout this chapter.

3. For further information and to view several images of BBC calibration cards, see the following http://www.videointerchange.com/color_correction1.htm.

4. There is neither scholarly nor colloquial consensus on the popular meaning of terms like "ethnic," "multicultural," and "multiracial."

5. At the time that these innovations and interviews took place, Thomson Multimedia Broadcast Solutions headquartered in Breda, the Netherlands, was called Philips. To be respectful of and consistent with the historical period in which the design adjustments took place, I am choosing to refer to the company by its original name/ Thomson, as in Philips/Thomson.

6. In the latter series of photo cards, the age-old contrast issue is yet again duplicated with the black tuxedo against a white shirt and the white wedding gown against a contrasting background.

7. Kodak and Fuji are being used here as examples. I do not wish to imply that alternative film emulsion companies, such as Ilford and Agfa, among others, did anything differently in this regard.

8. Carmen Miranda (1909–1955) was a Portuguese-born Brazilian singer who became famous for performing while wearing a tropical fruit–laden hat. Her iconic image became associated with the United Fruit Company and the "Good Neighbor" policy of the U.S. in the 1930s. Images of Carmen Miranda can be found on Google Images.

http://images.google.ca/images?gbv=2&hl=en&q=carmen+miranda&btnG=Search+Images (viewed on February 27, 2009).

REFERENCES

PERSONAL INTERVIEWS

Connor, Kathy. (1995, August 16). Executive. Kodak. Rochester, NY.

Harris, Brian. (1997, July 3). Lighting engineer for Black Entertainment Television. Washington, DC.

Hasegawa, Toru (Tom). (1996, May 27). Video engineer from NHK, National Broadcasting System, Japan. New York City.

Kage, Earl. (1995, August 21). Former Manager of Kodak Research Studios. Rochester, NY.

Le Brun, Thierry. (2006, November 27; 2009, January 7). Independent cinematographer and sociologist. Montreal.

Rice, Jim. (1995, August 18). Former Technical Sales Representative and Marketing Manager of Kodak, and McGee Endowed Professor at Rochester Institute of Technology. Rochester, NY.

Wien, Richard. (1995, August 18). Executive, Kodak. Rochester, NY.

PUBLISHED WORKS

Kearney, Richard. "Walter Benjamin." In *Modern Movements in European Philosophy: Phenomenology, Critical Theory, Structuralism* (pp. 151–168). Manchester, UK: Manchester University Press, 1994.

Vaccari, Franco. *La photographie et l'inconscient technologique*. Paris, France: Créatis, 1981.

Visser, Margaret. *Much Depends on Dinner*. Toronto: McClelland & Stewart, 1986.

Wien, Richard. Executive, Kodak. Rochester, NY, August 18, 1995.

Winston, Brian. "A Whole Technology of Dyeing: A Note on Ideology and the Apparatus of the Chromatic Moving Image." *Daedalus, 114*(4)(1985), 105–123.

Winston, Brian. *Technologies of Seeing: Photography, Cinematography and Television*. London: British Film Institute (BFI Publishing), 1996.

10 THE CHINA GIRL ON THE MARGINS OF FILM (2015)

GENEVIEVE YUE

In 2005, at the Fogg Art Museum at Harvard University, Julie Buck and Karin Segal exhibited *Girls on Film*, a series of seventy photographs blown up from "China Girl" frames the two artists and archivists had collected while working at the Harvard Film Archive. The show gave visibility to an unusual type of technical image used in film laboratories: typically an image of a woman positioned next to color swatches and patches of white, gray, and black. The China Girl, sometimes called the China Doll, China Lady, girl head, or any number of lab-specific nicknames like Ullie, Marcie, Shirley, and Lilly, was used from the late 1920s until the early '90s, and continues in limited use today. It appears in every country with a major film industry, including the United States, France, Germany, Italy, China, Korea, Japan, and India. In Western nations, the China Girl is almost always female, young, conventionally attractive, and, despite the racial connotations of the name, white. In film laboratories, it is an essential part of quality-control processes, used to calibrate the desired exposure and color balance of film reels as well as the functionality of developing and printing machines. It has analogues in still photography and computer technologies as well.[1]

Buck and Segal hoped that former China Girl models might come forward and identify themselves, as China Girls were never credited for their roles. None did, however. As Buck explained, "We wanted to free these women, but we realize they're still trapped in their images."[2] Outside of highly specialized sectors of the film industry, namely, those workers who handle and scrutinize celluloid prints, the China Girl is not well known, and most China Girl models have typically remained anonymous.[3]

In the film laboratory the China Girl has been ubiquitous. Every film commercially produced on celluloid, and nearly every strip of film, contains the image of a woman posed next to swatches of color. As John Pytlak, a former Kodak engineer, said when he accepted an Academy Award for his work on the Laboratory Aim Density or LAD system, "She has probably been in more movies than any other actress in the world." The China Girl has numerous quality-control applications in the film laboratory. Typically, when an exposed reel of film is printed, a corresponding strip of China Girls, usually four to six frames, will accompany the film through its development process. These China Girls will be compared to "master" China Girls that have already been developed to check accuracy. China Girls are then used in the process of color timing, where the China Girl acts as a kind of shorthand for the rest of the film, indicating whether the color processes occur as expected or deviate in any way. In the final printing of the film reels, China Girls are used again to make sure printing is consistent across the various reels of a feature-length film, each time tested against a master or control China Girl. In recent decades, despite the introduction of technological tools that have refined the process, the quality-control practices associated with the China Girl have remained largely unchanged.

Film viewers do not typically see the China Girl. In any given reel of a commercially produced 16mm or 35mm film, three to six China Girl frames are usually cut into the countdown leader, normally between the numbers 10 and 3. Yet this part of the filmstrip is not meant to be projected. The only way a China Girl may be glimpsed in a theater would be if a projectionist failed to switch reels at the correct time, allowing the ends of the film to run out. Even then, the short duration of the China Girl's appearance would be just as easy to miss, lost in the blink of an eye.[4]

The China Girl has been overlooked by film historiography as well.[5] Though many within the industry, particularly those who physically handle celluloid or acetate film, express familiarity or even longtime fascination regarding the China Girl, there is not, to my knowledge, a confirmed written account that describes the first usage of a China Girl nor the origins of the name.[6] Varying accounts have explained the term as relating to china dolls, or porcelain, a material that was sometimes used, in the case of posed dolls, and more broadly was suggestive of a delicate and refined quality that models aspired to in their appearance or dress. A more likely origin is the orientalist connotation of the term, expressed in the makeup, mandarin-collared

FIGURE 10.1
China Girl, circa early 1930s. Source unknown.

clothing, and tightly drawn hairstyles of the women, which privilege a woman's subordinate, submissive behavior, qualities that would be consistent with the technological function the image serves. The name "China Girl" and its variants circulate as part of laboratory vernacular and are only infrequently mentioned in the technical literature. As a result, it forms a kind of shadow history to the well-documented accounts of film stocks and the development of standardization methods for the film industry.

In this chapter, I address the theoretical implications of the China Girl as a consistent technical feature of film production and, significantly, as a subject imbued with what we might call a "reception history." Both aspects are expressed in and through a strand of experimental films and digital works that call into question the constitution of each medium in which she appears.[7] First, the China Girl links a notion of filmic materiality with femininity, particularly one that is racialized or orientalized. Second, the China Girl's enduring presence in increasingly digital contexts, despite its technological obsolescence, problematizes the supposed immateriality of these new technologies and offers an opportunity to reflect on the conditions of post-filmic materiality. Finally, the China Girl offers a way to think about the survival of filmic materiality through digital means—an issue made all the more urgent given the recent bankruptcy filings and lists of discontinued film stocks at Kodak, Fuji, Technicolor, as well as the closing of most commercial film laboratories around the world.

Although the China Girl denotes an image that does not appear to move, it is not photographic but cinematic: an image created by shooting a roll of 16mm or 35mm film, with the model sitting still in her pose. For much of the twentieth century, individual film laboratories produced their own China Girls, sometimes hiring models especially for that purpose or persuading their female employees to pose. Quality-control technicians would expose a roll of film and duplicate it in batches many times over.[8] The images would then be stored in a freezer to preserve their chemical composition, with individual frames periodically snipped to be used as test images in the various applications described above. In this way, the image of a given China Girl could linger in a laboratory for years, even decades. China

FIGURE 10.2
Kodachrome test strips, circa 1935.

FIGURE 10.3
BKSTS reference leader picture. 1970. Image courtesy of the British Kinematograph, Sound and Television Society (BKSTS).

Girls were given nicknames and absorbed into the workplace culture. Their images appeared everywhere, not only on every filmstrip but hanging from walls, glowing from light boxes, or sometimes blown up and admired on their own.

There are several industrial conditions that support the contention that the practice was introduced in the early 1920s and became widespread by the end of the decade. In the context of the first decade of the film industry, a period of rapid expansion and standardization of production practices, the research that led to the development of the China Girl came out of the Kodak Research Laboratories, which was closely affiliated with the industry's chief technical professional organization, the Society of Motion Picture Engineers (SMPE, later SMPTE, to include television). Kodak researchers set the standards and authored the procedures for quality control in processing and printing film. This included the use of instruments like the densitometer, which was used to measure the amount of light blocked by the silver grains in a filmstrip, or the density of an image. These researchers also established test cards and images by which such instruments were operated. As the industry grew, two major technological advances demanded more precise quality-control measures: sound in the early 1930s, and color in the early 1950s. Sound required a host of infrastructural changes, and in the laboratory, researchers began using the densitometer and other sensitometry instruments to more closely monitor the variations that could arise in printing and processing. Though densitometers and scene testers were invented in the early 1920s, it was only with the adoption of sensitometry for sound in the latter part of the decade and early '30s that these instruments, and the test materials associated with them—chiefly China Girls—became widely used.[9]

With the broad adoption of color in the early 1950s, there arose the demand for still more stringent quality-control measures. While color technology existed prior to the 1950s, the use of color in mainstream film had been monopolized by Technicolor since the late 1910s.[10] In 1949 Kodak introduced Eastmancolor negative color stock, its first three-color, single-strip color negative and printing film stock.[11] This was significant because it meant that any lab, and not just those operated by Technicolor, could develop and print color film. A host of other color stocks made by a number of rival manufacturers soon followed. Laboratories in the '50s once again faced pressure to maintain consistent control over variations, now in color as well as image density and sound. To address these increased demands in quality control, individual labs began producing their own China Girls. By the '60s, film manufacturers were also selling China Girls to accompany each film stock.

Another major industrial development requiring the use of a China Girl relating to color was the conversion of film for television broadcast in the 1960s and '70s. Accurate color reproduction was especially crucial in the representation of flesh tones, and SMPTE publications grappled with the challenges such representation posed. Predictably, these articles abstracted the body into a range of skin colors, couching implicit issues of race and racial differentiation within technical jargon and ostensibly scientific, objective rhetoric.

"Flesh tone" and "skin tone" do not appear in SMPTE publications until 1955 and 1956, respectively, and both occur in discussions of particular problems arising from tonal and color variables of broadcast television.[12]

In color television, Salvatore J. Bonsignore confirmed the primacy of the representation of flesh tones. "Skin tone is regarded as the most important standard in the color picture," he writes. "As long as skin tones appear natural to the public, the color picture will be successful."[13] Color effects in lighting, makeup, and staging were permissible so long as they did not interfere with the perceived natural color of a (white) actor's face.

The discrepancy between supposedly objective registration and its subjective apprehension persists in attempts during the 1970s to standardize values for skin tone with China Girl reference cards. In 1970, the British film manufacturer BKSTS produced its own China Girl, the BKSTS Reference Leader Picture, to "assist the object and subjective colour balance control of 16mm and 35mm prints in both the film laboratory and the television studio."[14] While the scrutiny of a China Girl by a technician's eye was the chief means of determining image consistency in the early twentieth century, its utility was diminished as technological developments took over its original functions. From a technological standpoint, the China Girl, by the

FIGURE 10.4
Morgan Fisher. *Standard Gauge*. 1984. Film still. Image courtesy of Morgan Fisher and the Bortolami Gallery.

late 1970s especially, was all but unnecessary to the procedures to which it had previously been essential. It became a visual reference, a check, and an unusually durable vestige of traditions past.

The BKSTS article touches on another ambiguity concerning subjective and objective apprehension: the presumption that a single, and implicitly white, skin tone could stand in for all others. The flesh-tone issue here speaks to the racial bias in film and television equipment, manifest as the degree to which white skin is perceived as natural, ideal, and uniquely correlative to instruments valued for their precision and accuracy.

This racial bias, however, is only implicit. In practice, laboratories avoided not only race but even bodies. In their place, we find "flesh tone" chips or dummy heads fitted with wigs made from human hair.[15] In a 1978 discussion of the BBC's Test Card 61, for example, E. W. Taylor and S. J. Lent note that in the printing of film, "the portrayal of a live model enables the subjective flesh tone balance to be readily made." This subjective apprehension is aided in large part by the "naturalness" of the China Girl.[16] Rather than accede to the perceived need for a human model, as other laboratory technicians maintained, Taylor and Lent attempt to reproduce the conditions of "naturalness" by making a test card representative of flesh tone.

The BKSTS article, along with Taylor and Lent's experiments, indicates that the China Girl was valued for the woman's flesh alone, which they agree provides an index of ideal skin tone. Given the fact that nearly all China Girls in Western countries are racially white, the woman's idealized appearance means an index based on her white skin. The notion of "ideal," of course, is conflated with whiteness, a (color) value presumed to be natural, ubiquitous, neutral, and, to a great extent, invisible or inconspicuous, like the China Girl herself.[17] The aim for an ideal, which is also measured as a norm, produces a fiction of a single color against which all others can and must be measured. Taylor and Lent's use of the word "natural" to designate a single, white subject signals the artificial and arbitrary premise of racial categorization.[18]

The presumed neutrality of white skin tones often presents a problem for technical analysis. When, for example, David L. MacAdam pursued printing tests of "a young lady" in the 1950s, he concluded that "[o]ptimum reproduction of skin color is not 'exact' reproduction . . . '[E]xact reproduction' is rejected almost unanimously as 'beefy.' On the other hand, when the print of highest acceptance is masked and compared with the original subject, it seems quite pale."[19] Lighting and exposing a film based on a white face often means that the values for everything else—sets, props, costumes, and other actors—are thrown off. A correct exposure for a pale white face, then, might mean that a darker complexion will be underexposed. In films where two actors with strongly contrasting skin tones appear in the same frames together, Hollywood cinematographers have tended to favor exposures that compliment the white actor.[20] In these examples, the objective neutrality of the white face is not only revealed to be a highly subjective and unreliable visual referent, but it can also undo the work of verisimilitude it is meant to uphold.

Although the China Girl is demonstrably racialized, the "China" modifier is innocent in this regard, or at least misleading where whiteness (and not "Chinese-ness") is apparently in question. "China," then, signals colorful embellishment or the model's femininity on its own.[21] The otherness signaled by the orientalist designation of the China Girl is deflected by adornment and other strategies that aim to minimalize and marginalize the oriental, the feminine, and the decorative. Rosalind Galt has argued that these three qualities form the denigrated category of the pretty, which, in its opposition to masculine, heroic, and Western aesthetics, can productively form a politics of the excluded, namely, that of race, gender, and sexuality.[22] Starting from appropriations of the China Girl in experimental film, the pretty can act as a disruptive shadow aesthetic, troubling the homogeneity of dominant forms of representation. In the context of Galt's theorization of the pretty, then, Knight's remark about a man's superior "colour fidelity" might accord with a desire for rationality and objectivity rather than with physiological traits. And it is this desire that indicates the degree to which arbitrary preferences for *certain* skin tones and a *particular* gender are made to conform to supposedly objective and neutral technological procedures.

In 1976, John Pytlak introduced the LAD system, which gives a more precise reading of density based on an 18 percent gray patch that forms part of the China Girl image. This became and remains the industry standard. If the technological justification for the China Girl had been on the wane before, this was its death knell: The LAD patch effectively rendered the China Girl, the face of a woman, functionally useless. Perhaps unsurprisingly, however, the China Girl survived this technological development and became a companion to the LAD patch, even sharing a name with it: the LAD Girl or, more affectionately, the LAD Lassie.

Digital technologies, of course, have developed their own variations of LAD controls. Like the telecine procedures in film-to-television conversions, which are workflows that involve analogue film at some level, processes of digital conversion like digital film recorders depend on digital LADs, or DLADs. There are also numerous faces used in reference charts for printer settings as well as monitor-calibration reference sheets. Despite the increasingly sophisticated technological means, the image of the China Girl persists even when—especially *after* the innovation of the LAD patch—it is in many ways unnecessary to the process for which it was originally intended. Many of the technicians I've spoken to in both traditional film and digital facilities have admitted that there is no longer any real need for the China Girl—as a white, female face or as any human face at all. As David Corley, a former film laboratory technician and founder of DSC Labs, a company that produces digital test patterns for television broadcast, remarked: "Personally, I found [China Girls] totally useless. Because there is an enormous range of skin tones, how do you say one skin tone is right and another is wrong? The only thing a China Girl would be useful for would be a subjective evaluation of the image."[23]

Given the apparent nonfunctionality of the China Girl, the image's enduring presence can seem puzzling. Some technicians insist that it offers a visual check, however subjective, of densitometer readings, and that it provides a recognizable form to confirm sensitometric measurements. As a face, however, the China Girl suggests something more, an attachment that approaches a kind of pathos or sentimentality, which is evident in the many private collections of clipped China Girl frames. As film facilities shrink or shut down, the China Girl has become one of the more poignant symbols of an outmoded industrial model in an era of profound technological change. Its ephemerality, the anonymity of the women who posed as models, the abstractions and rare mentions of the image in industrial manuals, and the literally marginal position the figure occupies in relation to the history of cinema may be, somewhat ironically, the traits that best express a sense of nostalgia for a vanishing medium.

The China Girl poses a problem of critical looking: How do we, or how should we, look at this image, or at the woman in the image? Though China Girls are rarely viewed in the theater, they are the most carefully scrutinized image in film laboratories. The way a China Girl is regarded in a laboratory, however, is highly particular: Technicians rely on the woman's familiarity, memorizing the ideal amount of detail and contrast in her hair, the color of her eyes and lips, and comparing them against the test China Girls that travel with other images. In the lab, a China Girl presents an image of an ideal appearance, but one that is also ordinary. It is most effective when it ceases to be a distinctive image and registers instead as a common, ubiquitous one, noticeable only when something appears out of place.

When we happen to see a China Girl image in the theater, it offers another view of cinema. Its visibility is a kind of glitch, a momentary breakdown in the cinematic machine. It brings to light, to visibility, all the previously unseen processes, both cultural and technical, that attend its image. The China Girl is prohibited because it reveals the inner workings of the processes that bring films into being, disrupting a movie's narrative reverie by exposing the constitutive seams of the film's construction. When we see it, we realize there is a different system of signification at work behind the images we typically see: For every *leading* lady, there is an attendant *leader* lady.[24]

While the China Girl is rooted in practices associated with film, especially in its linking of femininity and materiality, [Cécile Fontaine's digital video *China Girl* (2010)] demonstrates how the figure can also function

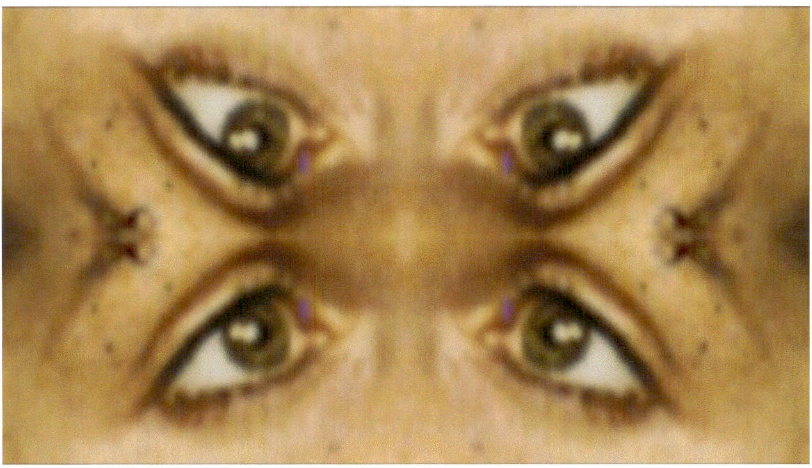

FIGURE 10.5
Cécile Fontaine. *China Girl*. 2010. Film still. Image courtesy of Light Cone and Cécile Fontaine.

emblematically for digital technologies as well. If China Girls, in their celluloid incarnations, told us something about the physical and chemical processes that attend the processing and printing of film images, then the digital China Girl, as a structural unit, offers a way of looking at the history of cinema from the perspective of new media. . . . In film, the China Girl highlights the disjunction between the women onscreen, whose images are given to be seen, and the women behind the scenes who are not meant to be seen at all. She is constitutive of cinema's system of representation, but largely excluded from it. In digital media, the China Girl is similarly obscured by the processes that depend on her image—chiefly monitor calibration and other forms of image adjustment—and in these ways the China Girl is subsumed into software, immaterial to the extent that there is no longer a physical artifact to grasp. With digital production and projection, the incidental view of a China Girl during a screening is no longer possible. Fontaine's continually morphing image, a familiar face made strange, reminds us of the transmutation not only of the image onscreen but the entire process by which films are made and exhibited in a digital context. Yet the China Girl need not be allied to either film or digital media. Rather, she offers a compelling opportunity to examine the two media as they are related historically, materially, and technically as a form of intermedial archaeology.

One of the most recent films to feature the China Girl returns to celluloid and to the tradition of China Girl uses in experimental film, *Releasing Human Energies*, made by filmmaker and archivist Mark Toscano.[25] For a little over five minutes, the China Girl sits in real time, visibly uncomfortable and blinking . . . The stretch of uninterrupted time provides a clear view of the China Girl's movement. It is very unlike the brief glimpses and frozen bodies more commonly associated with the China Girl as it has been rendered filmically and digitally.[26] While on one level the experience of watching the reel is similar to that of a Warhol Screen Test, the knowledge that this found fragment originally came from an industrial source restores a sense of labor and endurance, of life itself, to the China Girl. This woman is there not merely to look pretty but must suppress her movements, doll-like, and become a picture for our prolonged pleasure. The film preserves a persistent temporality, along with the China Girl's sense of twitchy struggle, and we might consider it the insistent wholeness of the China Girl before she is adopted for technical use. As Lili Young, a former China Girl model I interviewed, described her

FIGURE 10.6
Lili Young. China Girl at DuArt. C. 1968.

experience posing, "It was not a very exciting thing to do, you know. Just sit and be very still. Expressionless . . . They just wanted my flesh tone . . . It was like being in a long still shot, forever."

NOTES

1. For the "Shirley" image used in still photography, see Roth, "Looking at Shirley, the Ultimate Norm" and "The Fade-Out of Shirley, a Once-Ultimate Norm." In computer technologies, the Lenna or Lena image, based on a November 1972 *Playboy* cover image of model Lena Sjööblom, has been used as a reference in image processing since 1973. See Hutchinson, "Culture, Communication, and an Information Age Madonna"

2. Julie Buck and Karin Segal, interviewed by Gerald Peary, "Girls on Film."

3. There are rare examples of well-known actresses, including Romy Schneider and Elizabeth Taylor, posing as China Girls. These, however, are the exception. In a few cases, I have been able to identify women who posed as China Girls. Emme Marcy Rylan, the woman who posed as the current Kodak LAD and Digital LAD, has acted in numerous soap operas. See https://sites.google.com/site/donutscience/Home/kodakmarciefinallyidentified (accessed December 1, 2013). In addition to interviewing Lili Young, the China Girl filmed in the late 1960s at DuArt in New York, I have also spoken to Ullie (who refused to give her last name), the technician who also posed as a model at Cineric in New York, and identified the son of the woman who posed as the BKSTS reference leader picture.

4. Becca Hall, projectionist at the Gene Siskel Film Center in Chicago and director of the Northwest Chicago Film Society, described such accidental occurrences as "the face of the ashamed projectionist." Phone interview with author, February 2, 2012.

5. There are few discussions of China Girls in film scholarship, although work has begun to emerge in recent years. See Doane, "Screening the Avant-Garde Face," 222–224, for a discussion of Julie Buck and Karin Segal's exhibition, as well as Silva's *China Girls* (2006). For a discussion of Gabi Horndasch's *Versteckte Catherine* (1999/2000), see Hüser, "Finding Openings with Opening Credits."

6. The first mention of a "close-up of a girl" doesn't occur in *Journal of the SMPE* until 1933, although the densitometer first appeared in the journal in 1923. The term "gray lily card" first appears in a 1958 article. "China Girl" and "girl head," although they had long been used in film-laboratory vernacular, appear in the journal only decades later.

7. In this chapter, I toggle between the pronouns "it" and "she" in describing the China Girl. Generally "it" will designate the China Girl as an image, a technical object, used in laboratories. "She" will be about the figure, the woman, the subject, who appears within that image. These are somewhat slippery designations, of course, but the intent is to clarify the duality of the China Girl as both an image and a subject.

8. A roll would be 400 feet of 16mm film or 1,000 feet of 35mm film.

9. J. I. Crabtree, the co-author of *Motion Picture Laboratory Practice* (1936), the industry standard for laboratory procedures, writes: "After sensitometry had proved its worth in the control of variable-density sound tracks, the next step was to apply it to the control processing of picture negatives and prints. . . . Provided with a sensitometer and densitometer and other refined instrumental aids, the laboratory was able to measure not only variations in the activity of the developer and variations in the degree of uniformity of processing resulting from changes in temperature and degree of agitation, but likewise variations in uniformity of the photographic characteristics of the film emulsions. This had the beneficial effect of compelling the film manufacturers to improve the uniformity of their product, the lack of which it was previously possible to blame on the laboratory." See C. E. Ives and J. I. Crabtree, "A Trial and Error Method of Preparing a Motion Picture Sensitometer Tablet"; and J. I. Crabtree, C. E. Ives, and Fordyce Tuttle, "A Semi-Automatic Timing Device for Motion Picture Negatives."

10. Technicolor dominated the color market from the late 1910s through the early 1950s, in large part owing to the vast resources the company poured into its color processes and the protectiveness with which it guarded its photochemical secrets. Although Technicolor opened up its operations to licensing services and its equipment to other producers in the 1930s, it maintained a significant degree of control and secrecy over the cinematography, developing, and printing of its film stocks, which remained in place throughout the first half of the twentieth century.

11. Previously, Kodak had done significant research into color film stocks including Leopold Mannes's and Leopold Godowsky's development of the second Kodachrome in the early 1930s. Collins, *The Story of Kodak*, 206. The image of the woman in a headscarf in the Kodachrome test strips above also appears as the China Girl for Eastman Color Negative Film Type 5251, which can be seen in Morgan Fisher's *Standard Gauge* (1984). Type 5251 was used from 1962 to 1968.

12. "Flesh tone" is used in Fink, "Color Television vs. Color Motion Pictures" (1955), and "skin tone" appears in Wright, "Television Studio Practices Relative to Kinescope Recording" (1956).

13. Bonsignore, "The Art of Color TV," citation on p. 435.

14. Knight, "The BKSTS Reference Leader Picture," citation on p. 248. The BKSTS image was manufactured by Studios and Engineering at Thames Television Limited, in collaboration with Technicolor and Rank Film Processing Limited. Additionally, Knight notes that Kodak survey control strips were used to "establish standard conditions" (p. 252).

15. The most extreme example of the color abstracting of the body can be found in the "skin" signal used for television in the late 1970s, produced by the WGBH Educational Foundation and the 3M Corporation. The signal's screen of pale orange was intended to match white skin tone, and, in a semantically loaded turn, it replaced the televisual

process of scanning with its own terminology of "skinning." As Richard Dyer notes, "The whole process centered on blank images representing nothing, and yet founded in the most explicit way on a particular human flesh colour." Dyer, *White*, 94.

16. Taylor and Lent, "BBC Test Card No. 61 (Flesh Tone Reference)," citation on p. 76.

17. Jean-Louis Comolli and Paul Narboni, Brian Winston, Lorna Roth, and Richard Dyer have all investigated the racial qua ideological biases within image-technology apparatuses that otherwise present themselves as based on neutral scientific principles. See Comolli and Narboni, "Cinema/Ideology/Criticism"; Comolli, "Technique and Ideology"; Winston, "A Whole Technology of Dyeing"; Winston, *Technologies of Seeing*; and Roth, "Looking at Shirley"

18. In his study of whiteness and photographic media, Dyer observes an inherent "racial character of technologies" that reveals itself in the use of white skin tone as a measure for others. It is important to note that this "racial character" is not tantamount to racism (e.g., the notion that technologies themselves are or can be racist), but rather is designed according to perceived markers of racial differentiation. Thus, as a racial trait, whiteness may be *ideologically* neutral or invisible, but in color analysis it nevertheless registers a range of chromatic values. Dyer, *White*, 83.

19. David L. MacAdam, cited in Winston, "A Whole Technology of Dyeing," 120.

20. Dyer discusses this embedded photochemical anti-miscegenation in Hollywood films that have had at least one black and one white lead, including the *Lethal Weapon* series, *In the Heat of the Night*, *Rising Sun*, and *A Few Good Men*. Dyer, *White*, 98–103.

21. The names for China Girls in other languages support this orientalism. Beyond "Lili," they are known in France as "la chinoise," "Kodakette," and "la bridée," with this last referring to the slanted shape of Asian eyes. In Thailand, "Muay" or "Ar Muay" is used, also referring to the look of a Chinese girl. "China Girl" is also employed in Japan. In other countries, China Girl names emphasize their doll-like characteristics: Mexico uses *muñeca* or *mona* to refer to dolls, and Argentina similarly uses *muñeca de porcelana*. Laboratories in Austria refer to them as "Conchita."

22. Galt, *Pretty: Film and the Decorative Image*, 20.

23. David Corley, phone interview with author, January 26, 2011.

24. I borrow the term "leader lady" from Becca Hall.

25. I owe a debt of gratitude to Toscano, since he was the one who first brought the China Girl to my attention when, during a workshop on moving images organized by Tom Gunning at the Getty Research Institute in 2009, he showed a 16mm camera original reel of China Girl footage that had been donated to the Academy Film Archive in Los Angeles. Toscano later used this same footage for *Releasing Human Energies*.

26. See especially Silva's *China Girls* (2006), in which whole lengths of leader containing momentary glimpses of China Girl frames are spliced together. Silva sought

to depict China Girls as they appear "in the wild," which is to say in brief, three- to six-frame bursts.

BIBLIOGRAPHY

Bonsignore, Salvatore J. "The Art of Color TV." *Journal of the SMPTE* 65, no. 8 (1956): 435–436.

Buck, Julia, and Karin Segal, interviewed by Gerald Peary. "Girls on Film." http://geraldpeary.com/essays/ghi/girlsonfilm.html. Accessed February 10, 2012.

Collins, Douglas. *The Story of Kodak*. New York: Abrams, 1990.

Comolli, Jean Louis. "Technique and Ideology: Camera, Perspective, Depth of Field," translated by Diana Matias. In *Narrative, Apparatus, Ideology: A Film Theory Reader*, edited by Philip Rosen, 421–443. New York: Columbia University Press, 1986.

Comolli, Jean-Louis, and Paul Narboni. "Cinema/Ideology/Criticism," translated by Susan Bennett. *Screen* 12, no. 1 (Spring 1971): 145–55.

Corley, David. Phone interview with author, January 26, 2011.

Crabtree, J. I., C. E. Ives, and Fordyce Tuttle. "A Semi-Automatic Timing Device for Motion Picture Negatives." *Journal of the SMPTE* 15, no. 5 (1930): 587–597.

Crabtree, John, and Glen Matthews, *Motion Picture Laboratory Practice and Characteristics of Eastman Kodak Films*. Rochester, NY: Eastman Kodak Company, 1936.

Doane, Mary Ann. "Screening the Avant-Garde Face." In *The Question of Gender: Joan W. Scott's Critical Feminism*, edited by Judith Butler and Elizabeth Weed, 206–229. Indianapolis: Indiana University Press, 2011.

Dyer, Richard. *White: Essays on Race and Culture*. London: Routledge, 1997.

Fink, Donald G. "Color Television vs. Color Motion Pictures." *Journal of the SMPTE* 64, no. 6 (June 1955): 281–290.

Fisher, Morgan. *Standard Gauge*. 1984. 16-mm film.

Galt, Rosalind. *Pretty: Film and the Decorative Image*. New York: Columbia University Press, 2011.

Hall, Becca. Phone interview with author, February 2, 2012.

Horndasch, Gabi. *Versteckte Catherine*. 1999/2000. Video.

Hüser, Rembert. "Finding Openings with Opening Credits." In *Media, Culture, and Mediality: New Insights into the Current State of Research*, edited by Ludwig Jäger, Erika Linz, and Irmela Schneider. Bielefeld, Germany: Transcript verlag, 2010.

Hutchinson, Jamie. "Culture, Communication, and an Information Age Madonna." *IEEE Professional Communication Society Newsletter* 45, no. 3 (May/June 2001): 1, 5–7.

Ives, C. E., and J. I. Crabtree. "A Trial and Error Method of Preparing a Motion Picture Sensitometer Tablet." *Transactions of the Society of Motion Picture Engineers* 11, no. 32 (1927): 740–749.

Knight, Ray E. "The BKSTS Reference Leader Picture." *British Kinematography Sound and Television* (September 1970): 248–252.

Roth, Lorna. "The Fade-Out of Shirley, a Once-Ultimate Norm." In *The Melanin Millennium*, edited by Ronald E. Hall, 273–286. Dordrecht, Netherlands: Springer, 2013.

Roth, Lorna. "Looking at Shirley, the Ultimate Norm: Colour Balance, Image Technologies, and Cognitive Equity." *Canadian Journal of Communication* 34, no. 1 (2009): 111–136.

Silva, Michelle. *China Girls*. 2006. Video.

Taylor, E. W., and S. J. Lent. "BBC Test Card No. 61 (Flesh Tone Reference): Colorimetric and Other Optical Considerations." *Journal of the SMPTE* 87, no. 2 (1978): 76–78.

Winston, Brian. *Technologies of Seeing: Photography, Cinematography and Television*. London: British Film Institute, 1996.

Winston, Brian. "A Whole Technology of Dyeing: A Note on Ideology and the Apparatus of the Chromatic Moving Image." *Daedalus* 114, no. 4 (Fall 1985): 105–123.

Wright, Harold. "Television Studio Practices Relative to Kinescope Recording." *Journal of the SMPTE* 65, no. 1 (1956): 1–6.

11 MAKING FANTASY MATERIAL: KODACHROME, EUGENICS, AND THE CHEMOAESTHETICS OF RACIAL CAPITALISM

ALI FESER

I SILVER, SALT, GELATIN, DYES

Kodak film[1] started with pine trees and cotton, harvested by Black workers in the U.S. south.[2] The cotton was bleached, dried, purified, turned to cellulose acetate at the Eastman Chemical factory in Tennessee, and shipped to Kodak Park in Rochester, New York.[3] There, workers made emulsion, the part of film that reacts to light. They threw silver into cauldrons of acid and broke down its bonds to remake them. With bromide salts, they coaxed it into crystals and suspended them in gelatin, that fleshy part of film, with just enough sulfur in it to spark a chemical reaction when it was exposed to light.

To make film base, they boiled down cellulose acetate into a golden, honey-like substance called dope and extruded it on a giant steel casting wheel, measuring 25 feet wide. The wheel turned and under the cover of darkness, solvents—mostly methylene chloride, now classified as a "probable" carcinogen[4] in the US—wafted from the film into the air. Machines coated the dope with emulsion. Early black-and-white film took only one layer of emulsion. Kodachrome took three, each sensitized to a different part of the visible light spectrum. The top layer formed a yellow image, the middle layer formed a magenta image, and the bottom formed a cyan image. By separating the different colors of light for processing, and then combining them, the film created the impression of a full-color spectrum.[5]

FIGURE 11.1
Department 51 Picnic, 1948. The images that follow were made by Hymen Meisel (1898–1985), a lifelong resident of Rochester and employee at Kodak's Camera Works factory. All images in Chapter 11 are provided by and courtesy of the Visual Studies Workshop, Rochester, NY.

Once dry, the dope and emulsion became film, that "prima materia" that does not just represent but remakes the world, transmutes it into spectacle and surface. Workers next perforated it to fit in the teeth of a camera. The women in the Packaging Department sliced, wound, stuffed it in yellow boxes, and sent it to distributers. Consumers dropped it in their cameras and pressed the button. Kodak did the rest. Light hit the surface, the silver decomposed, the crystal dissociated, and the molecules rearranged in the pattern of the impact of light. When the roll was through, consumers sealed it in yellow mailers, pre-addressed to the nearest Kodak processing plant.

The developing process was toxic and long. It took three and a half hours and three rounds of processing—once for each layer—to develop Kodachrome. In addition to the usual chemical components of developing solution, this developer contained special chemical couplers for each layer. When injected into the emulsion, they reacted with the oxidation product of the developer to form a color dye in the film itself. With the couplers, silver, salt, and gelatin became blue skies, greens lawns, white picket fences, and ads for everything imaginable.

FIGURE 11.2
Camera Works Department 51, 1945. Meisel was a serious amateur and member of the Kodak Camera Club. The largest in the world, it hosted lectures, contests, and classes and, for pocket change, offered members all the paper and chemistry they might need. This distribution of photographic materials to workers was indeed crucial to product development. But it also helped generate a collective identity.

While it took until the 1970s for the sale of color film to outpace that of black and white, Kodachrome was quickly taken up as a medium for print advertising. In this incarnation, it mediated every commodity, from cars to kitchen appliances. Through Kodachrome, we learned what to desire: scenes of nuclear abundance, made possible through the magic of synthetic chemistry and the racialized division of labor it obscured. Not everyone dreamt of the Kodachrome "good life," and not everyone was allowed in. As specified in advertisements and other mass media, this fantasy was straight, white, patriarchal, and middle class. But nonetheless, it became recognizable as a norm and aspiration.

This chapter proposes that this fantasy, the American Dream, issued from the material practices of synthetic chemistry, from the particular mode of production at Kodak's factories, and from the company founder, George Eastman's utopian attempts to remake the world in Rochester and beyond into a Kodak-dreamworld. Just as Kodak scientists had learned to "fix" in place the molecules that make up the photographic image, Eastman was trying to "fix" social life and ensure the smooth reproduction of Kodak as a brand and lifestyle. Locally, this meant transforming Rochester into "a place worth living" in, as Eastman often quipped. On a national level, this utopian impulse meant for Eastman funding scientific and policy research on eugenics.

Neither Kodak's association with eugenics nor the racialized division of labor at its factories were unique to the company. Racialization is a foundational violence of capitalist accumulation.[6] It is used to justify the disproportionate extraction of value from Black and other workers of color,[7] and it is embedded in the genealogy of the corporation.[8] But Kodachrome uniquely materialized this racialized division of labor in both chemical and visual form, and because of the ubiquity of Kodak film, consumers learned to see the world through this filter.

As Lorna Roth and Genevieve Yue have theorized through their research on the Kodak "Shirley,"[9] a preference for whiteness has become inherent in photographic technologies. This chapter offers a backstory to the Shirley card and contextualizes the chemoaesthetics of film within a broader atmosphere of racial capitalism. Thinking with Karen Barad's performative account of materialization, I consider Kodachrome as a "condensation[s] . . . of multiple practices of engagement."[10] It carries traces of the racialized division of labor from which it emerged, and it inscribed this history on the world as it remade it in emulsion. Film is not a passive object, and by "intra-acting" with the phenomena it observes, it creates the reality[11] it claims to

FIGURE 11.3
Untitled, 1959. Employees at all levels of corporate hierarchy rubbed elbows in the darkrooms, and even though labor was divided up, employees could grasp the sum of their work in the practice of photography. Photography mattered to them. They took courses, collected cameras, and submitted portfolios when applying for promotions.

represent. As such, this chapter narrates the process by which a fantasy of the white good life materialized as a lived social reality in Rochester, and how this lived reality of whiteness was transposed into the technologies that underlie the image and consumer culture of the American Century.

II YELLOW FATHERS, YELLOW BLOOD

In Rochester, Kodak functioned as a total institution and proxy and precursor to the welfare state. As was the case with many twentieth-century monopoly corporations, its paternalistic style of capitalism was anchored in an ideology of kinship.[12] Within the company, family figured as a metaphor for power relations. Internal documents referred to the "Kodak family," and workers said they were like brothers. Many joked that the company was "the Great Yellow father,"[13] while one woman, referring to environmental contamination at Kodak Park, derided Kodak as an "abusive spouse."

Through its generous employee welfare programs, Kodak also made the white, middle-class, nuclear family a sociological phenomenon in Rochester. It offered employees and their families health insurance, life insurance, bonuses, mortgages, vacation time, tuition subsidies, thirty years of employment, a pension after that, and a bevy of other perks and benefits. Kodak also tacitly promised to hire from within workers' families. While originally implemented to prevent workers from unionizing,[14] these measures materialized an iconic brand of white subjectivity in Rochester. They generated a shared social reality for workers and, to an extent, standardized their life experiences and aesthetic dispositions.

This fantasy of the Kodak family also circulated in Kodak advertisements, which so often showed families on vacation, on holiday, and in snapshots of ordinary, nuclear pleasures. As Bourdieu reported in a study funded by Kodak's French subsidiary, Kodak made photography into "a ritual of the domestic cult in which the family is both subject and object."[15]

There were some Black workers at Kodak; they too passed down jobs and said that their "blood ran yellow." But this soft nepotism—a free gift of capital[16] if you're on the right side—reinforced the white homogeneity of Kodak's workforce. As evidenced by the founding of the Memorial AME Zion Church in 1827, Rochester had long been home to a sizable Black community, which grew in the early twentieth century with migration from the U.S. South and Caribbean. Yet, in 1939, Kodak directly employed only one Black worker.[17] While the company began to hire Black PhDs to work in the Research Labs during World War II, it was not until the sixties—and the interventions of a local racial and economic justice organization[18]—that there were multiple Black chemists in a single cohort. Even then, within Kodak's racialized division of labor, Black, Jewish, and immigrant workers usually made hardware (cameras and projectors) at the Camera Works factory downtown.[19]

It was white men, for the most part, who made film at Kodak Park. They and their families became the experimental substratum of Kodak film. They were the perfect test subjects, as the presumed consumer was also the white, industrial worker. The scientists recruited women they thought were pretty,

Kodakers told me time and again, to model for film tests in the Research Labs, and they brought new products home to test on family members in the front yards and front rooms of company-built subdivisions. They studied how the emulsion reacted in *real* use. They manipulated the molecules of the emulsion to best image the objects of their desire, desires already informed by a mass media culture materialized in Kodak emulsion. These workers were consumers as well as producers; they thumbed through magazines, watched

FIGURE 11.4

Canandaigua Lake, 1964. Most of Meisel's photographs are Kodachromes, made between 1945 and 1979. He left behind tens of thousands of slides, carefully packaged and labeled with the names of the institutions where he intended them to be stored. Instead, by way of a dumpster, they ended up in the archives of the Visual Studies Workshop.

movies. It was their fantasies—conjured within George Eastman's vision of the "good life"—that made it into the emulsion and onto the screen.

Kodak may have been a family company, but Eastman himself had no children. When he died (a single gunshot to the heart), he willed the remainder of his fortune and his fifty-room mansion to the University of Rochester. The house boasted gardens, greenhouses, and darkrooms, but its heart was an organ. It sits in the conservatory, beneath a stuffed elephant head and among the palms. Its pipes ascend three stories. Every morning, beneath a stuffed elephant head and among the palms, with the skies outside at an average eighteen percent gray, Eastman spooned out his grapefruit, while Harold Gleason, his organist, turned the whole house into an instrument. After a decade on Rochester's Millionaires' Row, Eastman decided to improve the conservatory's acoustics. He would expand it, not by cutting into the dining room or the billiard room but by slicing the house in two and adding nine feet four inches exactly.

The world, according to Eastman, was meant to be cracked in half and made anew. This is the business of synthetic chemistry and what Eastman did for the city of Rochester. He built neighborhoods for workers to live in and parks for their days off. He funded dental clinics and hospitals and negotiated the construction of water and transportation infrastructures. He built a music school that bears his name, with the phrase "for the enrichment of community life" etched on the façade of the music hall he named for his mother. He personally approved the films that played downtown on weekends. He was trying to make a working class with middle-class desires.

Outside of Rochester, Eastman donated to MIT and Tuskegee. He built a school for the children of Black agricultural workers near his vacation home, a former plantation in North Carolina. Beginning in 1925, and along with John D. Rockefeller, Eastman also underwrote the American Eugenics Society (AES). When Rockefeller ceased his contributions, Eastman increased his own from $10,000 to $15,000 a year. Eastman gave away a total over $100,000,000 in his lifetime, but this generosity was also highly strategic. He understood that factories don't just make commodities. They make workers, too.[20]

At the same time, Eastman's attempts at world-making express a wish endemic to industrial chemical manufacturing. The wish is that by manipulating the bonds between molecules, it is possible to transform social bonds, too. Since its emergence in the nineteenth century, synthetic chemistry has made it possible to produce cheap substitutes for rare and costly natural materials, thus multiplying the bounty of the earth.[21] Cellulose substituted for the ivory of billiard balls, plastics for tortoiseshell, nylon for silk, and photographs for the three dimensional. According to the head of Kodak's Research Labs, Dr. C. K. Mees, these cheap substitutes were thought to contain a revolutionary potential. Mees, who had been socialist in his youth, changed his mind about politics when he moved to Kodak and saw all that could be made in the laboratory. It wasn't necessary to seize from the rich to provide for the poor. Rather, chemists could supersede natural limits and create—from coal and carbon chains—enough for everyone.[22] They could turn nature into something even better.

In this sense, synthetic chemistry bears a formal resemblance to eugenics. Whereas industrial chemistry promised a world of abundance by manipulating the molecules in oil and coal, eugenics claimed it could eliminate social ills by manipulating the material substance of bodies. Both promised to get to utopia without the costly and indeterminate interventions of democracy. Both turn women into vessels for reproduction, and both entail violence against those who are not part of that picture. Moreover, eugenics makes sense only in the world that synthetic chemistry made possible: whereas the latter enabled the growth of industry and an explosion of mass-produced consumer goods in the nineteenth century, eugenics expresses a capitalist logic by which the value of a person's life is reducible to the amount of capitalist value they produce.

III BLOOD, GERM CELL, GENE, PROTOPLASM

Marian Gleason, the wife of Eastman's personal organist, visited the businessman and benefactor often. Eastman would send her orchids, "pure white in beautiful box[es]."[23] She'd put them on ice to keep them fresh. "Sometimes that's all there'd be in the icebox," she joked. "If I could resell just one, I'd make enough for a month's groceries." But the Gleasons survived by Eastman's desires. Rather than selling them, Marian would put on her nicest dress, pin an orchid to it, and head to the conservatory. She claimed to have been there at least once, beneath the stuffed elephant and among the palms, when Irving Fischer was in town.[24]

Fisher was a Yale economist and founder of the American Eugenics Society (AES). Officially established in 1926, AES functioned as a national

FIGURE 11.5
Kodak Office Reception Center, 1961. The photographs are remarkable for both what they show and what is absent from them. While Kodak advertisements featured snapshot photography as an essential technology for the reproduction of family life, Hy's collection contains few weddings, babies, or graduations. According to census reports, Meisel was single.

association and clearinghouse for research and education on eugenics. Its central mission was to establish a scientific justification for the racial superiority of white people, or, in its own words, to "stand against the forces which work for racial deterioration, and for progressive improvement in the vigor, intelligence and moral fiber of the human race."[25]

AES aligned itself with Mendelian eugenics, which began to supplant earlier genetic models of classification in 1900.[26] When eugenics first emerged as a field of scholarship in the 1880s, it assumed genetic material to be malleable. By altering the environment itself, eugenicists could improve the genes passed down to future generations and thus eliminate antisocial behavior. This Progressive-era logic is evident in Eastman's attempts to restructure social life in Rochester. In contrast, Mendelians insisted that intelligence, health, and character were fixed according to ancestry. They concluded that the only way to improve social life was to remove genetic matter from "the national blood stream"[27] through a combination of negative and positive eugenics. Along these lines, to inhibit "unfit" subjects from having children, AES helped fund scientific projects at the Galton Society in New York and the Kellogg-funded Race Betterment Foundation in Battle Creek, Michigan. It also supported policy research regarding sterilization, immigration, and miscegenation laws, and its members campaigned against state welfare programs.[28]

At the same time, to prevent what Theodore Roosevelt called "race suicide," AES worked to encourage white couples to reproduce and to make eugenics—as both an ideology of race and a technical program of state led reform—more palatable to the white, American public. AES published a monthly journal in collaboration with other eugenics organizations and funded outreach programs, like the Fitter Family Contests held at state fairs around the Midwest. For these contests, families tracked the purity of their bloodlines in photo albums—photography has long been a technology of racialization[29]—and they were judged according how many children they had, how attractive, intelligent, and healthy those children were, and by their family histories.[30] The temporalities of eugenics extended towards both future and past.

In 1924, when AES founder, Irving Fischer, first began courting Eastman, he appealed to the philanthropist's capacity to see beyond the present moment. "My experience so far," Fischer wrote:

> has been that few business men try to look as far into the future as you do. Few will contribute to a project . . . beyond the confines of their own town and generation. Money for humanity in general, and posterity, at that, comes hard.[31]

Eugenics, Fisher implied—like the snapshot itself—was in the business of fixing the future. We often think of the act of making a photograph as one of preserving the past, but this is done with the future in mind, for the subject who will look back with pleasure. It is a wish for continuity, and the future that eugenicists wished for was as "pure white" as Eastman's orchids. To materialize this dream, they would apply the precision of mass manufacture

FIGURE 11.6
The Ogilvies, Mrs. Newman, and Hyman, undated. In contrast to the heteronormative domestic, his slides detail friendships, daily life in Rochester, and travels through the U.S., Europe, and Latin America—that is, the very possibility of middle-class flourishing afforded by postwar industrial capitalism.

and techniques of animal breeding to human sexuality. They would regulate desire and engineer a future that prioritized whiteness and the accumulation of value above all else. They would reorganize social bonds by manipulating the very substances of life: blood, germ cell, gene, and protoplasm.

Early in their correspondence, Fischer sent Eastman a copy of a pamphlet by Albert Wiggam, a psychiatrist whose syndicated columns helped install biological determinism as the vernacular of American racism. In his essay, Wiggam defines eugenics as "evolution made conscious and intelligent,"[32] and, like Fisher, he characterizes it as a framework for fixing the future. As a "projection of the golden rule down the stream of protoplasm,"[33] eugenics attempts to ensure the wellbeing of the "unborn" by selecting the heredity influences that will be passed down to the next generation. But while eugenics may be, as Wiggam writes, a "technic of righteousness"[34] ordained by a white God, the masses don't always know what is good for them. As such, the state,[35] with the help of science, must work to bring people's desires into alignment with eugenic principles. But, as "mate selection . . . is the supreme cause of both racial glory and decline,"[36] policy alone is not enough. Eugenics requires the regulation not just of behavior, but of consciousness and desire, too. This, Wiggam contends, is the work of art.

According to Wiggam, art shows us what we should want. If we learn to "worship . . . physical beauty" through art, he writes, we will seek the same characteristics in our partners. The images that surround us can literally "change the faces of men,"[37] bringing bodies more in line with hegemonic standards of beauty. While Wiggam is skeptical of mass education as a means for helping the poor, he endorses it, on the level of the senses, as a strategy for preserving racial purity.

Eastman was so enthralled by Wigwam's essay that he asked Fisher to find him 600 copies to distribute to friends and associates. Eastman also agreed, in early 1925, to contribute an annual $10,000 to the new organization. In 1929, when Rockefeller suspended his donations to AES, Eastman increased his own to $15,000.[38] Beyond AES, Eastman would also fund research into an intelligence test that correlated musical aptitude with race.[39]

Eastman did not speak publicly about eugenics, but years later, Marian Gleason described his views to Eastman's biographer. "Everyone should produce as much has he uses in the world," she quoted Eastman as saying, "if he can't do that he shouldn't live—shouldn't be born at all." She also reported that Eastman had planned to rescind his support.[40] "It did no good to promote [a] project that wasn't supported by the public," Eastman is said to have said to Fisher, "until people became so uncomfortable through huge taxes to keep up mental hospitals, homes for delinquents, penitentiaries and jails, until [they] felt unsafe themselves on street."[41] Yet, ten months before his death, Eastman was still funding AES, and correspondence between AES and Eastman's secretary after his death suggests that AES expected further contributions from the trust established in Eastman's will.[42] And even if, as Marian contended, Eastman had given up on a

FIGURE 11.7
Drake Sea, 1978. The notion of a tourist with a camera around their neck reminds us that film was part and parcel of American globalization. Kodak materialized racial capitalism and eugenic fantasies into a recognizable, aesthetic form that would circulate around the world.

state-led program of eugenic reform, what was not possible in flesh was still possible in emulsion.

IV CONCLUSION

When Kodachrome was announced to the public in 1936, Mees of the Research Labs called it a "revelation":

> When you see Kodachrome on the screen you realize how wonderfully colored the world is. An artist, of(?) course, knows this, but most of us are not artists

and we don't realize the subtle colors that occur in everyday scenes—flowers and foliage, and summer landscapes, when bright colors strike the eye.[43]

But rather than "depict[ing] what awaits"[44] in the world, Kodachrome *reorganized* the world. Through the manipulation of raw materials, Kodachrome remade it as a materially polychromatic, yet ideologically white world, structured by the patriarchal family. It was designed around the whiteness of Kodak's workforce and emerged from Eastman's eugenicist fantasies of what the world should look like. It transposed a racialized organization of labor and a political program into a chemical reality, order of representation,

FIGURE 11.8
Contr[a]st Reflection in War Memorial, 1962. There was something radically new and democratic about Kodak film. When it put cameras at the center of the world, it opened possibilities for self-raepresentation. It could also image forms of life beyond the norm.

and structures of feeling.[45] It was used to sell a vision of the good life that was only sustainable through patterns of racialized extraction and exclusion. As a mode of seeing, this chemoaesthetic was both a symptom of the organization of capital and the very structure of desire that held it in place.

Yet, the relationship between the chemoaesthetic of Kodachrome, the racialized organization of industrial capitalism, and the story of eugenics and utopian world-making is not deterministically causal. Nor was Kodak's visual regime totalizing. Film is a fantastically plastic media. By working with the ideological and material constraints of its manufacture, you can push and pull the emulsion to image almost anything, even critiques of the very system that created it. As Stuart Hall might contend, it is across such disjunctures[46] that Kodachrome articulated its racialized history and achieved its monopoly on vision.

ACKNOWLEDGMENTS

This chapter is based upon research supported by the National Science Foundation under Grant Number 1559245. Cameron Hu, Allyson Nadia Field, Rebecca Journey, Joseph Masco, Nicholas Shapiro, Salomé Aguilera Skvirsky, Christina Rosetti, and Michael Rossi all provided invaluable commentary on earlier drafts.

NOTES

1. This chapter is based on forty months of ethnographic fieldwork in Rochester, NY. From 2013 to 2017, I conducted oral histories and participant-observation with current and former Kodak workers and their families, photographers, scientists, environmentalists, and other city residents. In what follows, I weave conversations with my interlocutors together with archival research on the history of photochemical science and the many stories that circulate locally about Kodak and its founder, George Eastman. Though Kodak declared bankruptcy in 2012 and now functions as a shadow of its former self (a thousand or so employees versus 64,000 in 1983), the company still figures heavily in the local imaginary. The myths about Kodak are, as Kathleen Stewart writes, forms of "persuasion" and "social worlding" that make claims about the history of capitalism, the ethical obligations of the corporation, and what it means to exist in its aftermath. *Ordinary Affects*, 66. Following Stewart, I attempt to "[perform] the intensity and texture" of these myths and "fashion some form of address to their form" (4).

2. Fenton, *A Movie Trip through Filmland*, 1921.

3. This account of the manufacturing process draws from Shanebrook, *How Kodak Film Is Made* (Rochester, 2010); Fenton, *A Movie Trip through Filmland*; "How Film Is Made."

4. Agency for Toxic Substances and Disease Registry, "Toxicological Profile for Methylene Chloride."

5. L. D. Mannes et al., Color photography, U.S. Patent 2,113,329, filed February 27, 1935, and issued April 5, 1938; Mees, *The Theory of the Photographic Process.*

6. James, *The Black Jacobins;* Mintz, *Sweetness and Power;* Williams, *Capitalism and Slavery*; Robinson, *Black Marxism.*

7. Dubois, *Darkwater: Voices from Within the Veil.*

8. Mittman, *Empire of Rubber*; Rosenthal, *Accounting for Slavery: Masters and Management.*

9. In order to ensure that color photographs were printed at Kodak's standards, the company distributed reference cards to professional and amateur photo developers. These 2.5″×3.5″ photographs were accompanied by a negative and featured an image of a white woman model, framed by a grayscale and color chart. With the model positioned squarely towards the camera, the card resembles nineteenth-century anthropological images of "racial types" as well as the composite images produced in eugenics research. Both the reference card and the woman featured on it were called "Shirleys." Roth, "Looking at Shirley, the Ultimate Norm"; and Yue, *Girl Head: Feminism and Film Materiality*. For further elaboration on how Kodak film reproduced racial ideologies and how the materiality of film shaped visual culture see Dyer, *White.*

10. Barad. *Meeting the Universe Halfway*, 53.

11. "Realism, then, is not about representations of an independent reality but about the real consequences, interventions, creative possibilities, and responsibilities of intra-acting within and as a part of the world," Barad, *Meeting the Universe*, 37.

12. Kodak was exceptional among U.S. monopoly corporations in that its workers never unionized, but its labor management policies were characteristic of what Sanford Jacoby calls "welfare capitalism" in *Modern Manors: Welfare Capitalism Since the New Deal*, and what Gramsci summarizes as the "American methods of work." As Gramsci elaborates, within this mode of production, the nuclear family was developed as a disciplinary technique. Rather than maintaining its labor market through "brute coercion" (302), corporations worked through workers' desires and taught them to want forms of life that could only be produced through ever-intensifying, industrial labor (298).

13. *Editors' note:* The term "Great Yellow Father" stemmed from the Kodak Corporation's distinctive yellow packaging. It does not appear to tie into the racialized implications of yellow discussed in part I of this volume.

14. Correspondence from the Unemployment Committee to Frank Lovejoy, March 2, 1922, Box 39, Folder 1, Marion Bayard Folsom Papers, University of Rochester Rare Books and Special Collections.

15. Bourdieu, *Photography*, 19. See also Hirsch, *Family Frames;* Holland and Spence, eds., *Family Snaps*.

16. Marx describes the transmigration of value from the worker into the commodity on which they labor as a "gratuitous gift" *to* the capitalist. With a free gift from capital, I am referring to the ways in which white workers benefit materially from the hyper-extraction of value from workers of color. Marx, *Capital*, 315.

17. *Second Report of the New York State Temporary Commission on the Condition of the Colored Urban Population* (1939)

18. Wadhani, "Kodak, FIGHT, and the Definition of Civil Rights in Rochester, New York: 1966–1967," 62.

19. Oral history with Kodak executive, 1971, Lawrence Bachmann Papers, University of Rochester Rare Books and Special Collections.

20. Marx, *Capital*, 716.

21. Bensaude-Vincent and Stengers, *A History of Chemistry*; Bensaude-Vincent and Simon, *Chemistry: The Impure Science*. See also Leslie, *Synthetic Worlds*.

22. Mees, "An Essay on Conservativism," 1946. Box 10, C. E. Kenneth Mees Manuscript Collection, Richard and Ronay Menschel Library, George Eastman Museum, Rochester, New York.

23. Interview with Marian Gleason, Box 1, Folder 5, Eastman-Butterfield Collection, University of Rochester Rare Books and Special Collections, Rochester, New York.

24. Brayer, *George Eastman*, 475–476.

25. Correspondence from Irving Fisher to George Eastman, December 1922, George Eastman Legacy Collection, George Eastman Museum, Rochester, New York.

26. Roberts, *Killing the Black Body;* Tallbear, *Native American DNA;* Rossi, *The Republic of Color;* Stern, *Eugenic Nation*.

27. Wiggam, "The New Decalogue of Science," 645.

28. Forced sterilization, especially, had roots in the castration of Black men and the systematic rape of Black women under slavery. It became state practice in the early twentieth century, as Jim Crow laws were proliferating, the color line was calcifying, and workers were radicalizing. Roberts, *Killing the Black Body* (2017). The ascendency of negative eugenics in the 1920s and 1930s also speaks to anxieties about the presence of women in the public sphere. Any women—but poor women in particular—who appeared to transgress sexual norms could be cast as "feebleminded" and subjected to sterilization. Kline, *Building a Better Race*.

29. In the eighteenth and early nineteenth centuries and as part of the colonialism project, illustrations of racial types helped to establish race as a matrix of visible and

embodied signs, Poole, *Vision, Race, and Modernity*. The camera was quickly recruited into this taxonomical project and used to survey the surface of body for signs of hereditary inferiority. Ellenbogen, *Reasoned and Unreasoned Images*; Sekula, "The Body and the Archive"; Smith, *American Archives*.

30. Smith, *American Archives*, 113–135.

31. Correspondence from Irving Fisher to George Eastman, January 9, 1925, George Eastman Legacy Collection.

32. Wiggam, "The New Decalogue of Science," 647.

33. Wiggam, "The New Decalogue of Science," 648.

34. Wiggam, "The New Decalogue of Science," 643.

35. The "improvement of man's *inborn capacities* for happiness, health, sanity, and achievement," Wiggam writes, "must become the one living purpose of the state." Wiggam, "The New Decalogue of Science," 643.

36. Wiggam, "The New Decalogue of Science," 650.

37. Wiggam, "The New Decalogue of Science."

38. Brayer, *George Eastman*

39. Cowan, "Eugenics at the Eastman School: Music Psychology and the Racialization of Musical Talent."

40. This is a point that Brayer repeats in her biography of Eastman, *George Eastman*, 475.

41. Interview with Marian Gleason, Box 1, Folder 5, Eastman-Butterfield Collection.

42. Correspondence from Leon Whitney (Executive Secretary of AES) to Alice Whitney Hutchinson, May 18, 1932, George Eastman Legacy Collection.

43. Description of Kodachrome process, Eastman Kodak Public Relations Department, April 12, 1935, Box 72, Folder 9, Kodak Historical Collection #003.

44. Barad, *Meeting the Universe*, 53.

45. Williams, "Structures of Feeling"; Berlant, *The Female Complaint*.

46. Hall, "Culture, Media and the Ideological Effect" and "Race, Articulation, and Dominance."

BIBLIOGRAPHY

Barad, Karen. *Meeting the Universe Halfway: Quantum Physics and the Entanglement of Matter and Meaning*. Durham, NC: Duke University Press, 2007.

Berlant, Lauren. *The Female Complaint: The Unfinished Business of Sentimentality in American Culture*. Durham, NC: Duke University Press, 2008.

Bensaude-Vincent, Bernadette, and Jonathon Simon. *Chemistry: The Impure Science*. London: Imperial College Press, 2008.

Bensaude-Vincent, Bernadette, and Isabelle Stengers. *A History of Chemistry*. Cambridge, MA: Harvard University Press, 1996.

Bourdieu, Pierre. *Photography: A Middle-Brow Art*. Stanford, CA: Stanford University Press, 1990.

Brayer, Elizabeth. *George Eastman: A Biography*. Rochester, NY: University of Rochester Press, 2006.

C. E. Kenneth Mees Manuscript Collection, Richard and Ronay Menschel Library, George Eastman Museum, Rochester, New York.

Cowan, Alexander. "Eugenics at the Eastman School: Music Psychology and the Racialization of Musical Talent." Presented at the 2017 American Musicological Society conference in Rochester, NY. https://cowanaw.wordpress.com/2017/11/07/eugenics-at-the-eastman-school-music-psychology-and-the-racialization-of-musical-talent/.

Dubois, W. E. B. *Darkwater: Voices from within the Veil*. Mineola, NY: Dover Publications, 1999.

Dyer, Richard. *White: Essays on Race and Culture*. New York: Routledge, 1997.

Eastman-Butterfield Collection, University of Rochester Rare Books and Special Collections, Rochester, New York.

Ellenbogen, Josh. *Reasoned and Unreasoned Images: The Photography of Bertillon, Galton, and Marey*. University Park: Pennsylvania State University Press, 2012.

Fenton, Paul, dir. *A Movie Trip through Filmland*. 1921; Rochester, NY: DeBergerac Productions, 2006.

George Eastman Legacy Collection, George Eastman Museum, Rochester, New York.

Hirsch, Marianne. *Family Frames: Photography, Narrative, and Postmemory*. Cambridge, MA: Harvard University Press, 1997.

Holland, Patricia, and Jo Spence, eds. *Family Snaps: The Meanings of Domestic Photography*. London: Virago, 1991.

How Film Is Made. 1958; Juanjo Plaza, accessed at YouTube Video, https://www.youtube.com/ watch?v=3Kd580wxWko.

Jacoby, Sanford M. *Modern Manors: Welfare Capitalism since the New Deal*. Princeton, NJ: Princeton University Press, 1997.

James, C. L. R. *The Black Jacobins: Toussaint L'Ouverture and the San Domingo Revolution*. 2nd ed. New York: Vintage Books, 1989.

Kline, Wendy. *Building a Better Race: Gender, Sexuality, and Eugenics*. Berkeley: University of California Press, 2005.

Kodak Historical Collection #003, University of Rochester Rare Books and Special Collections, Rochester, New York.

Lawrence Bachmann Papers, University of Rochester Rare Books and Special Collections, Rochester, New York.

Leslie, Esther. *Synthetic Worlds: Nature, Art and the Chemical Industry*. London: Reaktion Books, 2005.

Mannes, L. D., et al. Color photography, U.S. Patent 2,113,329, filed February 27, 1935, and issued April 5, 1938.

Marion Bayard Folsom Papers, University of Rochester Rare Books and Special Collections, Rochester, New York.

Marx, Karl. *Capital: A Critique of Political Economy, Vol. I.* New York: Penguin, 1990.

Mees, C. E. Kenneth. *The Theory of the Photographic Process*, rev. ed. New York: Macmillan, 1954.

Mees, C. E. Kenneth. "An Essay on Conservativism," 1946 C. E. Kenneth Mees Manuscript Collection, Richard and Ronay Menschel Library, George Eastman Museum, Rochester, New York.Mintz, Sidney. *Sweetness and Power: The Place of Sugar in Modern History*. New York: Viking, 1985.

Mittman, Gregg. *Empire of Rubber: Firestone's Scramble for Land and Power in Liberia*. New York: New Press, 2021.

Poole, Deborah. *Vision, Race, and Modernity: A Visual Economy of the Andean Image World*. Princeton, NJ: Princeton University Press, 1997.

Roberts, Dorothy. *Killing the Black Body: Race, Reproduction, and the Meaning of Liberty*. 2nd ed. New York: Vintage Books, 2017.

Robinson, Cedric J. *Black Marxism: The Making of the Black Radical Tradition*. 3rd ed. Chapel Hill: University of North Carolina Press, 2020.

Rosenthal, Caitlin. *Accounting for Slavery: Masters and Management*. Cambridge, MA: Harvard University Press, 2018.

Rossi, Michael. *The Republic of Color: Science, Perception, and the Making of Modern America*. Chicago: University of Chicago Press, 2019.

Roth, Lorna. "Looking at Shirley, the Ultimate Norm: Colour Balance, Image Technologies, and Cognitive Equity." *Canadian Journal of Communication* 34(2009): 111–136.

Second Report of the New York State Temporary Commission on the Condition of the Colored Urban Population. Albany, NY: J. B. Lyon Company, 1939.

Sekula, Allan. "The Body and the Archive." In *The Contest of Meaning: Critical Histories of Photography*, edited by R. Bolton, 342–389. Cambridge, MA: MIT Press, 1989.

Shanebrook, Robert. *How Kodak Film Is Made*. Rochester, 2010.

Smith, Shawn Michelle. *American Archives: Gender, Race, and Class in Visual Culture*. Princeton, NJ: Princeton University Press, 1999.

Stern, Alexandra Mina. *Eugenic Nation*. Berkeley: University of California Press, 2005.

Stewart, Kathleen. *Ordinary Affects*. Durham, NC: Duke University Press, 2007.

Tallbear, Kim. *Native American DNA*. Minneapolis: University of Minnesota Press, 2013.

Wadhani, R. D. G. "Kodak, FIGHT, and the Definition of Civil Rights in Rochester, New York: 1966–1967." *The Historian* 60, no. 1 (1997).

Wiggam, Albert. "The New Decalogue of Science." *The Century Magazine* 3, no. 81 (1922): 643–650.

Williams, Eric. *Capitalism and Slavery*. Chapel Hill: North Carolina University Press, 1994.

William, Raymond. "Structures of Feeling." In *Marxism and Literature,* 128–135. New York: Oxford University Press, 1977.

Yue, Genevieve. *Girl Head: Feminism and Film Materiality*. New York: Fordham Universiy Press, 202.

12 ADVANTAGE BLACK: HYPERBLACKNESS IN PERFORMATIVE SELF-PORTRAITURE

TINA POST

1

In August of 1967, Arts/Canada convened a "simultaneous conversation" between seven artists and intellectuals, largely but not exclusively white, on the color black. As the magazine's description suggests, the conversation was far-ranging and hard to wrangle, covering "[b]lack as spatial concept, symbol, paint quality; the socio-political implications of black; black as stasis, negation, nothingness and black as change, impermanence and potentiality."[1] The published transcript has the feeling of several confident men speaking over and past one another, with a couple of other gentlemen trying to get a word in, largely without success. But whatever the challenge posed by position and personality, the multiplicity of black itself presented an underlying difficulty. A father's blindness informed one man's sense of what blackness meant; for another, the link between blindness and blackness was an unproductive sentiment. For one man the socio-political struggles of black people had to be foregrounded; for yet another blackness was man's primal encounter with the beginning and the end of the cosmos, the fullness of everything and the emptiness of nothing. One man would bring up blackness's material qualities—glossy or matte—another would fixate on blackness's largely negative religious connotations. Black, it seems, contains too much—the material, moral, and metaphysical are hopelessly entangled.

Indeed, a much more recent *New York Times* article about ultra-blacks in science and nature could not avoid a nod at blackness's other associations. The author of the 2019 article included a brief art history lesson in monochrome, as well as a color psychology study which found that even in children, black has stronger emotional associations than any color except red. "Black was the color of cheating, and black was the color of cleverness," the author tells us. "A black tie was the mark of a boss, a black collar the sign of a pit bull. Black was the color of strength and of winning. Black was the color of rage."[2] These strong yet multivalent associations haunt the persistent entanglement of the color black and the racialization of certain bodies, which a few participants in the 1967 conversation highlighted even as others ignored or denied its the significance. In the words of Jamaican art critic David Boxer, "the changing fortunes of black, the color, and how it is viewed, how it is used, how it is accepted, can act as a barometer of our own attitudes, the attitudes of Black peoples and people of Black ancestry toward ourselves."[3] Though Boxer notes the link between the color and racial category, he does not do so in order to break the imbrication of the two (one possible desire in the face of the strong and complicated associations mentioned above).[4] Instead, Boxer urges his reader's attention to the ways that black people take up chromatic blackness as a site of aesthetic assertion.[5]

Though Boxer's observation occurred in 1997, the urge toward chromatic blackness has exploded in the work of contemporary black visual artists. Kerry James Marshall is most famous for this in the US, but an extremely partial list of other contemporary artists working in this vein could include Lina Iris Viktor, a black British artist of Liberian descent; South African artist Mohau Modisakeng; Zanele Muholi, also South African; Toyin Ojih Odutola, an American of Nigerian descent, and Mary Simbande, another South African. These artists vary in gender and sexual orientation, in continental home, in medium and artistic tradition. Yet a few things unite them. First, of course, is the fact that deep black hues are central to each of their oeuvres, and in each stark blacks stand for racialized blackness. Moreover, this stand-in does not feel like mere substitution, but rather like an assertion. Second, the hyperblackness of these works triggers the haptic sense—that is, the sense of touch as it is informed by sight. Why is a complex matter, and variable by artist and artistic medium. In Marshall's work, for example, haptic desire comes from the sheer number of blacks he uses to render what at first glance seems a monochromatic figure; for Lina Iris Viktor, discussed below, the interplay between light absorption and sheen entices a viewer's sense of touch. Third, the excessive blackness of these works is *not* the hyperblackness of minstrelsy, even when, as is often the case, these images index racist legacies of visuality (and even though, as scholars remind us, blackface and minstrelsy traveled international circuits.)[6] If not an allusion to minstrelsy, why use hyperblackness, with all its freighted associations? Why use deeply black pigments as racial metonymic? What does this hyperblack palette allow that more "natural" pigmentation does not?

By way of forcing coherence through generic specificity, this chapter focuses on those artists working in performative self-portraiture to pose some theorizations about why and how hyperblackness appears. I also examine some other cultural instantiations of hyperblackness—namely scientifically engineered blacks—to see what insights their adjacent discourses might afford. Thinking across these visual archives, and building in particular on the work of theorist Erin Manning, I propose we might think of the embodiment of enhanced blackness in contemporary performative self-portraiture as a "minor gesture" against minstrelsy's major legacies. "In the field of art," writes Manning, "the artwork, the object, even the effect created by an ephemeral composition is not, in itself, a minor gesture. The minor gesture is what activates the work under precise conditions, what makes the attunements of an emerging ecology felt, what makes the work work."[7] In other words, the minor gesture occurs (or, perhaps more accurately, becomes minor) in the interstice between itself and its specific context. It is as at once a political gesture, insofar as it is recognizable within its current surroundings, and an indeterminate one, to the extent that is future-opening. In what follows, I argue that the discourses surrounding recently engineered black hues reveal how hyperblackness itself unfolds in something of a minor key. Its discourses repeat old tropes, but with critical opportunities for difference. Artists working in performative self-portraiture, I argue, extrapolate these possibilities into aesthetic worlds. But there, the possibilities remain minor—full of the freighted past and possible otherwise, unsettled and awaiting.

2

In 2014, Surrey NanoSystems unveiled Vantablack, the darkest material the UK's National Physical Laboratory had ever measured, reflecting only

0.036% of light. The effect of so little light reflection is an uncanny one. "In fact, it could be said that it is almost impossible to 'see' Vantablack," Surrey NanoSystems's FAQ page explains. "However, the observer's brain of course tries to make sense of what it is seeing, with the result that some people describe it like looking into a hole!" Unable to register what they are seeing, people have an urge to touch Vantablack. "Perception of depth and dimensionality disappears into a scotoma of darkness," reports an article in *Wired*. "You look at Vantablack, but nothing looks back at you."[8]

Vantablack's name derives from its chemical composition: "Vanta" is an acronym for "Vertically Aligned Nanotube Array." It is, in the shortest description, a chemically engineered substance comprised of millions upon millions of carbon nanotubes—incredibly long, thin tubes of carbon, each about 3,500 times narrower than the average human hair.[9] The FAQ page of Surrey NanoSystems describes the engineering behind Vantablack's optics in the following way:

> Light energy striking the Vantablack surface enters the space between the nanotubes and is rapidly absorbed as it 'bounces' from tube to tube and simply cannot escape as the tubes are so long in relation to their diameter and the space between them. The near total lack of reflectance creates an almost perfect black surface. To understand this effect, try to visualize walking through a forest in which the trees are around 3km tall instead of the usual 10 to 20 meters. It's easy to imagine just how little light, if any, would reach you.[10]

In order to help their virtual visitors grasp the visual and haptic qualities of Vantablack, Surrey NanoSystems also offers a gallery of high-resolution images. One series of these images presents a piece of slightly crumpled aluminum foil partially coated with Vantablack. The Vantablackened section loses all surface detail, appearing as flat blackness or, as the associated copy declares, a "void." A second series of images showcases a miniature facial likeness of the rather Elizabethan-looking British science commentator Marty Jopson, completely coated in Vantablack. Displayed alone, this mask intrigues through its uniform hue and matte surface. Finer details are lost except at the mask's edges, where, for example, the ridges representing Jopson's facial hair can be seen. Surrey NanoSystems also presents images of the Vantablackened mask alongside an original bronze counterpart. As in the images of foil, Vantablack's light-absorbing qualities are contrasted with a reflective metallic sheen. On the one hand, we see a detailed mask, on the other a black hole lacking contour or texture, consisting merely in

FIGURE 12.1
A Vantablackened square of foil. Image courtesy of Surrey NanoSystems.

outline. Lest this seem a mere photographic trick, other views of the mask are offered as well: three quarter view, profile.

The metaphorical and visual examples Surrey NanoSystems uses to explain the profundity of this blackness are revealing of our cultural imaginaries and associations. Said differently, to illustrate a material that most of us are unlikely to experience firsthand, Surrey NanoSystems attempts to build on something we think we already know—or at least in which we already have an imaginative foothold. Mike Michael, a sociologist of science and technology at the University of Exeter, calls attention to the fact that Surrey NanoSystem's metaphor for understanding the pigment's structure asks us to inhabit an impossible forest of nanotube-like trees. "Strangely," he writes:

> the logic of forest analogy seems to be at odds with the technical operation of VANTAblack. If we were among the nanotubes, would we not witness some light as it "'bounces' from tube to tube," that is, tree to tree? The lack of light reflectance is surely best observed from outside the forest. The point is that even in the process of explicating the apparently basic technical properties of VANTAblack, there is recourse to the aesthetic, to experience.[11]

FIGURE 12.2
Two miniature likenesses of British science commentator Marty Jopson. Image courtesy of Surrey NanoSystems.

Michael points out that Surrey NanoSystem's explainer is perspectivally sub-par: it asks us to place ourselves amongst the imagined trees even though, scientifically, this metaphor would work better if we were outside the forest. But what is one to make of the *particular* experience called upon: the impossible forest whose closest analogues might be found in the Amazon Basin, the Congo Basin, and the extensive rainforests of Southeast Asia, which are, in order, the world's three densest canopies? While for many these locations might conjure the term "jungle" rather than "forest," (rain) forest is indeed the correct naming of their trees. It turns out there is no scientific definition for the term "jungle"—something you might have implicitly gleaned if you try to recall the last time you heard Sir David Attenborough utter the word—and presumably the staff of Surrey NanoSystems are aware of this technical distinction.[12] To be clear, it's not my assertion that Surrey NanoSystems's technical writers mean to conjure, explicitly and exactly, the terrifying bush of *Heart of Darkness*. After all, that's not—or at least is not exactly—a contemporary cultural imaginary. Nevertheless, I would assert that there's still something particular in Surrey NanoSystems's suggestion that the best way to imagine Vantablack is through the act of conjuring oneself into the darkness of uncivilized wilds—and then make them even darker. After all, they might have instead suggested their readers place themselves amongst cement walls, or stone pillars, or any number of other objects. In truth, fewer and fewer of us are likely to have extensive experience of dense forests, suggesting that most of us call to mind an imagined forest rather than a real one. Whether pre-colonial, colonial, or post-colonial, the imaginary of the dense, dark forest is, generally speaking, not the space of the rational western modern.

Relatedly, Surrey NanoSystems's curatorial image selection suggests that an optimal experience of Vantablack's phenomenology occurs when it is applied to a small facsimile of a human face. Again, I'm not suggesting any direct attempt at racialization through this application—especially given that the likeness is of a particular (white) person. Still, I feel there's something to say about the aesthetic experience of a dense black coating's application to a face—rather than, say, a kettle or a toy train or a boot. There is obviously something lively about hyperblackness applied to a visage, something that makes embodied hyperblackness especially compelling in our contemporary moment.

More recently, the MIT Center for Art, Science, and Technology bested Vantablack with their own nanotube-based black, this one absorbing 99.995 percent of light. Their illustration of their black's efficacy took the form of an artwork, *The Redemption of Vanity,* created in a collaboration between artist Diemut Strebe and professor of aeronautics and astronautics Brian Wardle. According to *MIT News,* the artwork "features a 16.78-carat natural yellow diamond from LJ West Diamonds, estimated to be worth $2 million, which the team coated with the new, ultrablack CNT material. The effect is arresting: The gem, normally brilliantly faceted, appears as a flat, black void."[13] I suspect I am not alone in connecting "blackness" and "diamonds" through the conflict-laden history of the gemstone's artificial scarcity—which is to say: Africa again.

In each of these examples, something harkens to racial blackness, but this invocation is subtly askew of previously prevalent discourses. There is an echo of an older racial imaginary, but faint enough to be missed—at least by the conscious mind, though all kinds of things might register at more subliminal levels. If the world of science suggests that hyperblackness might sounds itself in a minor key, it's my contention that the world of contemporary art suggests the same.

3

Consider, briefly, the example of visual artist and LGBTQ activist Zanele Muholi, especially the works that comprise *Somnyama Ngonyama*. Muholi's self-portraiture presents fantastic selves styled through the quotidian materials of culture: clothes pins, gloves, the inner tubes of bicycle wheels, scissors, wool, belts. The promotional copy of an exhibition at Spelman College Museum of Fine Art is a fine example of the discourses that surround this work: "Muholi playfully employs the conventions of classical painting, fashion photography, and the familiar tropes of ethnographic imagery to rearticulate contemporary identity politics. Each black and white self-portrait asks critical questions about social injustice, human rights, and contested representations of the Black body."[14] Certainly this much is true, but while the text mentions that Muholi's works are self-portraits, this description—and so many like it—insufficiently acknowledges the extent to which Muholi's project develops self-stylized avatars in order in order to engage external political, social, and historical structures.[15] By calling these selves avatars, I want to suggest this self-styling is in conversation with, but not predetermined by, the decolonial South African context. In other words, Muholi's *internal* or individual symbol structures are engaged as well as external or cultural ones. "This is my thesis," Muholi says in a video posted by the Aperture Foundation. "This is almost a whole self."[16]

Excessive blackness is central to the realization of these avatars—which are not simply a means of speaking back to power but also of finding the black self—here finding it through visual excess. "[Their] photographs return 'blackness' to the texture of [their] skin," writes Ellen Agnew, "the refraction of light on her dark body, the 'luminescent explosion of melanin.'"[17] While multiple articles quote Muholi saying—almost word for word—"I don't paint my face black because I am black anyway," it was only the article that took *exception* to their playing with hyperblackness that explicitly commented on the fact that they do digitally heighten the contrast of their images.[18] Muholi bumps the black in the creation of the avatars that represent self and that speak back to colonial legacies, even as they do so in part by reproducing them.

In the works of fellow South African Mohau Modisakeng, the (self-) subjects at the center similarly live at the intersection of the political and the mysterious or sublime, and again their capacity to do so rests in part on

FIGURE 12.3
Zanele Muholi, *Ntozabantu VI* (2016), from Somnyama Ngonyama (Hail the Dark Lioness). © Zanele Muholi. Courtesy of the artist and Yancey Richardson, New York.

hyperblackness and the way it fosters slippage between race and hue. Modisakeng originally trained as a sculptor before moving into photography and film, and his works exhibit the sparseness and high attention to materiality that one might expect of such a lineage. He heightens blackness in several ways, including, I suspect, a digital manipulation light consonant with that of Muholi's works and in harmony with the workings of Vantablack.[19] But Modisakeng also surrounds his body with backdrops and materials that amplify the differential workings of light absorption. In *Lefa*, meaning "Inheritance," the central figure merges with a bed of coal. In *Metamorphosis*, it is white powder rather than black powder that obscures the subject—he is behind its sheer white curtain, receding behind its light-reflective sheen. In his Untitled (Frames) series, Modisakeng wears a back shift and horse blinders in front of a stark white background. At times he turns his back to the camera, at others he arranges the blinders so that they cover his face. Most commonly, he is obscured through deep shadow. In any case, the result is that the black subject—in part or in whole—disappears into blackness. In his black and white film *Inzilo* (2013) an additional highly textured blackness has accrued to a figure's hands. He is able to away peel this waxy, powdery crust—but it magically drifts and adheres instead to his feet.

From the coal that fueled Soweto's fires to the blinders worn by the bony horses that pulled it, the traditional women's rituals of morning to the otherworldly black asphalt fields—once a site of plague quarantine—that form the backdrop of yet another series, Modisakeng's photographs do not abandon their particular South African context or its histories of colonial oppression. Yet neither are they limited to it, in content or symbol structure. Ruth Simbao reads *Inzilo* as a critical act of hovering, an intervention into colonially informed structures of time.[20] Though she reads this as enacted by movement, we might think about the supporting role the film's deep black hues play in fostering a sense of timelessness. Ashraf Jamal sees a "between times" at play in this work, quoting John Berger: "The true content of a photograph is invisible, for it derives not from a play with form, but with time."[21] He, too, invokes the word "hover"; he invokes the "enigmatic" and "sublime," "mystique" and "mysticism." And he quotes T. S. Eliot:

Between the idea
And the reality
Between the motion

FIGURE 12.4
Mohau Modisakeng, *Lefa La Ntate* (2016). Image courtesy of Mohau Modisakeng and Galerie Ron Mandos, Amsterdam. Copyright Mohau Modisakeng.

FIGURE 12.5
Mohau Modisakeng, *Untitled (Metamorphosis 1)* (2015). Image courtesy of Mohau Modisakeng and Galerie Ron Mandos, Amsterdam. Copyright Mohau Modisakeng.

And the act
Falls the Shadow
. . .
Between the conception
And the creation
Between the emotion
And the response
Falls the Shadow
. . .
Between the desire
And the spasm

Between the potency
And the existence
Between the essence
And the descent
    Falls the Shadow

Jamal introduces Eliot's inclusion by saying, "[Modisakeng's] signature—his enigmatic pull, as it were—stems from some infinitely gentle grasp of what it means to be human in Africa. If he understands the power of the epiphanic he also knows the inescapability of shadows. And here the Catholic poet T. S. Eliot, arrives upon the crux of Modisakeng's vision." In other words, Jamal's invocation of the Shadow is meant to signify spiritually and metaphorically. But I think he's also right *literally*: it is literally the black shadow that here facilitates the negotiation between emotion and response, between essence and descent.

Taken together, these artists' performative works suggest that heightened blackness might amount to a kind of archetypal hack—a shortcut into the world of mythos or spirit at odds with the legacies of racial blacking—even while (or especially when) it's used to signify on the legacies of racist visual legacies and/or ethnographic image making. This is all to reiterate that the minor gesture requires the counterpoint of the major, and that the major is what creates the possibility of the minor contrast. To the extent that hyperblackness works as a minor gesture, it requires the linkage between race and hue *and* our profound cultural memory of this linkage's horrific historical—and, Calvin Warren would say, metaphysical—consequences, so that it can perform its repetition with a difference.

4

As one might imagine given the monies involved in its development, Vantablack's provenance was scientific: ultra-blackness has application for telescopes and cameras, solar and thermal technology, military-grade weaponry, and space travel. But its creation also has artistic implications. Vantablack's use has been licensed exclusively to Sir Anish Kapoor, who has also attempted to describe Vantablack to the public. In an interview with the BBC's Radio 4, after employing the same technical descriptions as Surrey NanoSystems (picture yourself in a forest where the trees are one meter around and 300 meters tall and so forth), Kapoor proposes an alternative imaginative scenario. He says: "Imagine a space that's so dark that as you walk in you lose all sense of where you are, what you are, and especially all sense of time. Because when you lose orientation you also lose time. Something happens to your emotional self and in disorientation one has to reach in for other resources." "What we're looking to do," he says later in the short interview, "is to make real void objects."[22] Intriguingly, "void" seems to take on multiple parts of speech: is Kapoor making *real void* objects, or making real *void objects*? Moreover, this extreme blackness "mak[es] void" the subject who loses not only where they are, but also *what* they are, and when.

Kapoor also emphasizes the interaction of hue, haptics, and affects in absorptive blackness. "Well, it's blacker than anything you can imagine," he says, pushing excessive blackness beyond the boundaries of the thinkable. "It's so black you almost can't see it. It has a kind of unreal quality and I've always been drawn to rather exotic materials because of what they make you feel. So, that relationship between the materiality of the thing and the way it makes you feel is terribly important." Kapoor asserts that it is the materiality of excessive blackness, alongside the fact of its hue, that allows it to convey what Mike Michael summarizes as the "disorienting but valued aesthetic experiences" of "mystery, disorientation, the sublime (of the black hole) even."[23] This disorientation was emphasized when an Italian museum goer quite literally fell into the 8-foot pit comprising the Kapoor artwork *Descent Into Limbo* at the Museum of Contemporary Art in Porto, Portugal.[24]

Kapoor's discursive slide from familiar but exaggerated materiality—extremely thin hair, extremely tall trees—into a cosmic non-space outside of self and time is remarkably consonant with the ways that the artist Lina Iris Viktor envisions the use of black paint. Viktor uses her body as performative object, first painting her body, then taking a photograph, digitally enhancing it and printing it on canvas, where it is ultimately painted and gilded.[25] Viktor's work is striking for several reasons, but perhaps the most profound is her limited color palette: for a long time she adhered to a palette of black, white, 24-karat gold, and Yves Klein blue (to which she added red around 2017). Viktor believes her "purist palette" helps create a mythical world in which "natural laws, hermetic philosophies, mathematic & scientific principles," are all in play. "The world she imagines," says her website, "is perfectly curated, puristic, and minimalist—it tends to cancel out materiality, to transfer it to the realm of the spiritual & the alchemical."[26] The work "seeks to instill a divine order," and to "create immersive environments that transport the viewer into other worlds." Though the works feature Viktor, she doesn't consider them self-portraits. She says:

> The portraits are about abstraction, disidentification, transformation. I use myself because I am always present and it is a performance. But I feel no particular affinity or ownership over the final image as it is not about me or "self." Different bodies of work are about different aspects of these motifs. . . . *The Body Black: Iron, Gold, Ether is* about the unity on an elemental or subatomic level between our form, our bodies and the greater universe—the oneness of it all. We are all made up of the same elements; we are simply star matter.[27]

Yet for all her talk of spiritualism and cosmology, Viktor is not advocating an abandonment of race, as the patterns and landscapes that surround her embodied avatar are often meant to reference Africa, if also an Afro-Futurist not-yet-here.[28]

Though it is never mentioned, the brown of Viktor's skin comprises one further color within her limited palate—brown as a hue is elided into the category of *racial* blackness, which is represented, as well, through paint and/as makeup. Viktor herself seems to endorse this interpretive collapse (or, at the very least, the journalistic art world comes to that conclusion after interviewing her). Consider an example from *The Cut*:

> "[The word black] has a lot of negative associations and synonyms attached to it," she says, "along with the fact that it is a word used to describe . . . the black race. So, think about the psychological connotations of that." So in her paintings, Viktor challenges this conventional understanding of blackness, and the idea that "void is black to the human eye." "You look at the night sky, and it looks black, like a void, but is actually full," she says. Her work tells the story of blackness as "the saturation, the full, of the entire color spectrum."[29]

Viktor evokes the color black only to yoke it to racial blackness, *then* in quick turn, weds *both* to the cosmos. Ironically Viktor here suggests that blackness's cosmic resonance resides in its manifestation of the *entire* color spectrum, even though the text of her "About" page suggests the cosmic resides in *delimited* color. Black is everything and nothing. The language of the void returns. This is true even though the black of Viktor's work is not solely light-absorbing. Instead, as rendered in black paint or makeup, Viktor's body is a *mixture* of light absorption and light reflection—of void and sheen. If shine obscures the black subject, as art historian Krista Thompson has argued, the rhetorics around excessive blackness suggest that the optics and haptics of absorption *also* dislocate the black subject.

FIGURE 12.6
Lina Iris Viktor, *The Body Black*. Iron, Gold, Ether (2015–16).

If I were to name what seems *distinct* about Viktor's invocations of hyper-blackness, it would be her use of makeup in general, and the incompletion and smearing that can accompany her use of blacking in particular. In *Yaa Asantewaa* though the blacking around Viktor's face has clean edges, the top of the black circle on her breast is smudged. In *Materia Prima*, Viktor's thighs reveal streaks and smudges. Similarly, while the main interaction of skin and paint in *The Body Black. Iron, Gold, Ether* is arguably the dappled

FIGURE 12.7
Lina Iris Viktor, *Yaa Asantewaa* (2016).

meeting point of black and gold, the bottom edge of the picture once again iterates the line between makeup and skin.

This aesthetic gesture works on at least three levels. The first is technical: smearing and smudging are especially good indications of makeup's material presence, signaling this blacking *as* blacking (and not, or not only, as digital manipulation)—and therefore invoking the histories and legacies of other cultural blacking practices (including minstrelsy). The second is philosophical: if indeed Viktor is trying to promote the collapse of the hue and the racial structure in order to lend *both* kinds of blackness a kind of cosmic significance, then a certain imprecision of line, a certain bleeding of color, is supportive of this collapse. I'm fairly confident that Viktor is intuitively or intellectually aware of these consequences of her image-making practice. But I also wonder if there isn't some tension in Viktor's commitment to the cosmic, spiritual plane, in that the smeared and smudged border also enacts a haptic and visual cleaving, revealing the ways that literal blackness (hue) and figurative blackness (race) *both* split apart *and* stick together. The smudgy borders of Viktor's blacking invite questions about the "truth" of blackness as racial category and/as hue. And even as it attempts to signal void, it invites a haptic sensorium in its edges (whereas the void, of course, could not be touched).

Ultimately, the possibility that the gestures of hyperblackness can be used as a shortcut to the mysterious and sublime acknowledges, and yet turns to different ends, the assumptions of racial hyperblackness. That is, the supposed incomprehensibility of blackest Africa is here embraced and employed by peoples of African descent to convey a degree of incomprehensibility employed in the name of mystical otherworldliness *and* geopolitical sophistication. If there is an ontological terror in this blackness, they seem to say, it is the terror of the divine sublime.

This allusion to ontological terror is, of course, a reference to the influential work of Calvin Warren. Warren takes as a point of departure the philosopher Alain David's question, "Why are Negroes Black?" In answer, Warren argues that black personhood developed out of metaphysical necessity, and that black people are a scapegoat for the Western terror of nothingness. In his own words:

> colonialism and Eurocentrism are antiblack strategies for attempting to obliterate, and to forget, dreaded nothing—since black bodies, cultures, and existence are assigned this unbearable formlessness within modernity. . . . The Negro is

FIGURE 12.8
Lina Iris Viktor, *Materia Prima* (2016).

> invented precisely to absorb the terror of this nothing, of the interruption of time and space, within modernity. . . . Antiblackness and its technologies of destruction are designed to obliterate nothing: nothing as formlessness, nothing as interruption, nothing as black, and ultimately, nothing as the Negro.[30]

This passage is an instantiation of the afro-pessimism that periodically ripples through the scholarship of blackness and has seismically increased in the wake of police shootings, racist attacks, Neo-Nazi rallies, and the recent US presidency of a racist demagogue. Still, Warren's book isn't thoroughly hopeless. Rather he ends with a coda that suggests that blackness must "get over the human and its humanism fantasies." He writes, "The important task for black thinking (philosophizing, theorizing, theologizing, poeticizing) is to imagine black existence without Being, humanism, or the human. Such thinking would lead us unto an abyss. But we must face this abyss—its terror and majesty. I would suggest that this thinking leads us into the spirit, something exceeding and preceding the metaphysical world."[31] When it comes to correcting for the nonbeing of blackness, perhaps the minor aesthetics of the void join what Sarah Cervenak calls the "cosmopoetics" of blackbody radiation, and the gravitational pull that Andrew Sofer locates in the dark matter of theater and performance, in asking what aesthetic and agential properties necessarily flow from abyss, and what the abyss can claim as its own.[32]

NOTES

Kyle Frisina, Vivian L. Huang, and Sarah Stefana Smith thought alongside me at the American Studies Association's panel "Visualizing In/Emergence: Race, Aesthetics and Ethics of Obfuscation." Denise Ryner and Yaniya Lee invited this work to be part of the "Bodies, Borders, Fields" symposium, where I was privileged to share a panel with keyon gaskin and M. NourbeSe Philip. Maria DeSimone invited my participation in Brown University's Pembroke Seminar, convened by Leslie Bostrom and Evelyn Lincoln. Kareem Khubchandani, Vanatta Ford, Anna Storti, and Tao Leigh Goffe were my invaluable cohort members. Finally, my gratitude to Lida Zeitlin-Wu and Carolyn Kane for improving this work through their feedback and engagement.

1. "Black," *Arts/Canada.*

2. Angier, "Ultra-Black Is the New Black."

3. Boxer, "Notes for Black as Colour," 4.

4. For one theorization of the consequences of this imbrication, see Warren, *Ontological Terror.*

5. Boxer uses a capital letter "B" to refer to the racial category and a lowercase "b" to refer to the color. I prefer a lowercase "b" for both. La Marr Jurelle Bruce has already beautifully articulated the philosophy that I, too, hold. See Bruce, *How to Go Mad without Losing Your Mind.*

6. Roper, "Blackface at the Andean Fiesta; Hoxworth, "The Jim Crow Global South"; Thelwell, *Exporting Jim Crow.*

7. Manning, *The Minor Gesture*, 65.

8. Rogers, "Art Fight! The Pinkest Pink Versus the Blackest Black."

9. "FAQs | Surrey NanoSystems."

10. "FAQs | Surrey NanoSystems."

11. Michael, "On 'Aesthetic Publics': The Case of VANTAblack®."

12. Bouchard, "Is There Really a Jungle Out There?"

13. Chu, "MIT Engineers Develop 'Blackest Black' Material to Date."

14. "Zanele Muholi Art Papers Live."

15. My use of the term "avatar" is indebted to McMillan, *Embodied Avatars.*

16. *Zanele Muholi's Visual Activism.*

17. Agnew, "HAIL! The Dark Lioness: Part 1."

18. Joja, "New Work of Zanele Muholi"

19. There is more to say about hyperblackness that occurs as a function of light-based technology (and? versus?) that which is achieved through pigmentation. Here, I will simply flag this difference in the technological workings of hyperblacknesses as something that deserves greater thought. My gratitude to Lida Zeitlin-Wu for tugging at this thread.

20. Simbao, "Performing Stillness In Order to Move."

21. Jamal, "Mohau Modisakeng: Occupation."

22. "How Black Can Black Be?"

23. Michael, "On 'Aesthetic Publics': The Case of VANTAblack®."

24. Cascone, "A Man Fell Into Anish Kapoor's Installation of a Bottomless Pit at a Portugal Museum"; Deron, "A Man Fell into an Art Installation Called Descent into Limbo."

25. Frank, "Artist's Dazzling Self-Portraits Explore The Cosmic Power Of Gold"

26. Viktor, "About."

27. Harris, "Lina Viktor Makes Otherworldly Paintings from Star Matter."

28. McMillan, *Embodied Avatars*; Muñoz, *Cruising Utopia.*

29. Dellas, "This British-Liberian Artist Paints with 24-Karat Gold."

30. Warren, *Ontological Terror*, 36.

31. Warren, 170–171.

32. Cervenak, *Black Gathering*, 16; Sofer, *Dark Matter*.

BIBLIOGRAPHY

Agnew, Ellen. "HAIL! The Dark Lioness: Part 1." *Art Africa* (blog), December 4, 2017. http://artafricamagazine.org/hail-dark-lioness/.

Angier, Natalie. "Ultra-Black Is the New Black." *The New York Times*, November 11, 2019, sec. Science. https://www.nytimes.com/2019/11/11/science/black-fashion-physics-animals.html.

Autograph. "Zanele Muholi Art Papers Live." Accessed November 5, 2018. https://autograph.org.uk/events/zanele-muholi-artpapers-live.

BBC News. "How Black Can Black Be?" Accessed November 2, 2018. https://www.bbc.com/news/av/entertainment-arts-29326916/how-black-can-black-be.

"Black." *Arts/Canada*, no. 113 (October 1967): 3–18.

Bouchard, R. Philip. "Is There Really a Jungle Out There?" *The Philipendium* (blog), November 28, 2018. https://medium.com/the-philipendium/is-there-really-a-jungle-out-there-17d50f45b8aa.

Boxer, David. "Notes for Black as Colour." In *Black as Colour*, Exhibition Catalogue. Kingston: National Gallery of Jamaica, 1997.

Bruce, La Marr Jurelle. *How to Go Mad without Losing Your Mind: Madness and Black Radical Creativity*. Black Outdoors: Innovations in the Poetics of Study. Durham, NC: Duke University Press, 2021.

Cascone, Sarah. "A Man Fell into Anish Kapoor's Installation of a Bottomless Pit at a Portugal Museum." *Artnet*, August 20, 2018, News. https://news.artnet.com/art-world/man-injured-falling-into-anish-kapoor-hole-1335176.

Cervenak, Sarah Jane. *Black Gathering: Art, Ecology, Ungiven Life*. Black Outdoors: Innovations in the Poetics of Study. Durham, NC: Duke University Press, 2021.

Chu, Jennifer. "MIT Engineers Develop 'Blackest Black' Material to Date." *MIT News* | Massachusetts Institute of Technology, September 12, 2019. https://news.mit.edu/2019/blackest-black-material-cnt-0913.

Dellas, Mary. "This British-Liberian Artist Paints with 24-Karat Gold." *The Cut*, March 5, 2018. https://www.thecut.com/2018/03/24-karat-gold-painting-lina-iris-viktor.html.

Deron, Bernadette. "A Man Fell into an Art Installation Called Descent into Limbo." Edited by John Kuroski. *Ati*, August 21, 2018. https://allthatsinteresting.com/anish-kapoor-descent-into-limbo.

"FAQs | Surrey NanoSystems." Accessed October 25, 2018. https://www.surreynanosystems.com/about/vantablack

Frank, Priscilla. "Artist's Dazzling Self-Portraits Explore The Cosmic Power Of Gold." *Huffington Post*, May 11, 2016, sec. Culture & Arts. https://www.huffingtonpost.com/entry/artist-explores-the-mystical-powers-of-gold-in-stunning-sculptural-works_us_57322536e4b016f37897483e.

Harris, Marquita. "Lina Viktor Makes Otherworldly Paintings from Star Matter." *I-D*, May 4, 2016. https://i-d.vice.com/en_us/article/pab4mb/lina-viktor-makes-otherworldly-paintings-from-star-matter.

Hoxworth, Kellen. "The Jim Crow Global South." *Theatre Journal* 72, no. 4 (2020): 443–467. https://doi.org/doi:10.1353/tj.2020.0099.

Jamal, Ashraf. "Mohau Modisakeng: Occupation." In *Mohau Modisakeng*, 100–105. Cape Town, South Africa: What If The World, 2016.

Joja, Athi Mongezeleli. "New Work of Zanele Muholi." *Africanah.Org: Arena for Contemporary African, African-American and Caribbean Art* (blog), March 2, 2016. https://africanah.org/new-work-of-zanele-muholi/.

Manning, Erin. *The Minor Gesture*. Durham, NC: Duke University Press, 2016.

McMillan, Uri. *Embodied Avatars: The Art of Black Female Performance*. New York: NYU Press, 2015.

Michael, Mike. "On 'Aesthetic Publics': The Case of VANTAblack®." *Science, Technology, & Human Values* 43, no. 6 (May 2018): 1098–1121.

Muñoz, José Esteban. *Cruising Utopia: The Then and There of Queer Futurity*. New York: New York University Press, 2009.

Rogers, Adam. "Art Fight! The Pinkest Pink Versus the Blackest Black." *Wired*, June 22, 2017. https://www.wired.com/story/vantablack-anish-kapoor-stuart-semple/.

Roper, Danielle. "Blackface at the Andean Fiesta: Performing Blackness in the Danza de Caporales." *Latin American Research Review* 54, no. 2 (June 2019): 381–397.

Simbao, Ruth. "Performing Stillness in Order to Move: Mohau Modisakeng's Becoming." In *Mohau Modisakeng*, 68–73. Cape Town, South Africa: What If The World, 2016.

Sofer, Andrew. *Dark Matter: Invisibility in Drama, Theater, and Performance*. Theater: Theory/Text/Performance. Ann Arbor: University of Michigan Press, 2013.

Thelwell, Chinua. *Exporting Jim Crow: Blackface Minstrelsy in South Africa and Beyond*. Amherst: University of Massachusetts Press, 2020.

Viktor, Lina Iris. "About." Artist's Website. Lina Iris Viktor. Accessed October 17, 2018. http://www.linaviktor.com/about/.

Vimeo. "Zanele Muholi's Visual Activism." 2018. https://vimeo.com/279552605.

Warren, Calvin L. *Ontological Terror: Blackness, Nihilism, and Emancipation*. Durham, NC: Duke University Press, 2018.

III ALGORITHMS

INTRODUCTION TO PART III: RECODING COLOR IN AN AGE OF INFORMATIC BIAS

It's hardly news that algorithms intensify preexisting stereotypes by classifying people in ways that draw on much longer legacies of racist grammars and protocols, as contoured in the previous chapters in this volume. From the 2015 object recognition algorithm of Google Images sorting Black people into a folder labeled "Gorillas"[1] to a Nikon camera's facial recognition system asking a Taiwanese-American woman and her family members if "someone blink[ed]?" (despite everyone's eyes being fully open),[2] these appalling but ultimately unsurprising technological errors shed a glaring light on only a few of the myriad ways in which algorithmic racial bias is never merely technical. Rather, informatic bias is always already deeply entrenched in a culture's history, social and political beliefs, and, most crucially, unspoken prejudices and assumptions about who gets to be considered human.

Unlike previous, arguably less mediated forms of human-to-human animosity, there is something especially insidious about the violence that results when human bias is embedded in otherwise opaque computational technologies that are perceived to be divorced from human agency as a result of developments in automation and engineering. Several recent racialized artificial intelligence (AI) pitfalls from Google, Facebook, and Instagram (including those mentioned here) all serve as a case in point.[3] The implementation of these digital racial protocols has fatal, real-world consequences, as with, for example, the predictive policing algorithms that make a Black person more than twice as likely to be arrested than a white person, or a Muslim person forty-two times more likely to be searched or detained at the airport.[4]

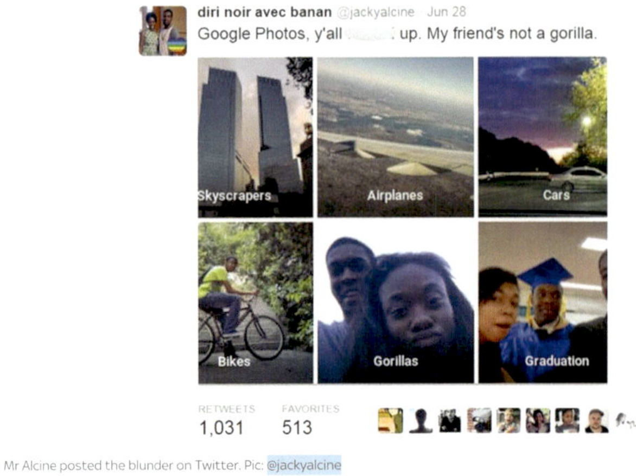

FIGURE PIII.1
Jacky Alciné's 2015 tweet about the infamous Google Photos "Gorilla" incident.

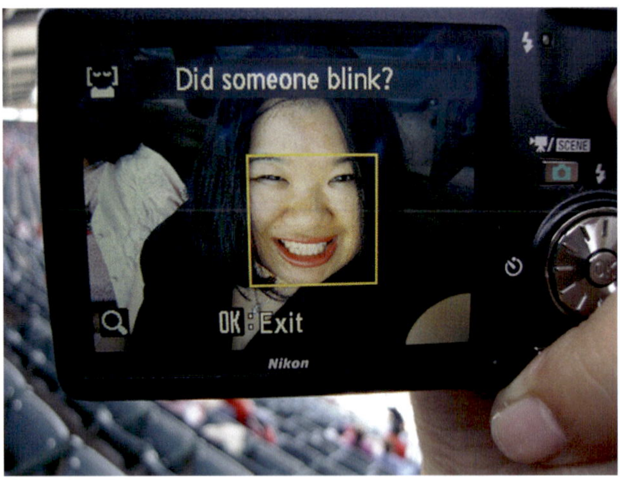

FIGURE PIII.2
Screenshot of Nikon interface (2015). Nikon's facial detection software asked a Taiwanese-American woman, "Did someone blink?"

Machine learning systems rely on neural networks paired with datasets that "train" algorithms to behave a certain way. However, these training sets are almost always built on the preexisting assumptions of those (primarily computer scientists and engineers) who input the data and create the rules (protocols) for them—technicians who have historically been white, male, and well educated. The result, unsurprisingly, is that informatic racial protocols necessarily exacerbate existential threats to those already coded as nonwhite (and therefore deviant or substandard). Consider Samsung mobile phone users who may observe (and then internalize) the way in which their phone's "Beauty Mode" filter automatically lightens their skin tone to bring out their "best features." If the biased assumptions already deeply engrained in *both* the histories of colorism and white supremacy and the politics of pattern recognition in AI today are not brought to light, they will only recede further into the darkened recesses of an already obfuscated black box.[5]

How did so-called advanced hypertechnology land us in this predicament, where racial bias is embedded everywhere, from police databases to selfie beautification filters? Will it ever be possible to re-encode the choices and decisions that inform how pattern recognition systems function in a way that encompasses the nuances of lived experience and identity?[6] A handful of engineers, scholars, activists, and artists have tried, collectively

FIGURE PIII.3
Joy Buolamwini, "The Coded Gaze: Unmasking Algorithmic Bias" (2016).

FIGURE PIII.4
Joy Buolamwini, *AI, Ain't I a Woman?* (2018).

maintaining that racist algorithmic protocols are not set in stone. A much-cited example is Nigerian-American Massachusetts Institute of Technology (MIT) researcher Joy Buolamwini, who while still a graduate student found that the facial-recognition software that she was programming was unable to detect her face unless she donned a white mask, exhibiting an encoded color bias that she would eventually label "the coded gaze."[7] She also noticed that similar facial recognition systems have greater difficulty discerning the gender of Black women, illustrated in her spoken-word video *AI, Ain't I a Woman?* (2018), in which Buolamwini juxtaposes the faces of historical figures like Sojourner Truth and Shirley Chisholm with those of living Black women (including Oprah Winfrey, Serena Williams, and Michelle Obama). The resulting message after uploading each photo is the same: "Appears to be male."[8] As a result of her findings, Buolamwini founded the Algorithmic Justice League in 2016, a nonprofit that raises awareness about algorithmic bias and advocates for inclusive datasets and more equitable hiring practices in the tech industry.

While the Algorithmic Justice League, alongside other initiatives focusing on equitable AI, such as Data for Black Lives and the AI Now Institute, are undoubtedly doing important policy and advocacy work, it is still crucial to understand these efforts alongside broader histories of the development of algorithmic bias. Simone Browne's 2015 *Dark Matters: On the Surveillance of Blackness* charts some of these parallels by illustrating how contemporary surveillance and policing practices uphold centuries-old racist ideologies of

European colonialism and slavery. By analyzing the physical branding of enslaved people's bodies as an early form of biometric technologies, Browne excavates a protoalgorithmic form of producing and tracking Blackness. Both AI and biometrics, she concludes, derive from the epidermal (and therefore chromatic) protocols of the transatlantic slave trade and its aftermath.[9] By extension, surveillance practices across a multitude of technologies likewise create and reinforce analogous racialization.

CV Dazzle Look 1. 2010. Hair by Pia Vivas Model: Jen Jaffe Original look for CV Dazzle thesis at NYU. Photo: ©Adam Harvey

CV Dazzle Look 2. For DIS Magazine Creative direction by Lauren Boyle and Marco Roso. Model: Irina. 2010. Hair by Pia Vivas. Photo: ©Adam Harvey

CV Dazzle Look 3. For DIS Magazine. Creative direction by Lauren Boyle and Marco Roso. Model: Jude. 2010. Hair by Pia Vivas. Photo: © Adam Harvey

CV Dazzle Look 4. For DIS Magazine Creative direction by Lauren Boyle and Marco Roso. 2010. Hair by Pia Vivas. Photo: ©Adam Harvey

FIGURE PIII.5
Adam Harvey, *CV Dazzle* (2010–ongoing).

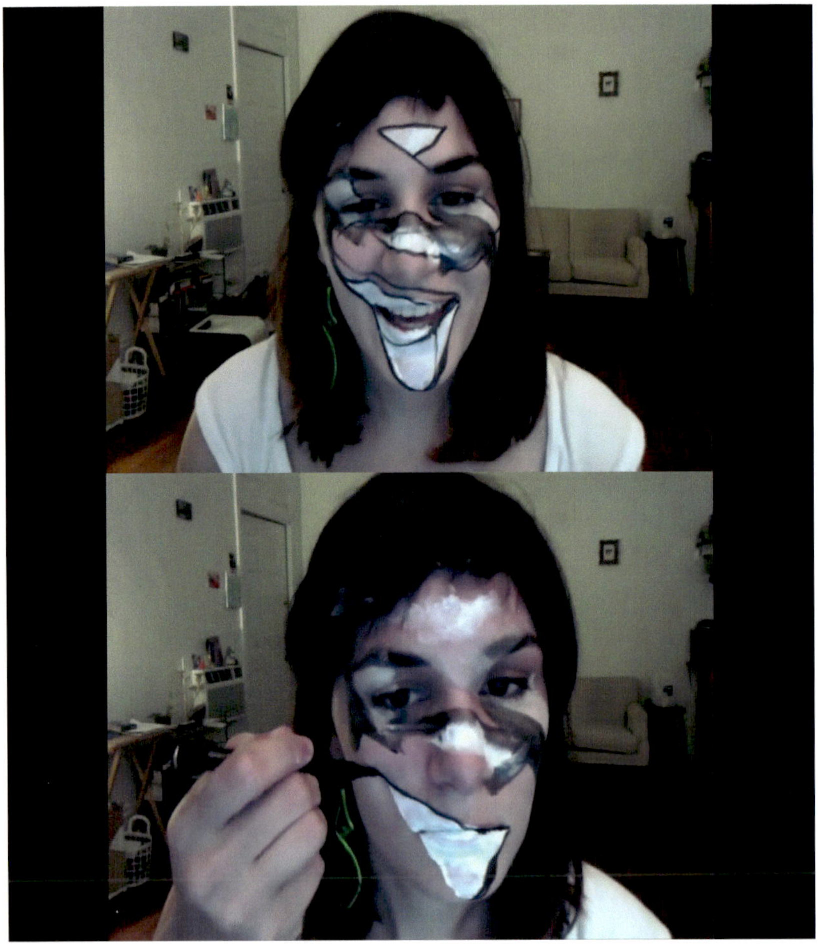

FIGURE PIII.6
Jillian Mayer, *How to Hide from Cameras* (2013).

By the same token, the same algorithmic biases that fail to detect darker skin tones provide an opportunity for the formation of counterstrategies. One common tactic has been for people of color to make themselves invisible, what Browne calls "dark sousveillance." In contemporary media art—as reflected in the recent Guggenheim exhibition *Going Dark: The Contemporary Figure at the Edge of Visibility* (October 2023–April 2024)—this has been taken up through creative uses of pigment and illuminated color to thwart and confuse algorithmic identification systems. Projects such as Adam Harvey's

FIGURE PIII.7
Merchant ship in dazzle camouflage (left) and the same ship uncamouflaged (right), *Encyclopædia Britannica*, 1922.

CV Dazzle and Jillian Mayer's *How to Hide from Cameras*, for instance, apply a tactic used during World War I known as "dazzle camouflage" or "reverse camouflage" to human faces. Reverse camouflage, which was primarily used on naval ships, involved painting a ship's exterior with contrasting colors and elaborate patterns to make it more difficult to detect its distance and speed.[10] In the case of the "Razzle Dazzle ships," as they were sometimes called, as well in as Harvey and Mayer's work, the application of color is not intended to blend in, but instead to stand out—to become unreadable by pattern recognition systems, whether human or machine. "Apply makeup that contrasts with your skin tone," Harvey instructs. "[Use] unusual tones and directions: light colors on dark skin, dark colors on light skin."[11]

Another set of digital artists have, for decades now, been experimenting with strategies of AI opacity. These artists include Zach Blas and Hito Steyerl, both of whom have PhDs and write about contemporary surveillance culture. Following the work of philosopher and poet Édouard Glissant, Blas argues that everyone has a right to opacity, to be unknowable and undifferentiated as an ethical stance against imperial conquest and domination.[12] He calls this "Informatic Opacity," which, following Phil Agre (see the introduction to this book), holds that "capture" is the dominant mode of surveillance today,[13] marking a break from the optical-based panoptic paradigm of part II

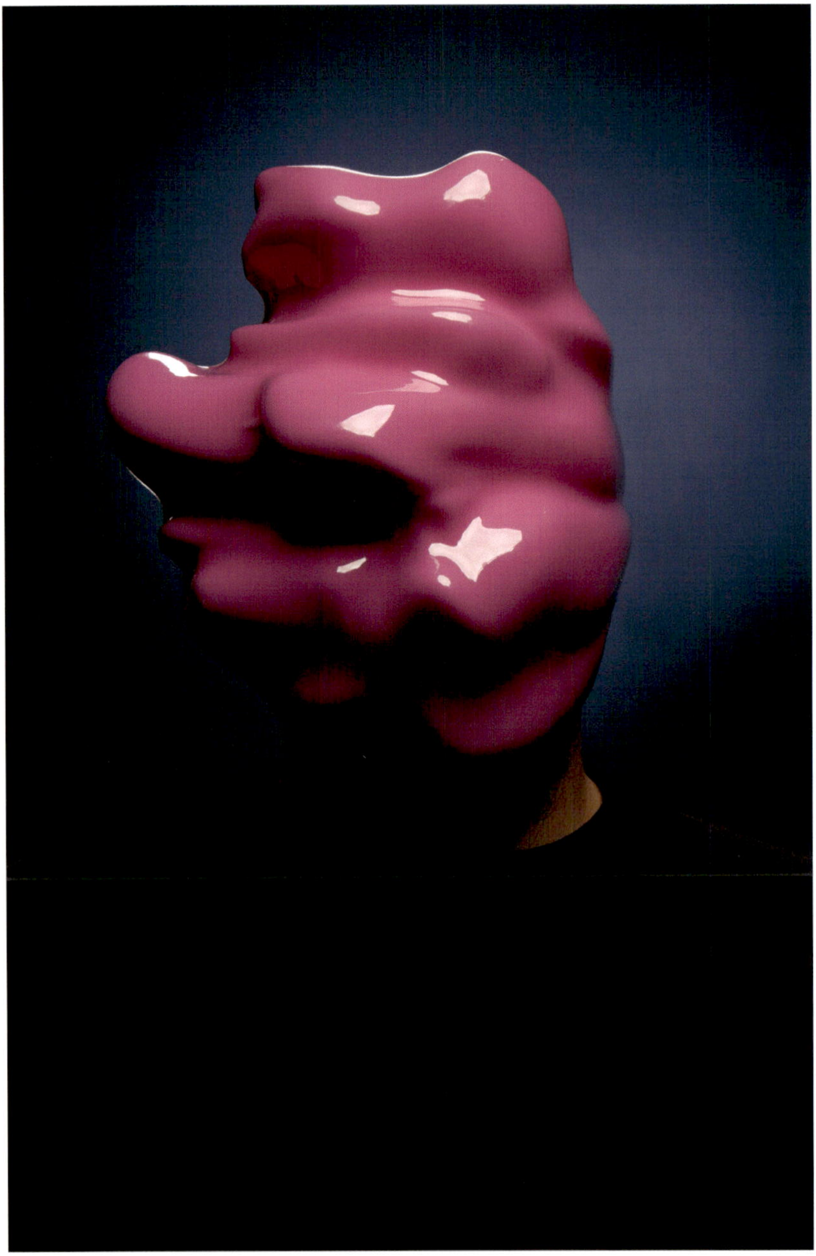

FIGURE PIII.8
Zach Blas. *Facial Weaponization Suite* (2011–).

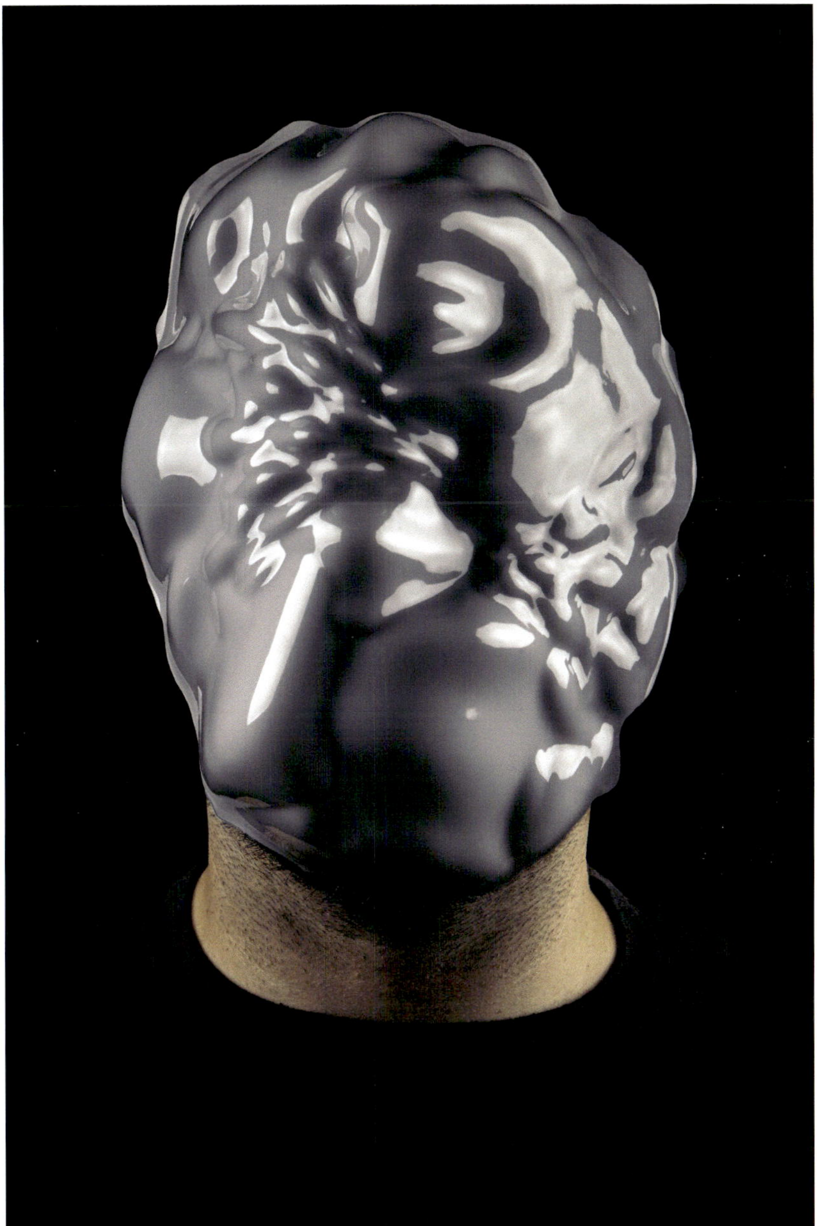

FIGURE PIII.9
Zach Blas. *Facial Weaponization Suite* (2011–).

FIGURE PIII.10
Hito Steyerl, *How Not to Be Seen: A Fucking Didactic Educational .MOV File* (2013).

and expanding the locus of surveillance to the "computational logics of data aggregation," which, as noted, are more readily bound to the numeric and linguistic realm than the visual.[14] Because human subjects in this paradigm are no longer coherent by optically available color, but rather "through data and information,"[15] Blas argues, an emergent politics of resistance must accordingly be rooted in strategies of data decoding and recoding.

For example, in the two images from *Facial Weaponization Suite* (2011–) shown here, Blas develops site-specific mask-making workshops to respond collectively to biometric control in a local context. By disallowing easy calculations and categorizations of the face, the masks generated in each workshop present new ways of being and seeing. In so doing, community members can actively resist the computational logic of data aggregation that otherwise profile and classify subjects as datafied individuals.[16]

Similarly, in her 2013 *How Not to Be Seen: A Fucking Didactic Educational .MOV File,* which is frequently assigned in critical media studies courses, Steyerl likewise explores the politics of invisibility through a series of playful but critical tactics (presented as a "how-to" video) for opting out of visual detection systems indigenous to the digital age.[17] Her premise is that on digital platforms, identity is now recognized and captured by way of *resolution* (versus optical color), which as we know is engineered through a sophisticated (black-boxed) numeric grid. "This is a resolution target," drones the automated voiceover as we see an image of Google Earth spinning and rotating at the start of the video. "It measures the resolution of the world as a picture. Resolution determines visibility. Whatever is not captured by resolution is invisible."[18] Blas has observed that Steyerl's video takes place at an abandoned resolution targeting location in the California desert, once used by the US Air Force to calibrate aerial photographic technologies and weapons in the 1950s and 1960s. The military resolution targeting field was replaced, however, "by a new resolution standard: the pixel."[19] In 1996, photographic resolution was approximately twelve meters per pixel (the video's narrator explains), while today it is one foot. Thus, the viewer is told, "to become invisible one has to become smaller or equal to one pixel."[20] Three of Steyerl's other noted techniques include related digital postproduction techniques such as "face painting" (in the video, we see Steyerl composite herself out of an image through paint as others wear green-screen bodysuits and burkas), "living in a gated community," and "living in a military zone." For Blas and Steyerl, as well as for Harvey and Mayer, the tactic of using color—whether digital or analog—to make one's face or being unreadable to AI pattern recognition protocols suggests that what has become increasingly desirable in a world of surveillance capitalism is the luxury of *not* being seen.

Yet there is also the luxury and freedom of being seen anew in displays of difference that do not coincide with dominant modes of governance, as Blas's work suggests. The question of who has the right to be seen and who has the right to hide once again falls along racial lines. If, as Wendy Chun argues, machine learning algorithms lump users with similar behaviors and preferences together in a way that mimics racial segregation in cities,[21] how do we find a way out of this reductive logic that reduces personhood to a series of checked boxes? Can algorithmic vision be rerouted for productive, collective, and communal recognition? In public space, now highly surveyed by closed-circuit television and related tracking software, informatic-era color encoding is contingent on AI that determines who counts as legitimate and valuable members of society and who constitutes threats to so-called normality. Such determinations are largely accomplished by way of individual (and therefore highly racialized) data profiling.

Today, we have arrived at a key juncture in algorithmic culture. Deepfakes, AI art generators, and chatbots—particularly Open AI's ChatGPT—have ushered in a new wave of techno-utopian and dystopian discourse about the increasingly blurred lines between humans and machines. As information becomes increasingly entangled with obfuscated notions of power and control, we need to remember that humanity is not something that is equally afforded to everyone.

THE ESSAYS

While algorithmic bias clearly permeates a wide range of cultural practices, the chapters in part III focus on biometric forms of skin-based racial-encoding between the 2000s and 2020s. Chapter 13, an excerpt from Lisa Nakamura's foundational *Cybertypes: Race, Ethnicity, and Identity on the Internet* (2002, republished here with archival images), offers an exception, as it does not focus on biometrics per se, but rather on "identity tourism" in late-1990s chat rooms—virtual spaces that attempted to decouple a user's given race from the customizable color of their avatar.

In response to the then-prevalent rhetoric that the Internet neutralized and erased social and bodily differences ("on the Internet, nobody knows you're a dog"), Nakamura shows how this misguided techno-utopianism is also inseparable from racial performativity and othering.[22] Far from disappearing online, she asserts, "the Internet is a place where race happens."[23] Nakamura thus coins the term "cybertypes" to describe how race is codified through visual online protocols.[24] She offers the example of her own experience creating an avatar in the virtual chat space Club Connect. Despite the prevalence of "identity tourism" on these platforms, where the predominantly white male users choose Orientalized stereotypes such as the "horny geisha" or "exotic samurai"[25] as their avatars, she observes how race is effectively severed from skin color in the digital realm. A fellow user laughs and tells her that her skin and face color are mismatched and helps her with the color settings on the platform—all without any mention of race.[26] Here, "phenotypical" color, denoted by customizable pixel settings, operates in what pretends to be a postracial space, precisely because it is customizable and no longer tied to the user's physical body. Customizing avatars through digital color thus embodies the precise form of racial colorblindness that Nakamura critiques: "This fantasy of skin color divorced from politics, oppression, or racism seems to also celebrate it as infinitely changeable and customizable: as entirely elective as well as apolitical."[27]

The next three chapters offer twenty-first century critiques of facial recognition software, which can arguably be seen as the apotheosis of a century-long tendency to privilege the prototypically white face as the norm across modern media platforms. These processes of racialization originating with analog photography and film (as described by Dyer, Roth, and Yue in part II) remain crucial for understanding racially encoded biases in algorithmic pattern recognition today, which likewise operate by a logic of default whiteness. In this sense, the Shirley Cards and China Girls from part II can be seen as datasets for "training" cameras to behave a certain way long before the advent of digital imaging and facial recognition technologies. Unlike the image-based technologies of photography and film, however, which have historically been linked to realism and visual truth, the emergence of digital color-grading processes and computer-generated imagery technologies now rely on an emergent rhetoric of "quantification," "numerical representation," "differentiation," and "discretization." This remediated form of default whiteness demands a new set of questions: How is digital color calibrated and formulated on digital platforms, and in whose interests? And how do supposedly customizable and individualized digital features like filters and camera settings actually reinforce a set of racialized standards or norms?

Where Nakamura's *Cybertypes* was arguably the first to articulate how racist standards operate within a supposedly postracial digital society, chapter 14, an excerpt from Ruha Benjamin's *Race after Technology: Abolitionist Tools for the New Jim Code* (2019), takes up this task for the post–Web 2.0 era. In "Coded Exposure," Benjamin defines and traces the multiple meanings of the word "exposure," from the amount of light contained in a photograph to the condition of unwanted visibility, bridging the visual paradigm of part II with the algorithmic one of part III. By tracing the ways that race has historically been exposed in photographic technologies perceived as neutral or objective, she shows how the visual color-coding of racialized imagery (namely, the Shirley Cards discussed in part II)[28] gains new meaning and value in today's milieu of racialized digital surveillance. "That new tools are coded in old biases," Benjamin concludes, "is surprising only if we equate technological innovation with social progress"[29](see chapter 14).

The next two chapters in part III focus on specific social media case studies in different geographical contexts. Each case study explores, albeit in very different ways, the technological and existential dilemmas posed by filters and the digital manipulation of whiteness in virtual spaces. Chapter 15, Amber Sweat's "Cyborg-Cute: Colors of 'Good' Mixed-Race Algorithms" analyzes how contemporary image-sharing platforms like TikTok foster so-called safe spaces for multiracial love, marriage, and reproduction under the auspices of the platform's assumed utopian racelessness. Focusing on the TikTok hashtag #swirl, primarily used by Black-white interracial couples, Sweat illustrates how the digitally enhanced image of the mixed-race child (which she dubs "cyborgian cuteness") circulates online in ways that are not only far from utopian, but also reinforce anti-Black and white supremacist logics in the age of social media.

Jianqing Chen's "Yellow Women's White Faces in Chinese Little-Fresh Photographic Self-Portraiture," in chapter 16, expands this book's largely American focus to examine digitally enhanced whiteness in twenty-first century beauty ideals in the East Asian context. Her contribution analyzes the use of "fair and flawless" selfies as ideal images for online circulation and the way that they have led to a crisis of recognition, authenticity, and subjectivity among Chinese women. Chen argues that the "Little-Fresh" (from the Mandarin *xiao qingxin*) selfie style that results from automated beauty filters and simplified photo editing tools popular in apps like Meitu operates as a form of whiteface, perpetuating gendered and raced ideals that simultaneously corroborate and contest white supremacist logic by normalizing pale and "flawless" skin as an aestheticized ideal among Chinese touchscreen users. Rather than a simple illustration of white supremacy's global spread, Chen concludes, the pale palette of Little-Fresh selfies can also be seen as a repudiation of the rosy-faced heroines of Cultural Revolution–era imagery, expressing a desire to return to the premodern Chinese philosophical and aesthetic ideal of diluted colors or "blandness" (*dan*).

The penultimate essay in *Color Protocols,* chapter 17, offers a pivotal contribution to part III by interrogating novel technologies that are being developed as we write this. In "Algorithmic Images and Recursive Epidermalization," Quran M. Karriem argues that, regardless of the good or ill intent of the organizations behind the design of chromatic technologies, when it comes to algorithmic images created with generative AI, we encounter a fundamental propagation and reification of color-based racial stereotypes. Putting media theorists such as Walter Benjamin and Vilém Flusser in dialogue with Franz Fanon's foundational concept of epidermalization, he argues that algorithmic images have lost the vestiges of their predecessors' links to physicality (i.e., traditional filmic images), while reproducing hegemonic ideologies in unprecedented ways. Using examples from generative adversarial networks (GANs)—deep learning models that train algorithms to discern real images from AI-generated ones—Karriem shows how the algorithmic image is fundamentally conceived as a counterfeit, designed to imitate the properties that we associate with the traditional photograph, while undermining its material connection to physical reality. The most likely and efficient use of algorithmic images, he concludes, is to exploit the lingering cultural belief in photographic indexicality by replacing it with the novel social authentication mechanism of "believability."

Where part III opens with an excerpt from Nakamura's *Cybertypes*, it concludes in chapter 18 with a follow-up interview with Lisa Nakamura conducted by co-editor Lida Zeitlin-Wu just after the text's twentieth anniversary in 2023. Here, Nakamura addresses how the book's central conceits have changed since its initial publication, particularly in relation to the digital encoding of race via the development of digital color. In a wide-ranging conversation touching on topics from emoji color customization to Crayola's 1990s "Multicultural" pack to tech companies' outsourcing of microwork to women of color in the Global South, Nakamura and Zeitlin-Wu reflect on digital color's entanglements with shifting identity politics and global infrastructures based on racialized and gendered labor.

Altogether, the chapters in part III analyze how the encoding of racism within color media persists alongside new developments in informatics and algorithmic abstraction. Whether these biases manifest in customizable color graphics, mixed-race TikTok communities, or built-in smartphone filters, it's clear that color standards need to be rethought to reach a more just and equitable future. Systemic inequality is baked into the very infrastructures of color technologies, and these infrastructures need reimagining. Throughout this book, our hope has been for scholars and practitioners alike to conscientiously develop a critique of modern and contemporary color across technological forms that situates them within broader histories and cultures of oppression. Only then can we mobilize algorithmic color protocols as a force for social change.

NOTES

1. The incident went viral in 2015, when Google software engineer Jacky Alciné, who is Black, posted an image of it to Twitter. As of 2018, Google's "solution" was to censor the words "gorilla," "chimpanzee," "chimp," and "monkey" from searches altogether—a woefully inefficient patchwork solution in lieu of any systemic overhauling of their algorithm. Vincent, "Google 'Fixed'."

2. Rose, "Are Face-Detection Cameras Racist?"

3. Other notable examples of algorithmic racial bias in biometrics include when an automatic soap dispenser failed to detect the hand of a Nigerian man in 2017, and more recently, during the push to online learning during the COVID-19 pandemic, when test-proctoring software was often unable to register the faces of people of color. See Goethe, "Bigotry Encoded: Racial Bias in Technology," and Feathers, "Proctorio Is Using Racist Algorithms to Detect Faces."

4. Heaven, "Predictive Policing."

5. See Apprich et al. *Pattern Discrimination,* and Chun, *Discriminating Data.*

6. As numerous histories of computing have shown, these computer scientists and engineers were overwhelmingly white men. However, counterhistories addressing the crucial role that women (and especially women of color) have played in its formation can be found in the work of Donna Haraway, Lisa Nakamura, Sadie Plant, and Tung-Hui Hu, among others.

7. In addition to her written work, Buolamwini explores these issues in the short documentary *Coded Bias* (Shalini Kantayya, 2020).

8. Buolamwini and Gebru, "Gender Shades."

9. Browne, *Dark Matters,* 91.

10. See Behrens, *False Colors,* and Shell, *Hide and Seek,* for longer histories of camouflage and visuality.

11. Harvey, *CV Dazzle.*

12. Blas, "Opacities"; Paulsen, "Rogue Pixels," and Monahan, *Crisis Vision.*

13. Blas, "Informatic Opacity," 5.

14. Blas, "Informatic Opacity," 5.

15. Blas, "Informatic Opacity," 21.

16. Blas, "Informatic Opacity," 80.

17. Steyerl, *How Not to Be Seen.*

18. Steyerl, *How Now to Be Seen.*

19. Blas, "Informatic Opacity," 24.

20. Steyerl, *How Now to Be Seen*; Blas, "Informatic Opacity," 24.

21. Chun, *Discriminating Data.*

22. Nakamura, *Cybertypes,* 5.

23. Nakamura, *Cybertypes,* xii.

24. Nakamura, *Cybertypes,* 3.

25. Nakamura, *Cybertypes,* 5.

26. Nakamura, *Cybertypes,* 52.

27. Nakamura, *Cybertypes,* 53.

28. Another example is "Lena," an image that originated as a Playboy centerfold in the 1970s and was digitized and circulated as the primary test card among computer engineers (Mulvin, *Proxies*).

29. Benjamin, *Race after Technology*, 74.

BIBLIOGRAPHY

Apprich, Clemens, Florian Cramer, Wendy Hui Kyong Chun, and Hito Steyerl. *Pattern Discrimination.* Minneapolis: University of Minnesota Press and meson press, 2018.

Behrens, Roy. *False Colors: Art, Design, and Modern Camouflage.* Dysart, IA: Bobolink Books, 2001.

Benjamin, Ruha. *Race after Technology: Abolitionist Tools for the New Jim Code.* Cambridge, UK: Polity Press, 2019.

Blas, Zachary. "Informatic Opacity: Biometric Facial Recognition and the Aesthetics and Politics of Defacement." PhD dissertation, Duke University, 2014.

Blas, Zach. "Opacities: An Introduction." *Camera Obscura: Feminism, Culture, and Media Studies* 31, no. 2 (September 2016): 149–153.

Browne, Simone. *Dark Matters: On the Surveillance of Blackness.* Durham, NC: Duke University Press, 2015.

Buolamwini, Joy, and Timnit Gebru. "Gender Shades: Intersectional Accuracy Disparities in Commercial Gender Classification." *Proceedings of the 1st Conference on Fairness, Accountability and Transparency* 81 (2018): 77–91.

Chun, Wendy Hui Kyong. *Discriminating Data: Correlation, Neighborhoods, and the New Politics of Recognition.* Cambridge, MA: MIT Press, 2021.

Feathers, Todd. "Proctorio Is Using Racist Algorithms to Detect Faces." *Vice,* April 8, 2021. https://www.vice.com/en/article/g5gxg3/proctorio-is-using-racist-algorithms-to-detect-faces.

Goethe, Taylor Sinclair. "Bigotry Encoded: Racial Bias in Technology." *Reporter,* March 2, 2019. https://reporter.rit.edu/tech/bigotry-encoded-racial-bias-technology.

Haraway, Donna. *Simians, Cyborgs, and Women: The Reinvention of Nature.* New York: Routledge, 1991.

Harvey, Adam. *CV Dazzle.* https://ahprojects.com/cvdazzle.

Heaven, Will Douglas. "Predictive Policing Algorithms Are Racist. They Need to Be Dismantled." *MIT Technology Review,* July 17, 2020. https://www.technologyreview.com/2020/07/17/1005396/predictive-policing-algorithms-racist-dismantled-machine-learning-bias-criminal-justice/.

Hu, Tung-Hui. *Digital Lethargy: Dispatches from an Age of Disconnection.* Cambridge, MA: MIT Press, 2022.

Kantayya, Shalini, dir. *Coded Bias.* Brooklyn, NY: 7th Empire Media, 2020.

Monahan, Torin. *Crisis Vision: Race and the Cultural Production of Surveillance.* Durham, NC: Duke University Press, 2022.

Mulvin, Dylan. *Proxies: The Cultural Work of Standing in.* Cambridge, MA: MIT Press, 2021.

Nakamura, Lisa. *Cybertypes: Race, Ethnicity, and Identity on the Internet.* New York: Routledge, 2002.

Nakamura, Lisa. "Indigenous Circuits: Navajo Women and the Racialization of Early Electronic Manufacture." *American Quarterly* 66, no. 4 (December 2014): 919–941.

Paulson, Kris. "Rogue Pixels: Indexicality and Algorithmic Camouflage." *Signs and Society* 6, no. 2 (2018): 412–434.

Plant, Sadie. *Zeroes and Ones: Digital Women and the New Technoculture.* New York: Doubleday, 1997.

Rose, Adam. "Are Face-Detection Cameras Racist?" *Time,* January 22, 2010. http://content.time.com/time/business/article/0,8599,1954643,00.html.

Shell, Hanna Rose. *Hide and Seek: Camouflage, Photography, and the Media of Reconnaissance.* Brooklyn, NY: Zone Books, 2012.

Steyerl, Hito. *How Not to Be Seen: A Fucking Didactic Educational .MOV File,* 2013. https://www.e-flux.com/video/384606/how-not-to-be-seen-a-fucking-didactic-educational-mov-file/.

Vincent, James. "Google 'Fixed' Its Racist Algorithm by Removing Gorillas from Its Image-Labeling Tech." *The Verge,* January 12, 2018. https://www.theverge.com/2018/1/12/16882408/google-racist-gorillas-photo-recognition-algorithm-ai.

13 HEAD-HUNTING ON THE INTERNET: IDENTITY TOURISM, AVATARS, AND RACIAL PASSING IN TEXTUAL AND GRAPHIC CHAT SPACES (2002)

LISA NAKAMURA

Is it accurate to say that on the Internet nobody can tell what race you are? For the large (and increasing) number of people who use the Internet as a social space via chat rooms and other forms of online interaction, this seemingly philosophical question has acquired increasing urgency. Despite claims by digital utopians that the Internet is an ideally democratic, discrimination-free space—without gender, race, age, or disability—an analysis of both textual and graphic chat spaces such as LambdaMOO [*Editors' note: a MOO is a multiuser online system that allows for object-oriented programming within its server*], Time Warner's The Palace, and Avaterra's Club Connect will reveal that these identity positions are still very much in evidence. Though it is true that users' physical bodies are hidden from other users, race has a way of asserting its presence in the language users employ, in the kinds of identities they construct, and in the ways they depict themselves online, both through language and through graphic images. These depictions of the self, or online identities, have been termed "avatars." Avatars are the embodiment, in text and/or graphic images, of a user's online presence in social spaces.

The Internet is a theater of sorts, a theater of performed identities. "Passing" is a cultural phenomenon that has the ability to call stable identities into question, and in that sense can be a progressive practice, but the fact remains that passing is often driven by harsh structural cultural inequities, a sense that it really *would* be safer, more powerful, and better to be of a different race or gender. Millions of computer users "pass" every day, and much scholarly work has been devoted to examining how, why, and what it means that this happens in relation to gender. Piles of articles have been written about cross-gender passing, or "computer cross-dressing," but very little has been done on the topic of cross-racial passing despite the fact it may be as common, or even more so. I read this notable *lack* of research on race in cyberspace springing in part from the "digital divide"—that is, the relative lack of computer and network access for racial minorities in comparison to whites.

The celebration of the Internet as a democratic, "raceless" place needs to be interrogated, both to put pressure on the assumption that race is something that *ought* to be left behind, in the best of all possible cyberworlds, and to examine the prevalence of racial representation in this supposedly unraced form of social and cultural interaction.

If race is indeed a cultural construct rather than a biological fact, as Anthony Appiah and others have asserted, then cyberspace is a particularly telling kind of example when we wish to look at the vexed and contested position of race in the digital age. The Internet is literally a "construct" (as the recent film *The Matrix* terms it); like race itself, it is a product of culture and its attendant power dynamics rather than an object that somehow existed prior to linguistic and cultural definition. Cyberspace is a place of wish fulfillments and myriad gratifications, material and otherwise, and nowhere is this more true than in chat spaces. Both textual and graphic chat spaces encourage users to build different identities, to take on new nicknames, and to describe themselves in any way they wish to appear. Digital avatars, or renditions of self, provide a pipeline into the phantasmatic world of identities, those conscious or not-so-conscious racial desires and narratives that users construct and inhabit during their interactions in cyberspace.

. . . The World Wide Web's dominance of the Internet has transformed cyberspace into a world of visual images, a world in which text has taken a backseat. Graphic avatars are *visible,* and thus race, which was invisible in textual avatars unless specifically put there, became visible as well. Websites that support graphic avatars like Time Warner's The Palace and Avaterra's Worlds Away Club Connect allow users to create images of themselves that they can move through cyberspace and customize. In this way, graphic chat

FIGURE 13.1
The Palace, "Palace in Wonderland," Time Warner. Circa 1997. The Palace was a 2-D graphical chat environment that provided users access to graphical chat room servers called "palaces." The software concept was created by Jim Bumgardner and produced by Time Warner in 1994. Archival screen grab from Cyberia.

resembles a video game more than it does live action e-mail in the sense that other users can see you and interact with you as an image. . . .

Club Connect, a graphic chat space run by Avaterra, a company specializing in web-based business meeting spaces, can be accessed via its direct link from NetNoir's website. NetNoir is probably the oldest and certainly one of the best-known examples of an ethnic-identity website, which makes it a rare and important example of a minority presence on the web and the Internet in general. NetNoir started out as a text-only bulletin board addressing African-American concerns, and though it is less than five years old, it is still about as venerable as a website can be. (It is well known that Internet years are like "dog years": one of them might equal

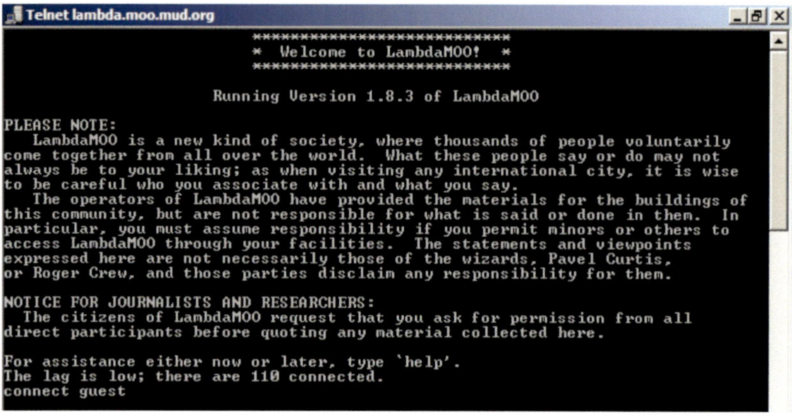

FIGURE 13.2

Interface of LambdaMoo January 1, 2009. Archival screenshot. LambdaMOO was an on online community founded in 1990 at Xerox PARC to expand the domain of Multi-User Dungeons (MUDs) in virtual communities. As noted by Nakamura, the text-based chatroom circa 2002 required gender designation but not ethnic or racial designation. Image courtesy of Gesche Joost.

seven in "normal" time.) I wanted to experience graphic chat through a minority-run website to see if this factor might prove an exception to the overwhelming whiteness I had seen on the Internet in general. I wish I could say that it did, but unfortunately it didn't, or rather only did in fairly limited ways. As Thomas Foster writes, "Virtual reality privileges vision as a mode of information processing, and visual perception remains inextricably linked to a history of racial stereotyping" (160). When contrasting graphic chat with earlier forms of text-only chat, such as LambdaMOO, it can be seen that the more image-rich "virtual reality" enabled by the superior bandwidth it employs only intensifies the "privileging of vision as a mode of information processing," and consequently produces racial and gendered cybertypes of the body that can come across as potentially and perhaps perniciously more "real" than their textual counterparts. . . .

LAMBDAMOO AND ORIENTALISM

A cute cartoon dog sits in front of a computer, gazing at the monitor and typing away busily. The cartoon's caption jubilantly proclaims, "On the Internet, nobody knows you're a dog!" This image resonates with particular intensity for those members of a rapidly expanding subculture that congregates within the consensual hallucination defined as cyberspace. Users define their presence within this textual and graphic space through a variety of different activities—commercial interaction, academic research, netsurfing, real-time interaction and chatting with interlocutors who are similarly "connected"—but all can see the humor in this image because it illustrates so graphically a common condition of being and self-definition within this space. Internet users represent themselves within it solely through the medium of keystrokes and mouse clicks, and through this medium they can describe themselves and their physical bodies any way they like; they perform their bodies as text.

"On the Internet, nobody knows you're a dog."

FIGURE 13.3
Peter Steiner, "On the Internet, nobody knows you're a dog." Cartoon, *The New Yorker* (1993).

On the Internet, nobody knows you're a dog; it is possible to "computer cross-dress" and represent yourself as a different gender, age, or race [Stone 1994, 84]. In millennium America, this supposedly radically democratic aspect of the Net is celebrated frequently and unconditionally. The cartoon celebrates access to the Internet as a social leveler that permits even dogs to freely express themselves in discourse to their masters, who are deceived into thinking that dogs are their peers rather than their property. The element of difference, in this cartoon the difference between species, is comically subverted in this image; in the medium of cyberspace, distinctions and imbalances in power between beings who perform themselves solely through writing seem to have been deferred, if not effaced.

This utopian vision of cyberspace as a promoter of a radically democratic form of discourse should not be underestimated. Yet the image can be read on several levels. The freedom of which the dog chooses to avail itself is the freedom to "pass" as part of a privileged group—human computer users with access to the Internet. This is possible because of the discursive dynamic of the Internet, particularly in chat spaces like LambdaMOO, where users are known to others by self-authored names they give their "characters" rather than the more revealing e-mail addresses that include domain names. Defining gender is a central part of the discourse-players who choose to present themselves as "neuter," one of the several genders available to players on LambdaMOO, are often asked to "set gender," as if the choice to have a neuter gender is not a choice at all, or at least one that other players choose to recognize. They are seen as having deferred a choice rather than having made an unpopular one. This is an example of the "informatics of domination" that Donna Haraway describes in *Simians, Cyborgs, and Women.*[1] Gender is an element of identity that must be defined by each player, though the creators of LambdaMOO try to contribute toward a reimagining of gender by offering four choices (two more than are acknowledged in "real life"); still, one must be chosen. Each player must enunciate a chosen gender, since this gender will be visible to all players who call up other players' physical descriptions on their screens. Race, however, is not an option that must be chosen. Although players can elect to write it into their descriptions, it is not required by the programming that they do so.

Nonetheless, race is "written" in role-playing cyberspaces as well as read by other players. It is crucial to direct critical attention toward the conditions under which race is enunciated, contested, and ultimately erased and suppressed online, and the ideological implications of these performative acts of writing and reading otherness. What does the way race is written in cyberspace reveal about the enunciation of difference in new electronic media? Have the rules of the game changed, and if so, how?

Role-playing sites on the Internet offer their participants programming features such as the ability to physically "set" one's gender, race, and physical appearance, through which one can, and indeed in many cases is required to, project a version of the self that is inherently theatrical. Since true identities of interlocutors at Lambda are unverifiable (except by crackers and hackers, whose outlaw manipulations of code are unanimously construed by the Internet's citizens as a violation of both privacy and personal freedom) it can be said that everyone who participates is "passing," since it impossible to tell if a character's description matches a player's physical characteristics. Some of the uses to which this infixed theatricality are put are benign and even funny. Descriptions of self as a human-size pickle or pot-bellied pig are not uncommon, and generally are received in a positive, amused, tolerant way by other players. In contrast to this, players who elect to describe themselves in racially "othered" terms, as Asian, African American, or Latino, are often seen as engaging in a form of hostile performance, since they introduce what many consider a real-life "divisive issue" into the phantasmatic world of cybernetic textual interaction. The borders and frontiers of cyberspace, which had previously seemed so amorphous, take on a keen sharpness when the enunciation of racial otherness is put into play as performance. While everyone is "passing," some forms of racial passing are practiced and condoned because they do not threaten the integrity of a national sense of self that is defined as white.

LambdaMOO requires that one choose a gender; though two of the choices are variations on the theme of "neuter," the choice cannot be deferred because the programming code requires it. It is impossible to receive authorization to create a character without making this choice. Race, on the other hand, is not only not a required choice, it is not even on the menu.[2] Players are given as many lines of text as they like to write any sort of textual description of themselves that they want. The architecture of belief that underpins social interaction in the MOO—that is, the belief that one's interlocutors possess distinctive human identities that coalesce through and vivify the glowing letters scrolling down the computer screen—is itself built upon this form of fantastic autobiographical writing. The majority of players in LambdaMOO do not mention race at all in their self-description, though most do include eye and hair color, build, age, and pronouns that indicate a male or a female gender. In those cases when race is not mentioned as such, but hair and eye color is, race is still being evoked—a character with blue eyes and blond hair will be assumed to be white. Yet while the textual conditions of self-definition and self-performance would seem to permit players total freedom, within the boundaries of the written word, to describe themselves in any way they choose, this choice is actually an illusion.

This is because the decision to leave race out of self-description does in fact constitute a choice: in the absence of racial description, all players are assumed to be white. This is partly due to the demographics of Internet users; most are white, male, highly educated, and middle class. It is also due to the utopian belief-system prevalent in the MOO. This system, which claims that the MOO should be a free space for play, strives toward policing and regulating racial discourse in the interest of social harmony. This system of regulation does permit racial role-playing when it fits within familiar discourses of racial stereotyping, and thus perpetuates these discourses. I am going to focus on Asian performance within the MOO because Asian personae are by far the most common nonwhite ones chosen by players and thus offer the greatest number of examples for study.

The vast majority of male Asian characters deployed in the MOO fit into familiar stereotypes from popular electronic media such as video games, television, and film, and popular literary genres such as science fiction and historical romance. Characters named Mr. Sulu, Chun Li, Hua Ling, Anjin San, Musashi, Bruce Lee, Little Dragon, Nunchaku, Hiroko, Miura Tetsuo, and Akira invoke their counterparts in the world of popular media. Mr. Sulu is the token Asian American in the television show *Star Trek;* Hua Ling and Hiroko are characters in the science-fiction novels *Eon* and *Red Mars;* Chun Li and Liu Kang are characters from the video games *Street Fighter* and *Mortal Kombat;* the movie star Bruce Lee was nicknamed "Little Dragon"; Miura Tetsuo and Anjin San are characters in James Clavell's popular novel and TV miniseries *Shogun;* Musashi is a medieval Japanese folklore hero; and Akira is the title of a Japanese anime film. The name Nunchaku refers to a weapon, as do, in a more oblique way, all of the names listed above. These names all adapt the samurai warrior fantasy for role-playing, and permit their users to perform a notion of the oriental warrior adopted from popular media. The effect of popular media in cyberspace has been to create a bricolage of figurations and simulations. The orientalized male persona, complete with sword, confirms the idea of the Asian man as potent, antique, exotic, and anachronistic.

This type of orientalized theatricality is a form of identity tourism; my research indicates that players who choose this type of racial play are almost always white, and their appropriation of stereotyped male Asian samurai figures allows them to indulge in a dream of crossing over racial boundaries temporarily and recreationally. Choosing these orientalized cybertypes tips their interlocutors off to the fact that they are not "really" Asian; they are instead playing in a familiar type of performance. Thus, the Orient is brought into the discourse, but only as a token or type. The idea of a nonstereotyped Asian male identity is so seldom enacted in LambdaMOO that its absence can only be read as a symptom of suppression.

DEFAULT WHITENESS

While many members of online social communities like LambdaMOO are stubbornly utopian in their attitudes toward the power dynamics and flows of information within the technologically mediated social spaces they inhabit, many of the academic theorists are pessimistic. Andrew Ross and Constance Penley introduce their collection *Technoculture* by asserting that "the odds are firmly stacked against the efforts of those committed to creating technological countercultures" [1991, xiii]. Chesher concedes, "In spite of the claims that everyone is the same in virtual worlds, access to technology and necessary skills will effectively replicate class divisions of the rest of reality in the virtual spaces" and "will tend to reinforce existing inequalities, and propagate already dominant ideologies" [1994, 28–29]. Indeed, the cost of Net access does contribute toward class as well as racial divisions. One of the dangers of identity tourism is that it takes this restriction across the axes of race/class in the "real world" to an even more subtle and complex degree by reducing nonwhite identity positions to part of a costume or masquerade to be used by curious vacationers in cyberspace. Asianness is co-opted as a passing fancy, an identity-prosthesis that signifies exotic sexual availability when female, and anachronistic dreams of combat when male. Self-identification as a samurai or geisha is diverting, reversible, and a privilege mainly used by white men. The paradigm of Asian transvestism on LambdaMOO itself works to suppress racial difference by setting the tone of the discourse in racist contours, which inevitably discourage "real life" Asian men and women from textual performance in that space, effectively driving race underground. As a result, a default whiteness covers the entire social space of LambdaMOO; race is whited out in the name of cybersocial hygiene.

Many Asian-American, African-American, and Latino users I interviewed commented on the reasons that they decide to exclude their "real life" race in their descriptions, and the consequences of their decisions. One player on LambdaMOO, a Korean-American woman, wrote,

> I never outright (in my desc) said I was Asian, because I felt that IRL [in real life] people already have stereotypes and felt that it would be at least as bad here, and I wanted to have a character that was free from that. Even before being experienced with net interactions, I pretty much feared what kind of people I would attract just by virtue of saying I'm an Asian woman. But then it bugs me that people just assume you're white if you don't say otherwise. Which is interesting because a lot of people here don't just assume you're heterosexual. They are careful to say SO instead of GF/BF and use [androgynous] pronouns or whatever.

She continued, "You know, though, people of color know exactly what it means to be limited in whatever way because they look the way they do. That is why, I'm sure people here don't take any pains to point out what they look like. . . ."

As this player explains, the few minorities on the Net are acutely aware of how their skin color "limits" them, and this—in conjunction with the informal "don't ask don't tell" policy of cyberspace—conspires to discourage minorities from presenting themselves *as* minorities.

The Berkeley Macintosh Users' Group, or BMUG for short, is a well-established community-based Internet bulletin board for Macintosh users, and it has an ethnically diverse membership. Players must use their real names and write resumes that any other player may read freely. After examining many of these, I noticed that users tended to describe their computers, their hair and eye color, their occupations, their pets—but race was very rarely mentioned. In this case of BMUG, a virtual space with strong ties to the University of California, Berkeley and progressive politics, the elision of race seems not to be motivated by fear of harassment or reprisal. Perhaps tolerance is so taken for granted there that users think that it doesn't need to be mentioned. Nonetheless, the effect is the same: if in cyberspace you are what you write, then the BMUGGERS, as they call themselves, have written race out too.

The dream of a new technology has always contained within it the fear of total control and the accompanying loss of individual autonomy. Perhaps the best way to subvert the hegemony of cyber- social hygiene is to use its own metaphors against itself. Racial and racist discourse in cyberspace are the unique products of machines and ideologies.

The erasure and domestication of Asianness on LambdaMOO perpetuates an Orientalist myth of social control and order. As Cornel West puts it, "race matters," and as Judith Butler puts it, "bodies matter." Programming language and Internet connectivity have made it possible for people to interact without putting into play any bodies but those they write for themselves. The temporary divorce that cyberdiscourse grants the mind from the body and text from the body also separates race and the body. Player scripts that eschew repressive versions of the oriental in favor of critical rearticulations and recombinations of race, gender, and class, and which also call the fixedness of these categories into question, have the power to turn the theatricality characteristic of MOO space into a truly innovative form of play, rather than a tired reiteration of old hierarchies. Role-playing is a feature of the MOO—not a bug—and it would be absurd to ask that everyone who plays within it hew literally to their "real life" gender, race, or condition of life. A diversification of the roles that are permitted and played can enable a thought-provoking detachment of race from the body, and a questioning of the essentialness of race as a category. Performing alternative versions of self and race jams the ideology-machine, and facilitates a desirable opening up of what Butler calls "the difficult future terrain of community" in cyberspace (242).

GRAPHIC CHAT: CYBERTYPING AND THE TYRANNY OF THE VISIBLE

According to Elizabeth Kolbert, in an article in the *New Yorker,* "Ultima Online, even with a quarter of a million subscribers, is two or three orders of magnitude larger than even the most populous MUD" [2001, 97]. This is so despite the fact that MUDs and MOOs are free, while most graphic chat spaces charge monthly fees. It seems that most users strongly prefer to "see" who they're interacting with, even if they are aware that what they are "seeing" may bear no relation to the actual appearance of the user on the other end. Race becomes part of the visual language of graphic chat in a way that it could not be in textual chat, a factor that creates less difference than one might expect in terms of how race is dealt with in these spaces.

Discourse about race is still elaborately "routed around," at least in terms of much interface design, avatar construction and deployment, and everyday discourse in Club Connect in spite of the fact that graphic avatars make race so visible. And even though the "club" is linked to an African American ethnic identity website, NetNoir, certain aspects of racial policing that obtain in LambdaMOO, albeit on an informal basis, are still in evidence at Club Connect.

My initial experiences with Club Connect did prove to inject race into graphic chat in ways that I had not seen in text-only environments and interfaces. The architects' dream of a multicultural virtual world is visible from the start. The log-on screen, which is the first thing that you see when you enter the website, features a little cartoon icon of a person's head, which greets you and asks you if you're having a problem with your password. Since it is practically guaranteed that most new users, especially ones who aren't computer-savvy, will have trouble with this step, viewers are given a great deal of exposure to this virtual greeter. The greeter is an icon of a fairly dark male with dreadlocks and African American features. Though it is generally a shaky move to make claims about which visual images look black, Asian, or in other ways "ethnic" since these judgments are always subjective and can tend to reduplicate the language of essentialism and racialism if not racism, this figure struck me as undeniably African American. This figure's position on this screen figures him as "tech support": both a servant of sorts and a technological expert. The figure paradoxically merges images of domestic laborers-like butlers and houseboys, images long associated with both blacks and Asians-and the computer geek figure, an image reserved for young white males. The figure looks young and "hip," as signified by the dreadlocks. These dreadlocks also gesture toward an ethnicity that is up front and center rather than elided or hidden, which is a real departure from the norm and a genuine innovation. (I can count the number of black icons I've seen on the Internet on *no* fingers.) This is a hopeful sign, particularly if one subscribes to the popular notion from media studies that models with color, images that "look like me," contribute to the inclusion of minorities offscreen as well.[3] Examining icons and avatars provides a wonderful occasion for an analysis of the principles of cybernetic iconography, a field that has yet to be born in relation to ethnic studies.[4] Analyzing the ways that icons and avatars are raced in cyberspace allows us to lay bare the principles of ethnic image-building on the Net that may seem transparent and "natural" to computer users, who are surrounded by icons every day. Thanks to Apple and Windows, icons are the matrix in which cyberspace users exist.

Once you've gotten past the password screen, Club Connect's software invites you to choose your gender, and then assigns you a "starter" or default body which you can further customize after getting acquainted with the site. The first time I logged on, I chose the gender "female" and was given the body and face of a young black woman, with fairly dark skin, African features, and short dark hair. I chose a name for my avatar, and began to explore the site. The first thing I noticed is that I was the only avatar among hundreds who was noticeably dark. While I saw many Asian-presenting female avatars, I saw no Asian-presenting male ones. This disparity can be accounted for, I am sure, by the fact that it isn't really a disparity; images of Asians in popular culture and the mainstream media do tend to be dominated by females, as in television newscasting, though this is slowly changing. On the other hand, I didn't see many strictly Anglo-presenting avatars, either. Many players chose light-skinned black female characters, Latino-presenting male characters, or Asian-presenting female characters, but there were relatively few black-presenting or Asian-presenting male characters.

I found other players to be extremely helpful and friendly. Perhaps this was because they could tell that I was a new player, but on a site this busy and populated, I suspect that their perceptions of me as "new" were conditioned by another factor. As I have mentioned earlier, I was the only truly dark character I could see anywhere. A friendly blond female, named mspiggy, offered to show me around, and when I expressed interest in exploring the ways I could customize my avatar, actually gave me a new head as a gift! This head, named "Rebecca," was one of dozens of models available via virtual vending machines that would take my virtual tokens and, in return, spit out a face for my avatar. Significantly, "Rebecca" was blonde and white-skinned; in fact, she greatly resembled my new online friend. While a great deal could be read into this psychologically, if one were so inclined, it seems to me indicative of the ways that beauty and race interact together in graphic chat spaces.

Mspiggy's assumption that I would want a head that looked less "ethnic" seemed to project a particular image of beauty that was less, well, dark. Mspiggy showed me how to detach my old head, which I tucked under my arm, and helped me insert the new one. She then pointed out that my skin color and head color didn't match, saying "lol; [laugh out loud] you look like you have a tan only on your arms!" and offered to help me to acquire some "body spray" so that I could change it. Throughout this exchange no mention of race was made.

Mspiggy led me to vending machines, where I viewed the multitude of different heads available to me. In an unconscious parody of cosmetic surgery and other technological image "enhancements," I noticed that these were priced across a wide range. I admit that I have not looked at every head available, but I did notice that none of them were as dark as my old one had been. This commodification of identity is reflected in the chat room's helpscreens on the topic of "changing and customizing your avatar's head," part of which reads,

> Each head comes with its own default hair and face colors, which you may be able to customize. If custom face and hair colors are available, look for vending machines that sell or dispense head spray. If your head is sprayable, the popup menus of your avatar and the spray can will give you the necessary options for using the spray. Go ahead and experiment; you'll be able to change and adjust the colors again and again until you find the combination you like.

As hilarious as this description is, it seems to be doing more than simply pointing out the extent to which race has become elective in cyberspace; indeed, the passage omits any mention of race altogether. Instead of inviting you to give your avatar a race, just as you had given it a gender, the passage invites you to "experiment; you'll be able to change and adjust the colors again and again until you find the combination you like." Here, race is constructed as a matter of aesthetics, or finding the color that you like, rather than as a matter of ethnic identity or shared cultural referents. This fantasy of skin color divorced from politics, oppression, or racism seems to also celebrate it as infinitely changeable and customizable: as entirely elective as well as apolitical. Clearly, we must look to the subtextual, to the omitted and repressed, to find the place of race in graphic chat.

Implicit statements about race are to be found by looking at the relative pricing of heads, and the avatars of those around me. The marketplace and traffic in heads and bodyspray work in a traditional capitalistic system of supply and demand; that is to say, the types of heads for sale are available because there is demand for them: avatars are market-driven. If more players wanted to buy, for instance, extremely dark or Asian-looking male

FIGURE 13.4
WorldsAway archival screenshot. "Your WorldsAway avatar always had a specific gender and body type, but you could change the heads easily (there were hundreds of options)." In 1995, the Tokyo-based virtual world developer Fujitsu Systems' Worlds-Away division spawned the American spin-off Avaterra.com, which, in turn, developed three virtual worlds: Club Connect, New Radio World, and Dreamscape. These platforms "promised a technicolor online world where you could be anything you wanted, and share a virtual city with people all over the globe." Image courtesy of Benj Edwards and PCWorld (2015).

heads, undoubtedly they would have been on display, but they were not. The prevalent type seemed to be ethnically hybrid along particular lines: Asian-white females and light-skinned black females were quite popular. There were many male-presenting avatars with animal heads such as tigers, cats, and wolves—a fact in which a more inquiring mind than mine could undoubtedly find much meaning. Perhaps players choosing this option wished to defer the question of race permanently.

I did not try to ascertain the "real" racial identity of the other players on the site. In any case, the veracity of this information would always be in doubt, and it hardly seemed to matter, in a way. This conundrum, and a conundrum it is, reminded me of a similar one from literary studies: Does

FIGURE 13.5
The VZones 3.0 client archival screenshot. "Essentially just a minor reskin of the WorldsAway 2.4 client, with extra web browser functionality built in. It was not very popular with the membership and eventually phased out in favor of WorldsAway." VZones 3.0, like Club Connect, also features male avatars with animal heads. © 2013–2023 RenoProject. All Rights Reserved.

race reside in the author or in the text? It seems silly to say that a text like Maxine Hong Kingston's *Woman Warrior* should be discussed without any reference to the author's ethnicity and race. On the other hand, would that text not be just as "Asian" if we did not know the race and ethnicity of the author? In any case, I knew that many of the players had come from NetNoir's website, which was a good indication that many of them were African American, but since there are other ways to gain access to the site I knew

FIGURE 13.6
Pride (or, "Queery Universe") was WorldsAway's first adult world, targeting the LGBTQ community. It was also the first to introduce the famous WorldsAway UFO (implicating a precarious alignment between non-heteronormative sexuality and aliens).[5] Image courtesy of Edd Coates (aka Metal Edd in VZones).

also that many were not. The more interesting aspect of this whole thing seemed to be the question, When users are allowed to choose their races in graphic chat spaces, what do they choose? What do their choices say about the place of race in cyberspace? Why do they create the personae that they do? Where does race go when Asian-American and African-American users log on? Does it disappear?

As in text-only chat, the identities users choose say more about what they want than who they are, or rather, since these are eminently social spaces, what they think *others* want. In some sense, Club Connect is a racially diverse space, since players are choosing Asian- and African-American-presenting avatars, but since it is impossible to tell the race and ethnicity of these avatars offline, one must ask what kind of authenticity, integrity, or political efficacy these communities can have. The site's persistent avoidance of race as any kind of factor at all in the buying and selling of avatars that clearly *are* raced indicates a radical repression of what is all too often coded as a "divisive" subject, one to be avoided in the name of social harmony. However, like LambdaMOO, Club Connect displays a plethora of raced and gendered bodies, Asian and otherwise, that gesture toward a

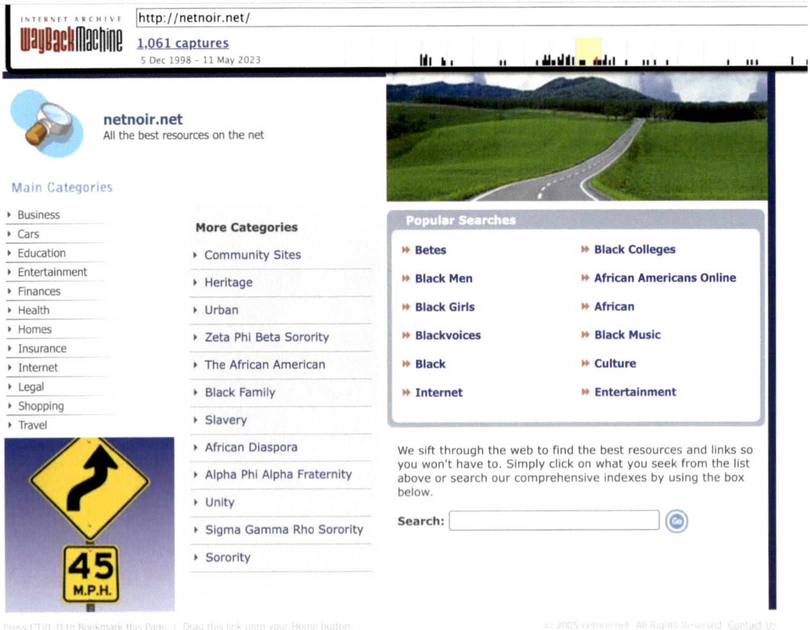

FIGURE 13.7

NetNoir archival interface. AOL (October 24, 2005). Image courtesy of Alexa Crawl EF Collection and the Wayback Machine.

complex, multifaceted, and sometimes conflicted awareness of racial diversity. The notion that race can be customized, changed, and taken on or off as easily as one pops the head off a Barbie doll invokes the distinctive features of identity tourism. The metaphor of headhunting in cyberspace is a powerful one, since it also locates identity tourism in a matrix of colonial trophy-getting, a way of "eating the Other," to use bell hooks's phrase. Taming and framing the other by buying its signs and signifiers with virtual tokens seems indeed to be a form of virtual tourism, a kind of souvenir acquisition. The tyranny of visibility, of cybernetic "ways of seeing" in regard to race, has yet to be challenged in spaces like these.

NOTES

Editors' note: This is an excerpt from *Cybertypes: Race, Ethnicity, and Identity on the Internet* (Routledge, 2002, pp. 31–39, 46–55); "Race In/For Cyberspace: Textual Performance and Racial Passing on the Internet," in *Works and Days* 25/26 (Fall 1995/

Winter 1996), and "Head-Hunting in Cyberspace: Identity Tourism, Asian (?) Avatars, and Racial Passing in Web-Based Chatspaces," in *The Women's Review of Books,* February 2001.

1. See Haraway, *Simians, Cyborgs, and Women,* 163.

2. Some MUDs, such as Diku and Phoenix, require players to select a race. These MUDs are patterned after the role-playing game *Dungeons and Dragons,* and unlike Lambda, which exists to provide a forum for social interaction and chatting, focus primarily on virtual combat and the accumulation of game points. The races available to players (orc, elf, dwarf, human, etc.) are familiar to readers of the "sword and sorcery" genre of science fiction and determine what sort of combat attributes a player can exploit. The combat metaphor, which is part of this genre of role-playing, reinforces the notion of racial differences.

3. Of course, the nature rather than merely the quantity of online representations of racial minorities is what needs to be considered here. As Mimi Nguyen writes, "If representational visibility equals power, then almost-naked young Asian women should be running a very big chunk of cyber-space. That is, whenever I type 'asian+-women' [into a search engine] I get almost nothing but them" ("Tales of an Asiatic Geek Girl," 183).

4. See Jeff Ow's article on Orientalist yellow-face imagery in video games entitled "The Revenge of the Yellow-Faced Cyborg" for a notable exception to this rule.

5. "VZ History: Pride/Queery Universe.".

BIBLIOGRAPHY

Butler, Judith. *Bodies That Matter: On the Discursive Limits of "Sex."* New York: Routledge, 1993.

Chesher, Chris. "Colonizing Virtual Reality: Construction of the Discourse of Virtual Reality, 1984–1992." *Cultronix* 1.1 (Summer 1994).

Foster, Thomas. "'The Souls of Cyber-Folk': Performativity, Virtual Embodiment, and Racial Histories." In *Cyberspace Textuality: Computer Technology and Literary Theory,* edited by Marie-Laure Ryan. Bloomington: Indiana University Press, 1999. 137–163.

Haraway, Donna. *Simians, Cyborgs, and Women.* New York: Routledge, 1991.

Kingston, Maxine Hong. *The Woman Warrior.* New York: Knopf, 1977.

Kolbert, Elizabeth. "Pimps and Dragons (Dept. of Gaming)." *New Yorker,* May 28, 2001: 88–98.

Nguyen, Mimi. "Tales of an Asiatic Geek Girl: *Slant* from Paper to Pixels." In *Technicolor: Race, Technology, and Everyday Life,* edited by Alondra Nelson and Thuy Linh N. Tu with Alicia Headlam Hines. New York: New York University Press, 2001.

Ow, Jeff. "The Revenge of the Yellow-Faced Cyborg: The Rape of Digital Geishas and the Colonization of Cyber-Coolies in 3D Realms' Shadow Warrior." In *Race in Cyberspace.* edited by Beth Kolko, Lisa Nakamura, and Gilbert Rodman. London and New York: Routledge, 2000. 51–68.

Penley, Constance, and Andrew Ross. *Technoculture.* Minneapolis: University of Minnesota Press, 1991.

Stone, Allucquere Rosanne. "Will the Real Body Please Stand Up? Boundary Stories about Virtual Cultures." In *Cyberspace: First Steps,* edited by Michael Benedikt. Cambridge, MA: MIT Press, 1994.

"VZ History: Pride/Queery Universe," http://vzn.eddcoates.com/worldpride.shtml.

14 CODED EXPOSURE: IS VISIBILITY A TRAP? (2019)

RUHA BENJAMIN

> I think my Blackness is interfering with the computer's ability to follow me.
> —Webcam user[1]

EXPOSURE

- the amount of light per unit area
- the disclosure of something secret the condition of being unprotected
- the condition of being at risk of financial loss
- the condition of being presented to view or made known.

In the short-lived TV sitcom *Better off Ted*, the writers parody the phenomena of biased technology in an episode titled "Racial Sensitivity." This episode presents the corporation where the show takes place installing a "new state of the art system that's gonna save money," but employees soon find there is a "glitch in the system that keeps it from recognizing people with dark skin."[2] When the show's protagonist confronts his boss, suggesting the sensors are racist, she insists otherwise:

> The company's position is that it's actually the opposite of racist because it's not targeting black people, it's just ignoring them. They insist that the worst people can call it is indifferent . . . In the meantime, they'd like to remind everyone to celebrate the fact that it does see Hispanics, Asians, Pacific Islanders, and Jews.[3]

The show brilliantly depicts how the default Whiteness of tech development, a superficial corporate diversity ethos, and the prioritization of efficiency over equity work together to ensure that innovation produces social containment.[4] The fact that Black employees are unable to use the elevators, doors, and water fountains or turn the lights on is treated as a minor inconvenience in service to a greater good. The absurdity goes further when, rather than removing the sensors, the company "blithely installs separate, manually operated drinking fountains for the convenience of the black employees,"[5] an incisive illustration of the New Jim Code wherein tech advancement, posed as a solution, conjures a prior racial regime in the form of separate water fountains.

Eventually the company sees the error of its ways and decides to hire minimum-wage earning White employees to follow Black employees around the building, so that the sensors will activate. But then the legal team determines that, for each new White worker, they must hire an equal number of Black workers, and on and on, in a spiraling quota that ends when the firm finally decides to reinstall the old sensors.

Playing off of the political anxieties around reverse discrimination and affirmative action, the episode title "Racial Sensitivity"—a formula that usually designates a charge brought against Black people who call attention to racism—is a commentary on the company's insensitivity and on the absurdity of its fixes. The writers seem to be telling us that more, not less, sensitivity is the solution to the technological and institutional dilemma of coded inequity. The episode also manages to illustrate how indifference to Blackness can be profitable within the logic of racial capitalism until the social costs become too high to maintain.[6]

MULTIPLY EXPOSED

Some technologies fail to see Blackness, while others render Black people hypervisible and expose them to systems of racial surveillance.[7] Exposure, in this sense, takes on multiple meanings.[8] Exposing film is a delicate process—artful, scientific, and entangled in forms of social and political vulnerability and risk. Who is seen and under what terms holds a mirror onto more far-reaching forms of power and inequality. Far from being neutral or simply aesthetic, images have been one of the primary weapons in reinforcing and opposing social oppression. From the development of photography in the Victorian era to the image-filtering techniques in social media apps today, visual technologies and racial taxonomies fashion each other.[9]

Photography was developed as a tool to capture visually and classify human difference; it also helped to construct and solidify existing technologies, namely the ideas of race and assertions of empire, which required visual evidence of stratified difference.[10] Unlike older school images, such as the paintings and engravings of exotic "others" that circulated widely before the Victorian period, photographs held an allure of objectivity, a sense that such images "were free from the bias of human imagination . . . a neutral reflection of the world."[11] Yet such reflections were fabricated according to the demands and desires of those who exercised power and control over others. Some photographs were staged, of course, to reflect White supremacist desires and anxieties. But race as a means of sorting people into groups on the basis of their presumed inferiority and superiority was staged in and of itself, long before becoming the object of photography.

What of the modern photographic industry? Is it more democratic and value-neutral than image was in previous eras? With the invention of color photography, the positive bias toward lighter skin tones was built into visual technologies and "presented to the public as neutral." Neutrality comes in the idea that "physics is physics," even though the very techniques of color-balancing an image reinforce a dominant White ideal.[12] And when it comes to the latest digital techniques, social and political factors continue to fashion computer-generated images. In this visual economy, race is not only digitized but heightened and accorded greater value.

This chapter traces the complex processes involved in "exposing" race in and through technology and the implications of presenting partial and distorted visions as neutral and universal. Linking historical precedents with contemporary techniques, the different forms of "exposure" noted in the epigraph serve as a touchstone for considering how the act of viewing something or someone may put the object of vision at risk. This kind of scopic vulnerability is central to the experience of being racialized.

In many ways, philosopher and psychiatrist Frantz Fanon's classic text *Black Skin, White Masks* is a meditation on scopic vulnerability. He describes the experience of being looked at, but not truly seen, by a White child on the streets of Paris:

> "Look, a Negro!"
> It was an external stimulus that flicked over me as I passed by.
> I made a tight smile.
> "Look, a Negro!" It was true. It amused me.

> "Look, a Negro!" The circle was drawing a bit tighter. I made no secret of my amusement.
> "Mama, see the Negro! I'm frightened!" Frightened! Frightened! Now they were beginning to be afraid of me. I made up my mind to laugh myself to tears, but laughter had become impossible.

This story reveals to us that a key feature of Black life in racist societies is the constant threat of exposure and of being misread; and that being exposed is also a process of enclosure, a form of suffocating social constriction.

In a beautiful essay titled "Skin Feeling," literary scholar Sofia Samatar reminds us: "The invisibility of a person is also the visibility of a race . . . to be constantly exposed as something you are not."[13] Yet, in the distorted funhouse reflection of racist conditioning, the White children are the ones who fancy themselves as being at risk. Fanon's experience on the streets of Paris foreshadows the technologically mediated forms of exposure that proliferate Black life today. Whether we are talking about the widespread surveillance systems built into urban landscapes or the green light sitting above your laptop screen, detection and recognition are easily conflated when the default settings are distorted by racist logics.[14]

Finally, as it circulates in the domain of finance, the term "exposure" quantifies how much one stands to lose in an investment. If, as legal scholar Cheryl I. Harris argues, Whiteness is a form of property and if there is a "possessive investment in whiteness" (as sociologist George Lipsitz describes it), then visual technologies offer a site where we can examine how the value of Whiteness is underwritten through multiple forms of exposure by which racialized others are forcibly and fictitiously observed but not seen. That said, photography has also been a powerful tool to invest in Blackness. . . .

EXPOSING WHITENESS

The most concrete technique through which Whiteness has fashioned photography is the Shirley Cards produced by Kodak from the 1950 to 1990s. The cards were an integral part of film exposure methods and used the image of a White woman to standardize the exposure process [see figures 9.1–9.3 in chapter 9]. Since the model's skin was set as the norm, darker skinned people in photographs would be routinely underexposed. In short, skin tone biases were embedded in the "actual apparatuses of visual reproduction."[15] As one photographer recently put it, "It turns out, film stock's failures to capture dark skin aren't a technical issue, they're a choice."[16] This also implies we can choose otherwise.

Photographers developed a range of "fixes" for underexposure in order to calibrate the color; for instance they could add more lighting to darker subjects. But these only worked for images containing a single variation. If more than one skin tone were represented in an image, such fixes were harder to employ. At least three social shifts propelled more fundamental changes to this form of discriminatory design. As public schools in the United States began desegregating and students of different skin tones were photographed for yearbooks in the same frame, the technical fixes that could be employed when a Black child was photographed alone were not useful. In particular, Black parents, objecting to the fact that their children's facial features were rendered blurry, demanded higher-quality images.[17] But the photographic industry did not fully take notice until companies that manufactured brown products like chocolate and wooden furniture began complaining that photographs did not depict their goods with enough subtlety, showcasing the varieties of chocolate and of grains in wood.

Finally, as U.S.-based visual technologies circulated in non-European countries, the bias toward lighter skin tones grew ever more apparent. Competition in Asian markets propelled Kodak to follow Fuji in "ethnicizing" Shirley Cards [see figures 9.7–9.8]. Kodak continued research on skin tone preferences in different countries. Roth describes a resultant "geography of emulsions," in which "film inventory is batched by region and distributed" according to the skin color biases of the various parts of the world.[18] The market and profitability imperative of tailoring technologies to different populations is an ongoing driver of innovation.

But the hegemony of Whiteness is exposed not only in the context of the global competition to capture different regional markets. It is also exposed through practices that are more explicitly political. In South Africa, for example, Polaroid's ID2 camera, with its added flash "boost button" was used to better capture Black citizens' images for the infamous passbooks that violently restricted the movement of Black people throughout the country.

Polaroid's profit from South African apartheid spurred widespread protest against the company. The protest was led by the Polaroid Revolutionary Workers' Movement, which was founded by several African American employees in the United States, most notably photographer Ken Williams and chemist Caroline Hunter. The Workers' Movement "pointed out that part of Polaroid's profit in South Africa was earned from the sale of photoidentification units" that were used

> To photograph Black South Africans for the much hated "reference" books or "passbooks," which control the movements of Blacks into urban areas in the Republic of South Africa. If a Black is without one of these documents he is subject to imprisonment, a fine, or deportation from the urban area in which he was found.[19]

One of the flyers that the Workers' Movement posted around the office summed up the problem: "Polaroid Imprisons Black People in 60 Seconds."[20]

After initial protests, Polaroid experimented with ways to improve the conditions of its Black South-African workers, but activists deemed the reforms inadequate and continued pressuring the company over a seven-year period, until finally Polaroid pulled out of South Africa completely. This move, in turn, propelled the broader movement for boycott, divestment, and sanctions throughout the 1980s, which prompted other companies to withdraw from South Africa too.[21] In short, the use of visual technologies "in systems set up to classify people is an important aspect of the history of photography,"[22] one that connects the design of technology and social policy.

Aiming to expose the Whiteness of photo technologies, London-based artists Adam Broomberg and Oliver Chanarin used decades-old film and Polaroid's ID-2 camera, which were the ones used for passbooks, to take

FIGURE 14.1
Sample South African apartheid passbook. Image courtesy of API/Time.

FIGURE 14.2
Adam Broomberg and Oliver Chanarin, "To Photograph the Details of a Dark Horse in Low Light." Courtesy of the Late Estate Broomberg & Chanarin and Goodman Gallery, Johannesburg.

pictures across South Africa for a month. Their exhibit, "To Photograph the Details of a Dark Horse in Low Light" (a line taken from a Kodak statement touting the ability of Kodak film to depict dark skin accurately), aims to explore "the radical notion that prejudice might be inherent in the medium of photography itself." Here the artists echo an assertion made by photographer Jean-Luc Godard who, in 1977, was invited on an assignment to Mozambique but refused to use Kodak film, saying that it was inherently "racist." According to Broomberg, the light range was so narrow that, "if you exposed film for a white kid, the black kid sitting next to him would be rendered invisible except for the whites of his eyes and teeth."[23] You might be thinking, *surely* this is no longer an issue.

In 2009, Hewlett Packard's MediaSmart webcam demonstrated how the camera would pan to follow a White face but would stop when individuals with dark skin entered the frame.[24] The issue, according to HP, was that "the software has difficulty recognizing facial features in lower light levels."[25] *What are we to make of such enduring invisibility?* That new tools are coded in old biases is surprising only if we equate technological innovation with social progress. The popular trope that technology is always one step ahead of society is not only misleading but incorrect, when viewed through the lens of enduring invisibility.

Just as Polaroid, Kodak, and others attempted to design differently so that their cameras could vividly represent a broader spectrum of skin tones, so too companies that manufacture digital cameras today are working to address bias in the design and marketing of their products. In developing smartphones for sub-Saharan Africa, China's Tecno Mobile made it a point to emphasize the quality of images for Black customers:

In one such ad, a wide screen smartphone is shown on a black page with the image of a black

> woman showing on the screen. The words "capture the beauty of darkness" are written in bold beneath the image, followed by the line "The phone is powered for lowlight shooting." The ad labels the phone, "Camon C8," as a solution for a commonly held frustration with mobile phone cameras that render poor quality photos of dark skinned subjects in low-light settings.[26]

In the United States, the default settings of photo technologies, both past and present, tend to cater to lighter-skinned subjects. This is not simply a benign reflection of designers' or technicians' unexamined biases, nor is it an inevitable result of technological development, as China's Tecno Mobile demonstrates. In the next section we observe how visual technologies expose Whiteness and regularly reinforce racist visions, although the way such images circulate and gain meaning is not always a direct reflection of their initial settings.

EXPOSING DIFFERENCE

In her study of colonial-era photography, visual anthropologist Deborah Poole argues that we must not assume a singular interpretation of the relationship between images and society, one that looks for an all-encompassing "gaze" to exercise domination and control. Rather it is important to investigate the social nature of vision, because, once "unleashed in society, an image can acquire myriad interpretations or meanings according to the different codes and referents brought to it by its diverse viewers."[27] In examining the complex coding of racial desire and derision, Poole's insights remind us that domination and surveillance typically go hand in hand with "the pleasure of looking." The seeming adoration poured on racialized others via visual technologies, however, does not in itself signal a decline in racism and domination.

The desire to see others in a derogatory or in an exotic light, just as much as the practice of invisibilizing them, reproduces long-standing forms of authority and hierarchy. When it comes to representing racialized others, the interplay between desire and derision, longing and loathing can get lost in the strict focus on how visual technologies fail to see darker skinned people.[28] We can overlook the technology—racism—that precedes technological innovations.

In 2015 Google Photo came under fire because its auto-labeling software tagged two Black friends as "gorillas"[29]—a racist depiction that goes back for centuries, being formalized through scientific racism and through the association of Black people with simians in the Great Chain of Being. It found its modern incarnation in cartoons of former First Lady Michelle Obama and in Roseanne Barr's racist tweets against Valerie Jarrett and was resuscitated in algorithms that codify representations used for generations to denigrate people of African descent . . . Similarly, in his examination of video game avatars, computer scientist and artist D. Fox Harrell found that, when he attempted to make an African-inspired avatar that looked like him, it "was automatically made less intelligent." He explains:

> Prejudice, bias, stereotyping, and stigma are built not only into many games, but other forms of identity representations in social networks, virtual worlds, and more. These have real world effects on how we see ourselves and each other. Even in systems that have very open identity creation options, like Second Life, there are still different valuations for skins, social groups and categories being formed.[30]

The impact of having seemingly "racist robots" as . . . players in video games may seem trivial, but similar biases are built into the technological systems used by police, where they have more lethal consequences.

In social theorist Michel Foucault's classic formulation, "visibility is a trap." Foucault explained how power is exercised through techniques of surveillance in which people are constantly watched and disciplined, "the object of information, never a subject of communication." The *less obvious* the mechanism, the *more powerful* the disciplinary function of surveillance. It is tempting to point to the smart recording devices we carry around in our pockets and exclaim that "we are all caught inside the digital dragnet!" But, the fact is, we do not all experience the dangers of exposure in equal measure. Consider the "If you *see* something, *say* something" signs that litter public spaces, the brigade of White women reporting Black people to the police, the broken windows policies that license law enforcement to discipline small infractions like vandalism and toll-jumping, allegedly in order to deter and prevent larger crimes, and police body cameras that supposedly capture what "really" happened when an officer harasses or kills someone: clearly people are exposed differently to the dangers of surveillance.

In the most comprehensive study of its kind, a group of researchers at Georgetown Law School obtained over 10,000 pages of information from more than 100 police departments across the country, to examine how the use of facial recognition software impacts different communities. They found that "[t]he databases they use are disproportionately African American, and the software is especially bad at recognizing Black faces, according to several studies." What's more, the different global settings in which AI is taught to "see" impacts the technical settings designed to identify individuals from various groups. It turns out that algorithms "developed in China, Japan, and South Korea recognized East Asian faces far more readily than Caucasians.The reverse was true for algorithms developed in France, Germany, and the United States, which were significantly better at recognizing Caucasian facial characteristics."[31] This suggests that the political–geographic setting augments the default setting of Whiteness. The ethnoracial makeup of the software design team, the test photo databases, and the larger population of users influence the algorithms' capacity for recognition, though not in any straightforward sense.

For instance, when it comes to datasets, a 2012 study found that an algorithm trained "exclusively on either African American or Caucasian faces recognized members of the race in its training set more readily than members of any other race."[32] Scholars at Georgetown University's Center on Privacy and Technology point out that the disparities in facial recognition across racial groups may be introduced "at a number of points in the process of designing and deploying a facial recognition system":

> The engineer that develops an algorithm may program it to focus on facial features that are more easily distinguishable in some race than in others—the shape of a person's eyes, the width of the nose, the size of the mouth or chin. This decision, in turn, might be based on preexisting biological research about face identification and past practices which themselves may contain bias. Or the engineer may rely on his or her own experience in distinguishing between faces—a process that is influenced by the engineer's own race.[33]

Now consider that these software programs are used by police departments all over the country; in those departments "digital eyes watch the public," comparing individual faces in real time to "hot lists" that are filled disproportionately with Black people—and these also happen to be the least recognizable figures in the world of facial recognition software.

The humor in the much circulated HP MediaSmart video of the Black user quoted in this chapter's epigraph saying "I think my blackness is interfering with the computer's ability to follow me" turns deadly in the context of digitally mediated police profiling, where suspects are caught in the crosshairs of being seen too much via surveillance practices and not enough via software; one's own body is called upon to testify against itself. Guilty, guilty, guilty. . . .

EXPOSING PRIVACY

Fanon's (2008) *Black Skin, White Masks* reverberates through the work of MIT Media Lab researcher Joy Buolamwini, who discovered that the facial recognition technology with which she was working could detect the contours of her face only when she put on a White mask. This is what she calls the "coded gaze." Buolamwini established the Algorithmic Justice League as part of her quest for "full spectrum inclusion," to counter the bias she experienced.[34] She asks: "If we do not improve the systems and they continue to be used, what are the implications of having innocent people identified as criminal suspects?"

While inclusion and accuracy are worthy goals in the abstract, given the encoding of longstanding racism in discriminatory design, what does it mean to be included, and hence more accurately identifiable, in an unjust set of social relations? Innocence and criminality are not objective states of being that can be detected by an algorithm but are *created* through the interaction of institutions and individuals against the backdrop of a deeply racialized history, in which Blackness is coded as a criminal. Inclusion in this context is more akin to possession, as in Fanon's plea that the "tool never possess the man," where possession alerts us to the way freedom is constrained.

Consider a population-wide facial recognition program in which the Zimbabwean government has contracted a China-based company to track millions of Zimbabwean citizens in order to make the Chinese database more comprehensive by "more clearly identify[ing] different ethnicities." The benefit for Zimbabwe is access to a suite of technologies that can be used by law enforcement and other public agencies, while positioning China to become "the world leader in artificial intelligence."[35] *Transnational algorithmic diversity training par excellence!* Perhaps. Or, better, neocolonial extraction for the digital age in which the people whose faces populate the database have no rights vis-à-vis the data or systems that are built with their biometric input.

Not only that. Since the biggest application of facial recognition is in the context of law enforcement and immigration control, Zimbabwe is helping Chinese officials to become more adept at criminalizing Black people within China and across the African diaspora.

Racist structures do not only marginalize but also forcibly center and surveil racialized groups that are "trapped between regimes of invisibility and spectacular hypervisibility,"[36] threatened by inclusion in science and technology as objects of inquiry. Inclusion is no straightforward good but is often a form of unwanted exposure. Jasmine Nichole Cobb's insight that "invisibility is . . . part of the social condition of blackness in modernity as well as an important representational tactic for people of African descent"—what Rusert describes as that "dialectic of calculated visibility and strategic invisibility"—is relevant to countering the New Jim Code.[37]

The figure of Saartjie ("Sara") Baartman illustrates the violent underside of being forcibly seen. Baartman, who was taken from South Africa to Europe in 1810, was publicly displayed for large audiences in London and Paris, photographed, studied, and eventually dissected in death by the leading scientist of the time, Georges Cuvier, and her skeleton, brain, and genitals were subsequently put on display until 1974. Baartman's horrific exposure in life and death illustrates the connection between visual and scientific technologies. While many people have heard some version of her story in scholarly texts and popular works, few know of Baartman's eventual repatriation to and burial in South Africa in 2002, through which the evidentiary politics surrounding her identity came to a climax. The protracted negotiations between South Africa and France for the return of Baartman's remains—her skeleton, brain, and genitals—were stalled by French claims that the remains had been lost and could not be identified among the museum's holdings. Consider that "Baartman was one of thousands from Europe's former colonial territories whose remains had been gathered in metropolitan museums." In 2002, "once the French government [finally] agreed to return them, a dispute arose about the veracity of the physical remains offered by the French."[38]

Despite this, the South African committee that negotiated her return declined to have the remains tested to verify whether they belonged to Baartman, or even whether the three sets of remains belonged to the same person. For the committee, to do so would amount to a replication of the violation, repeating once again the "great long insult" to which Baartman had been subjected during her life. Instead, on August 9, 2002, Baartman was given a ceremonial burial in Hankey, South Africa, near the place where she was born. This decision of not exposing Baartman's remains to scrutiny yet again was the South African committee's assertion and attempt to define a way of knowing differently, whereby it decided to accept without further DNA testing that the remains offered by France belonged to Baartman. This signaled an end to the invasive visibility to which Baartman was subjected during her lifetime and for 159 years after her death.

In "Baartman and the Private: How Can We Look at a Figure That Has Been Looked at Too Much," South African gender studies scholar Gabeba Baderoon explains how "dominated peoples have long crafted a way to exist and keep their histories outside of conventional archives." The politics of knowledge, in other words, is deeply entangled in a politics of the private and in who gets to lay claim to privacy and subjectivity. The assertion of "privacy" in this context is not the same as *privatizing*—an economic calculation "outside history," made for the purpose of maximizing profits; rather, in Baderoon's theory of the private, it draws attention to the "intimate, personal, closed, hidden, *coded*, secret, veiled, unknown, the apparently meaningless, the invisible, the ordinary, the in-between, the silent . . . the underside . . . unofficial . . . unpredictable, and unreliable in dominant views of history."[39]

What is privacy for already exposed people in the age of big data? For oppressed people, I think privacy is not only about protecting some things from view, but also about what is strategically exposed. This might look like Mamie Till-Mobley, mother of slain teenager Emmett Till, choosing to expose the mutilated body of her son because "I think everybody needs to know what happened." It could also look like the organization Stop LAPD Spying Coalition, exposing the lies of law enforcement officials who claim not to know about the very surveillance techniques that the organization records them using. Organizers participate in community events, public rallies, and protests and proceed to "watch the watchers." In one video, activists are shown interviewing police officers who act as if they had never heard of "Freedom on the Move" vans—a mobile video surveillance system "intended for use by military, border patrol, and law enforcement agencies"; and, just as they make these denials, the video cuts to the vans driving by the May Day rally.[40] Countering these forms of technological exposure, Stop LAPD Spying Coalition created "A People's Audit," which includes survey results of community members' experiences of preemptive policing and harassment.[41] If surveillance treats people like a "surface," then countering this form of violent exposure can entail listening deeply to the everyday encounters of those who are forcibly watched. This points to a different way of making sense and knowledge of the world, a theory of privacy predicated on mutual respect and dignity. . . .

NOTES

Chapter 14, "Coded Exposure: Is Visibility a Trap?," by Ruha Benjamin, originally published in *Race after Technology: Abolitionist Tools for the New Jim Code,* © 2019 Polity Books. Used with permission of Polity Books.

Editors' note: This excerpt is taken from the third chapter of *Race after Technology: Abolitionist Tools for the New Jim Code* (Cambridge, UK: Polity Press, 2019).

1. Bowles, "'I Think My Blackness Is Interfering'"; "wzamen0. 'HP Computers Are Racist'."

2. Merriam-Webster Online, "exposure."

3. "Racial Sensitivity."

4. "Racial Sensitivity."

5. "Top 100 Sitcom Episodes of All Time, No. 54."

6. The episode represents the ludicrous logic of racial capitalism, which Melamed aptly calls a "technology of antirelationality . . . Accumulation under capitalism is necessarily expropriation of labor, land, and resources. But it is also something else . . . the production of social separateness—the disjoining or deactiving of relations between human beings (and humans and nature) [and humans and technology!]—needed for capitalist expropriation of work" ("Racial Capitalism," 78).

7. See Buolamwini, "The Coded Gaze."

8. Samatar, "Skin Feeling."

9. Hamilton and Hargreaves, *The Beautiful and the Damned*; Poole, *Vision, Race, and Modernity.*

10. Gidley, *Representing Others.*

11. Wells, *Photography,* 103.

12. Lorna Roth, "Looking at Shirley," 131.

13. Samatar, "Skin Feeling"

14. Brown, *The Black Skyscraper.*

15. Roth, "Looking at Shirley," 168.

16. Roth, "Looking at Shirley," 169.

17. Roth, "Looking at Shirley," 119.

18. Roth, "Looking at Shirley," 121.

19. Unterhalter, "The Polaroid Experiment in South Africa," 109.

20. Massie, *Loosing the Bonds*, 272, quoted in Morgan, *American Slavery, American Freedom*, 525.

21. Morgan, *American Slavery.*

22. Sturken and Cartwright *Practices of Looking*, 357.

23. Quoted in Smith, "'Racism' of Early Colour Photography Explored in Art Exhibition."

24. Bowles, "'I Think My Blackness Is Interfering'."

25. Bowles, "'I Think My Blackness Is Interfering'."

26. Avle and Lindtner. "Design(ing) 'Here' and 'There'."

27. Poole, *Vision, Race, and Modernity*, 18.

28. Poole notes that "it is almost too easy to consign all images to the task of reproducing (or perhaps even producing) imperial, racial, and sexual ideologies. The more challenging task is to think about the ways in which aesthetics and the 'open code' of visual images occasionally disrupt the powerful hold that imperial discourse has over our imaginations" (*Vision, Race, and Modernity*, p. 17).

29. Plenke, "Google Just Misidentified 2 Black People in the Most Racist Way Possible."

30. Soep, "Chimerical Avatars and Other Identity Experiments from Prof. Fox Harrell."

31. Soep, "Chimerical Avatars and Other Identity Experiments from Prof. Fox Harrell."

32. Soep, "Chimerical Avatars and Other Identity Experiments from Prof. Fox Harrell."

33. Soep, "Chimerical Avatars and Other Identity Experiments from Prof. Fox Harrell."

34. Buolamwini, "The Algorithmic Justice League." See also Buolamwini and Gebru, "Gender Shades."

35. Chutel, "China Is Exporting Facial Recognition Software to Africa, Expanding Its Vast Database."

36. Rusert, *Fugitive Science*, 26.

37. Rusert, *Fugitive Science*.

38. Baderoon, "Baartman and the Private," 71.

39. Baderoon, "Baartman and the Private," 75 (emphasis added).

40. "Stop LAPD Spying: They Still Lie."

41. See https://stoplapdspying.org/wp-content/uploads/2013/04/PEOPLES-AUDIT UPDATED-APRIL-2–2013-A.pdf.

BIBLIOGRAPHY

Avle, Seyram, and Silvia Lindtner. "Design(ing) 'Here' and 'There': Tech Entrepreneurs, Global Markets, and Reflexivity in Design Processes." In *Proceedings of the 2016 CHI Conference on Human Factors in Computing Systems* (215), 2233–2245. https://doi.org/10.1145/2858036.285850.

Baderoon, Gabeba. "Baartman and the Private: How Can We Look at a Figure That Has Been Looked at Too Much?" In *Representation and Black Womanhood,* edited by N. Gordon-Chipembere. New York: Palgrave, 2011.

Bowles, Nellie. "'I Think My Blackness Is Interfering': Does Facial Recognition Show Racial Bias?" *The Guardian,* April 7, 2016. https://www.theguardian.com/technology/2016/apr/08/facial-recognition-technologyracial-bias-police.

Brown, Adrienne R. *The Black Skyscraper: Architecture and the Perception of Race.* Baltimore: Johns Hopkins University Press, 2017.

Buolamwini, Joy. "The Algorithmic Justice League." Medium Corporation, December 14, 2016. https://medium.com/mit-media-lab/the-algorithmic-justice-league-3cc4131c5148.

Buolamwini, Joy. "The Coded Gaze." Algorithmic Justice League, 2017. www.ajlunited.org/the-coded-gaze.

Buolamwini, Joy, and Timnit Gebru. "Gender Shades: Intersectional Accuracy Disparities in Commercial Gender Classification." *Proceedings of Machine Learning Research* 81 (2018), 1–15.

Chutel, Linsey. "China Is Exporting Facial Recognition Software to Africa, Expanding Its Vast Database." *Quartz,* May 25, 2018. https://qz.com/africa/1287675/china-is-exporting-facialrecognition-to-africa-ensuring-ai-dominance-through-diversity.

Gidley, Mark, ed. *Representing Others: White Views of Indigenous Peoples.* Exeter, UK: University of Exeter Press, 1992.

Hamilton, Peter and Hargreaves, Roger. *The Beautiful and the Damned: The Creation of Identity in Nineteenth Century Photography.* London: Lund Humphries, 2001.

Massie, Robert Kinloch. *Loosing the Bonds: The United States and South Africa in the Apartheid Years.* New York: Nan A. Taleze, 1972.

Melamed, Jodi. "Racial Capitalism." *Critical Ethnic Studies* 1, no. 1 (Spring 2015): 76–85.

Merriam-Webster Online, n.d., s.v. "exposure."

Morgan, Edmund S. *American Slavery, American Freedom: The Ordeal of Colonial Virginia.* New York: Norton, 2006.

Plenke, Max. "Google Just Misidentified 2 Black People in the Most Racist Way Possible." *Mic Network,* June 30, 2015. https://mic.com/articles/121555/google-photosmisidentifies-african-americans-as-gorillas.

Poole, Deborah. *Vision, Race, and Modernity: A Visual Economy of the Andean Image World*. Princeton, NJ: Princeton University Press, 1997.

"Racial Sensitivity," YouTube link, originally aired on April 8, 2009.

Roth, Lorna. "Looking at Shirley, the Ultimate Norm: Colour Balance, Image Technologies, and Cognitive Equity." *Canadian Journal of Communication* 34, no. 1 (2009): 111–136.

Rusert, Britt. *Fugitive Science: Empiricism and Freedom in Early African American Culture*. New York: NYU Press, 2017.

Samatar, Sofia. "Skin Feeling." *The New Inquiry*, September 25, 2015. Online.

Smith, David. "'Racism' of Early Colour Photography Explored in Art Exhibition." *The Guardian*, January 25, 2013. https://www.theguardian.com/artanddesign/2013/jan/25/racismcolour-photography-exhibition.

Soep, Elisabeth. "Chimerical Avatars and Other Identity Experiments from Prof. Fox Harrell." *Boing Boing*, April 19, 2010. https://boingboing.net/2010/04/19/chimerical-avatarsa.html.

"Stop LAPD Spying: They Still Lie," YouTube video, https://www.youtube.com/watch?v=tDYwvT_jPBo.

Sturken, Marita, and Cartwright Lisa. *Practices of Looking: An Introduction to Visual Culture*. Oxford: Oxford University Press, 2017.

"Top 100 Sitcom Episodes of All Time, No. 54: 'Racial Sensitivity,' Better off Ted," https://robertdavidsullivan.typepad.com/my_weblog/2012/04/top-100-sitcom-episodes-ofall-time-no-54-racial-sensitivity-better-off-ted.html.

Wells, Liz, ed. *Photography: A Critical Introduction*. New York: Routledge, 1996.

"wzamen0. 'HP Computers Are Racist,'" YouTube video, 2:15. December 10, 2009. https://www.youtube.com/watch?v=t4DT3tQqgRM&feature=youtube.

Unterhalter, Beryl. "The Polaroid Experiment in South Africa: A Progress Report." *Vanderbilt Journal of Transnational Law* 6 (1972): 109–120.

15 WHOSE WHITE BABY IS THAT? COLORS OF CYBORGIAN YOUTH AND "GOOD" MIXED-RACE ALGORITHMS

AMBER SWEAT

In the prison scene of Nick Cannon and Nile Evans' comedy, *School Dance* (2014), inmate Darren (Katt Williams) receives a visit from his significant other. Awash in a bluish mid-shot, Trina (Tiffany Haddish) arrives with their two children in tow; the teenage son, Darren Jr., shares a similar skin tone to his two black parents. However, contrasted against the sharp blue of the carceral scene, the younger unnamed infant is visibly lighter than his family members are. Such distinction prompts Darren to stare at the child in confusion before sharply asking, "Whose goddamn white baby is that?"[1] Trina affirms that he is the child's father in spite of the child's lighter-featured phenotype, only for Darren to launch into a string of refusals:

> [*Darren gestures to darker-skinned Darren Jr.*] That one might be mine, that one right there, I'm not totally sure. Just because you name him Darren, don't mean he belong to Darren. [*Darren turns to infant*] But I don't care what this baby's name is. This baby is rice-skinned, but not light-skinned. That is a white child. That is Caucasian from the mountains of Caucasus. That is a Slavic baby, a Viking from Iceland. Th-th-that baby got 730 as a credit ratin' right now as an infant.[2]

Darren's monologue has quickly become one of the film's most recognizable scenes, with just one upload boasting upward of twelve million YouTube views.[3] The scene's diegetic sound was also isolated from the video and uploaded as a viral TikTok "sound" by @tigerty_929, under which users upload videos that relate to the audio content in a humorous way. Of the

FIGURE 15.1
A scene from Nick Cannon and Nile Evans' *School Dance* displaying Darren's two children alongside their mother (2014).

School Dance sound's 22,000+ uploads, a vast majority of those most-liked depict comical situations like Darren's: real-life black parents (typically fathers) appear with their lighter-skinned children, making light of their physical differences and questioning how such a child could be a product of their genetics. Under a June 2021 video from famous family TikTokers @ fluellenfam where a darker-skinned father uses the sound to joke about his blonde-haired, blue-eyed son, @stormangel611 comments, "staaaaaap . . . each one got lighter tho . . . beautiful family . . ."[4] A similar interracial family, @ thewilliamsonfam, has twin girls—one is darker-featured and one lighter-featured. In a video featuring the lighter-featured twin, @__madssss enthuses, "Omg I love her blonde hair and blue eyes she is so beautiful."[5]

These feel-good family profiles have constructed much of the internet's belief in an interracial utopia, crafting a safe space for multi-racial love, marriage, and reproduction under the auspices of the platform's assumed race-lessness. Such assumptions receive critical pressure in Nakamura and Chow-White's 2012 *Race after the Internet,* which engages a "forceful critique of the utopian discourse that characterized early digital media studies" in response to the assertion that the internet age has crafted new information "societies" that are "radically equal."[6] Today, the #swirl phenomenon, a social media movement that champions post-raciality and assumed race neutrality via mixed-race couples' shared accounts of daily life in a post-Loving v. Virginia era, is a paragon example of TikTok's assumed racelessness.[7] While black and white couples make up most of the hashtag's landscape, couples of various racial assortments call for partnership beyond racial divisions, depicting their quotidian as one where racio-ethnic mixing is a pleasurable and comical part of the everyday. Of the various activities #swirl couples take part in, childbearing and childrearing draw the greatest influx of views, with videos vacillating between events as banal as lunch preparation and as intimate as live childbirth (the latter typically draws the highest views alongside pregnancy announcements). The #swirl couple—screened in all their saccharine glory and quotidian, marital bliss—reaches their most popular with the conception and birthing of a child.

The intrigue of the #swirl couple's mixed-race child is one of both phenotypic curiosity and color-evasive fantasy, where children exist as emblematic figures of universal goodness.[8] Because the assemblage of mixed-race features is an ungraspable, unpredictable, and overall hopeful image, the child is rendered a prism through which online users can find not only hope but also entertainment by envisioning a future that plays with expectations of human phenotype, distorts bodily hegemonies, and is racially heterogeneous and increasingly post-racial.

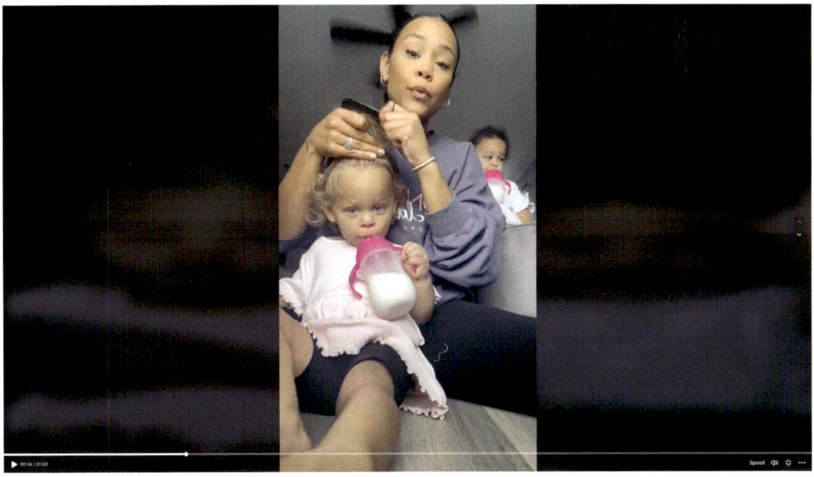

FIGURE 15.2
TikTok screengrab from @thewilliamsonfam, featuring their lighter-featured daughter (2022).

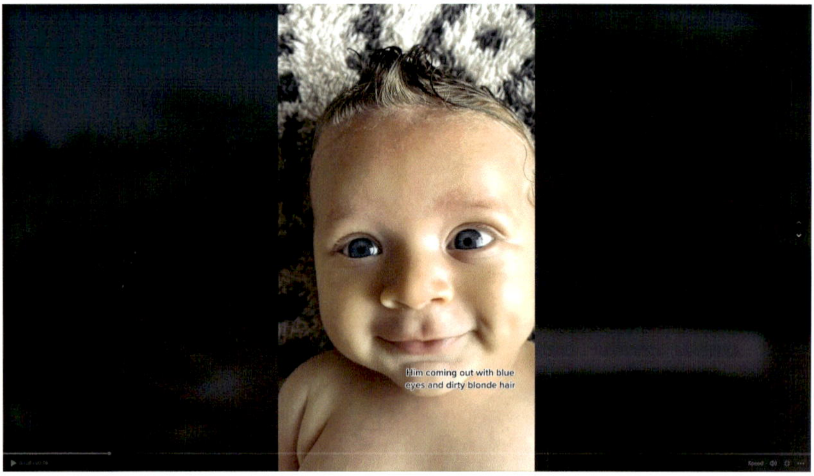

FIGURE 15.3
TikTok screengrab from @fluellenfam, featuring their light-featured son (2022).

However, while some contend that the mixed-race child symbolizes the post-racial nature of the internet, I posit that the opposite holds true. Using TikTok as a platform of study, this chapter addresses how the "cute" mixed-race child image instead exposits the ways that anti-blackness and white supremacy do not disappear in the net age. Instead, they are emboldened via the internet's human-generated interfaces which deploy equally human-generated algorithms to champion a "Good" type of mixed-raceness. When prototypically white features become increasingly evident in mixed black and white children, this "unexpected whiteness" becomes tantamount for attention generation. In turn, the attention that inflates viewership for white-looking children (later defined as *engagement*) nourishes an algorithm that feeds on the user's lingering gaze. Soliciting prolonged looking is especially precious in the internet-age attention economy where, with so much for people to look at with such rapid accessibility, human attention itself has become fragmented, short-term, and divided across unprecedented numbers of images.[9]

The result of this yearning for attention, I argue, is heightened colorism and predilections towards whiteness. And when parents seek to emphasize this whiteness for the children (largely through digital filters), we see the creation of a cyborgian aesthetic for images of the mixed-race child: what I term *Cyborgian cuteness*. I denote this cuteness as Cyborgian because it is contingent on images of mixed-race children that users (usually parents or guardians) technologically mediate to assume a digitally-manipulated whiteness. As a result, the child image is not merely a simulated image of the real human, but, like a cyborg, a hybrid of organic substance and augmented reality's inorganic matter.[10] Furthermore, as hinted, the force of the whitened, manipulated mixed-race child image does not end with its paradoxical supposition of post-racial fantasies; it is also an increasingly valuable commodity in the high-tech landscape. If the child's parents can "win the algorithm" by publicizing their child's image on the platform, they can boost financial gain via heightened viewership and the brand partnerships or advertisements that follow.

This piece tracks three specific phenomena that, like an algorithm, speak to, rely upon, and build upon one another. Firstly, the digital manipulation of children's images via filters and digital color manipulation has a specific effect for the mixed-race child, turning the child into a cyborg via the assemblage of their organic (material, untouched face) and the artificial (which takes place at the level of image manipulation). Although children of all races are subject to image manipulation, I reveal how mixed-race children are edited more intensely in order to emphasize their European features; parents modify the colors of their child's image towards a cuteness within the limitations of whiteness, but beyond the standard aesthetics of the organic human. Secondly, at the level of algorithm, I unravel how this digital manipulation stimulates the attention economy for these children—largely through increased participation via *staring*—which in turn may result in material wealth for families. Finally, this piece shows how this feedback loop has material consequences; parents in the new media landscape have openly discussed yearning for a child which fits this unattainable norm, expressing how Cyborgian cuteness has become a troubling corporeal expectation for mixed-race children IRL ("in real-life" as opposed to the digital URL—an acronym for an internet website's "Uniform Resource Locator"—which has come to metonymize computer-age practices that take place on-screen.)

UNEXPECTED WHITENESS: TIKTOK CUTENESS AND USER ENGAGEMENT

How do mixed-race children "win" the algorithm by their unexpected whiteness, or the surprising appearance of Eurocentric traits that first appear incongruent with Black parentage? Like the comedic shock surrounding

FIGURE 15.4
An image from Kim Kardashian's Instagram of her daughter, Chicago. Chicago, who is half-Armenian and half-Black, is one of the many Kardashian youth to have been filtered by their parental figures. This photo's filter adds anthropomorphic mouse ears, lightens the skin, enlarges the eyes, and lightens the eyes (2018).

Darren's baby in *School Dance,* that which is unanticipated or surprising functions as the "hook" for a user's bemused (and amused) engagement. A look into TikTok's algorithm helps explain how and why this happens. TikTok oft presents itself as a social media platform whose algorithm knows viewers better than viewers know themselves; as Twitter user @Pateezy says, "Your TikTok algorithm most probably knows you better than your closest friend."[11] This tweet points to the accuracy of the "For You Page" (FYP), which curates videos from creators who a user does not officially follow, but for whom the user's interests are calibrated (i.e., users who pass time

FIGURE 15.5
A TikTok screengrab from user @kat.lexii. This video of her daughter employs the "Big Eye Baby" filter, which lightens the skin, enlarges the eyes, and lightens the eyes (2022).

on "BeautyTok" are more likely to have an FYP of makeup videos, as would be the case for Motherhood hauls on a "MomTok" user's FYP). This configures TikTok as a platform whose site-specific exploration and interest-based entertainment (much like YouTube) opposes user-to-user social media based upon extended network formation or "friending" (such is the case with Facebook or Instagram). FYP's serendipity results from "stumbling across" something that TikTok knows will be fulfilling and calibrated towards a user's personal tastes.[12] This targeting happens through what is known as "collaborative filtering," wherein an algorithm, according to Google, "uses similarities between users and items simultaneously to provide recommendations. This allows for serendipitous recommendations."[13]

TikTok's algorithm is famously elusive. Officially, the platform contends that the content on a user's FYP is based on the following:

1. User interactions (videos that a user creates, likes, shares, or forwards; duration of engagement with certain videos or measured time spent viewing a certain video, completion of the video, or re-watches)
2. Video information (captions, or metadata tags such as hashtags or sounds)

3. Device settings (the device's geographical location, language settings, or device type).[14]

TikTok maintains that this triad exists in equal tandem with one another and "a number of [unnamed] factors," but my findings emphasize one of these practices over all others: "how long [one] linger[s] over a piece of content."[15] TikTok's short-form videos facilitate engagement in an era of decreased attention spans and increased want for concise narratives, leading to more video exposure in a reduced amount of scroll time.

However, the determination of which TikTokers will receive higher algorithmic engagement does not remove itself from long-held human biases. Human adoration towards that which is cute, anomalous, and (unexpectedly) white helps explain the reasons why mixed-race children have become critical figures in the production of net celebrity and economic gain. To start, child internet stars draw high user interaction due to their innate cuteness, regardless of race or ethnicity.[16] As Sianne Ngai shows, cuteness is a taste concept that "is firmly rooted in visual commodity culture," evidenced in the commodification of the net child image by #swirl parents. Further, cuteness catalyzes affects of care and attentiveness from adults and, by proxy, adult viewership, making the cute child a central figure for engagement.[17] Cute images also lead to increased user interactions, largely by way of their positive impacts on a user's brain chemistry that in turn conjure user desire for increased exposure to more cute images or videos. Given that increased user interactions such as lingering contribute to strengthening the algorithm, it is unsurprising that precious net avatars which incite looking (kittens, puppies, babies) are privileged in algorithmically-driven spaces. TikTok, thus, is a fertile space for the mixed-race child's entry into algorithmic fortune.

By the same token, pre-existent racial and racist tendencies also intensify online, largely orienting perceptions of cuteness, where racial or racist tendencies which proliferate through a user base harden in the algorithm.[18] Accordingly, if the fame of mixed-race children is contingent on cuteness, it is imperative to consider said cuteness in proximity to whiteness. The child of color—especially one that appears monoracially black—becomes distanced from notions of a largely white and Judeo-Christian cute and innocent child imaginary.[19] As a result, they receive less attention. Such trends crystallize in the label of proper or "Good" mixedness, with the word "Good" recurring in comments surrounding mixed-race net babies. Commenters often point out a child's "Good Hair" as the mother runs her fingers through it with various creams and serums and often define a child as a "Good type" due to their brighter skin tone or blue eye color. In short, to be a properly attention-earning, interracial child avatar, there is a vital necessity to be "Good" insofar as "Goodness" reflects colorist tendencies in global communities—both multi- and mono-racial—that largely favor white phenotypic traits such as hair texture or skin color. It is, above all, to centralize and exalt whiteness through the lip-service of "#swirl" and its discursive technophilia. Goodness drives the algorithm's central currency of increased attention, but only for children whose coloration merits said attention, exacerbating the racialized color encodings always-already in place.

Furthermore, it is not only whiteness itself, but an *unexpected* whiteness that reifies the gazing-based algorithm for TikTok. The unexpectedness is rooted in a collective scientific lexicon of what "should" show up in a particular genetic combination. Western science-based ordering systems oft categorize human features into either dominant or recessive traits; dominant traits are those whose colors tend to mark blackness (brown skin, brown eyes, and dark hair), and recessive traits (notably blonde hair and blue eyes) are those that mark whiteness.[20] These terms, dominant and recessive, imply the inherent subsuming of one by the other—the two genetic codes mixing cannot take place without the harsh effacing of white features and the continuation of black features. Commenters such as those on thewilliamsonfamily's account often point to the startling presence of "dormant" white-marked features as ones that have "defeated" genetically dominant black-marked features, with the white features returning in an almost messianic fashion. Thus, the apparition of white features immediately becomes intriguing, miraculous, or extraordinary, thereby catalyzing the want for these features in mixed kid images which, in turn, drive engagement.

This startling effect of unexpected whiteness—that "entices and disturbs" consumers—lends itself to staring and, by proxy, engagement. In 2022, a mixed-race Reddit user ignited a conversation by asking, "Anyone else get stared at in public?", receiving affirmative responses from fellow mixed-race individuals.[21] Similarly, in a comment for the now divorced #swirl couple, Jamie and Nikki (a white man and black woman respectively), a viewer mentions, "can't stop staring at that unique n beautiful face!"[22] in response to a video reel of their mixed-race daughters. The inability to categorize human faces into neatly determined phenotypical categories renders the anomalous face non-normative, increasing intrigue and, in turn, heightening the amount of time spent giving the unique face attention for appraisal. Taking this conclusion into account, the intrigue of lighter-featured mixed-race children is the result of this divergence from normative, (mono)racial expectations of phenotype. Accordingly, this intrigue is doubly impactful when mixedness produces the white features that it assumedly precludes.

Whilst studies on in-person interactions consider the stare as an unwelcoming or disquieting phenomenon,[23] the stare is reconfigured as a welcome occurrence in when mediated by the digital touchscreen. In targeted advertisements, typically included in the visual media themselves, a creator's higher engagement signals an increased likelihood for adjunct engagement with a sponsor, either by intentional or accidental clicking. And if this engagement is predicated on the amount of time a user spends absorbed in a particular image, then maintained looking—staring—is congruent with economic success by way of advertisers and brand deals. When it comes to conquering the algorithm, cute, mixed-race children with "unexpectedly" white features stimulate engagement, rendering them paragons for user captivation, media virality, and subsequent monetary gain for the parents and adults who moderate their images.

BIG-EYED DREAM GIRL

Scrolling through my own FYP, mixed-race children (due to my research) appear in a never-ending scroll of inhumanly large eyes filled with virtual sparkles, long eyelashes, and blushed noses. Most of the videos use the hashtags #swirl or #mixedbaby and utilize the sound from *School Dance* or one with a similar sentiment. A common comment on these profiles is that the children have ascended to an "unreal" cuteness or dimension of beauty. This is partially due to how the inability to place mixed-race children into pre-existing racial phenotypes renders them errant or impossible figures. However, there is a secondary dimension to this "unreality" of the child image; in the user's production of artificial whiteness, the child image (even if unknowingly) becomes distorted beyond reality.

At the bottom of many of these TikToks is a little gray box containing a yellow square and words such as "Glow," "Brightside," "Dreamy Girl," or "Big-eyed baby." Each signals that the TikToker has applied a filter to the raw image, employing augmented reality (AR) to change and distort the child's face at some point between recording and publication onto the platform.[24] The words listed above also denote and consolidate which filter the user has employed. Each of the aforementioned filters has a troubling effect on the videos' coloration; from adding blue filters to the eye region, to shrinking noses, to heightening exposure and brightness upon the child's skin, all enhance Eurocentric features. At the same time, the filters—made with the adult facial dimension as a template—are incongruent with the child's face. Making the already-large eyes a bit too large and the already-small nose a bit too small, AR manipulates the child face beyond artificial whiteness and into the realm of the uncanny.

However detrimental, this digital enhancement of whiteness is beneficial for the algorithm via three heightened metrics of engagement. For starters, it distorts the image towards the mixed-race child's unexpected whiteness as a key to algorithmic success. Secondly, it enhances engagement via increased staring beyond whiteness' intrigue, giving the child an uncannily non-human appearance that in turn startles and entices viewers to continue looking. Some users decry the act of filtering a child's image, often pointing to the racist implications of such an act by their assumed parent. Nevertheless, this decision to manipulate the child image towards a lactified combination of the human/non-human is increasingly frequent in net celebrity, and its effects are trickling down into the quotidian both URL and IRL (that is, online and offline), where increasingly narrowing beauty standards are impacting identity formation.[25] I turn here towards the impact of this algorithmic and economic predilection on real-life parents, children, and their relationship to "Cyborgian cuteness" as it constructs new corporeal expectations for generation α (generally considered those born between 2010 and 2024) and subsequent age groups.

The connection between child and cyborg has come into academic dialogue as of late, as in a 2014 study by May-Chahal.[26] Approximately a decade before, in 2001, Lammel first coined the "Cyborg-Child" as one whose life is largely contingent on a "massive" use of the internet to maintain a child's social existence and self-formation.[27] Whereas "cyborg" first designated a fictitious being comprised by both human and machine possessing the capabilities required to explore interstellar space, the twentieth century Cyborg-Child is an essentially biotechnical being. For Lammel, the modern child is a cyborg due to both their proximal relationship to the world of media as well as how media are innately embedded in their lived experience. I would argue that the child's development becomes reliant not just on their real-life subjectivity, however, but one that is mediated on sites such as TikTok. These virtual images—artificially white and uncanny—are now part of how a child should look. This strange, racially ambiguous cuteness is no longer bizarre or anomalous for the net-age child, but expected to come to corporeal fruition.

"MIXED W/ DELUSION"

As I have argued thus far, the algorithmic exaltation and economic potential of Good mixed-race children has created the "Cyborgian cuteness" on which parents and mixed-race children are modeling new ways of appearing in new media and, by proxy, in flesh. It hardly seems out of left field to claim that this white-centric Cyborgian cuteness is likely to result in immense psychic stress for mixed-race black-white children. Already presumed "unnatural" due to their multiracial existence, the secondary layer of re-coloration and distortion via technology doubles the shakiness of their selves in hopes of their parents' domination of the algorithm. Further, Cyborgian cuteness blurs the lines between how children physically can or cannot look in real life with the aid of tools like automated filters. Parents of mixed-race children begin to internalize that their children must adhere to an image standard that is simply unattainable for organic human features.

As AR filters reveal a propensity towards whiteness and increased engagement, TikTok parents have revealed their hopes for a child who will assume the Cyborgian aesthetic even while still in-utero. In July 2022, @yatziri-allison uploaded a video showing her ultrasound. In the video, which is now deleted, she wrote the caption, "Got pregnant and thought I would have an *exotic mixed* baby . . . ", only to cut to a photo of her child.[28] Stitching in an unfiltered photo of the child making a mess during a mealtime, the video reveals that that the child was not racially ambiguous nor 'unnaturally' cute, but "normal," striking a chord of disappointment with the mother-uploader. Whilst several comments laughed at the video or supported the mother's sentiments, many pointed out the concerning nature of the caption. "mixed w/ delusion," says a top comment, encapsulating the ways that Cyborgian cuteness has distorted the expectations for interracial human children in the TikTok age.[29] The mother, doubly disappointed

FIGURE 15.6
A TikTok screengrab from user @official.baby.names, which specializes in images of mixed-race children. All photos are overlaid with a filter which adds light and moving sparkle to the iris.

FIGURE 15.7
TikTok screengrab from user @mommydesi333. The filter she employs lightens and thickens the hair, lightens the eyes, and adds pink hues to the skin. The video was filtered, downloaded to secondary memory, and uploaded (bypassing the filter label).

that her child did not adhere to the ambiguous and super-human image that wins the mixed-race child algorithm, openly voices her discontent to an audience of both similarly minded and largely dissenting voices. Moreover, she is not the only one.

Jamie Perkins was once a pioneering force in the #swirl community with his South Sudanese wife, Nikki. After their divorce, he began attempting to crack the algorithmic code for TikTok fame, integrating hashtags such as #kidsoftiktok, #mixedkids, and #mixedfamily into his videos. When his daughter, Ava, was born in 2016, the first images of a Cyborgian cuteness were proliferating wildly on YouTube and Instagram. In the video of their daughter's birth, Jamie stares intently at her newborn face and smiles, saying, "Look mom, she has blue eyes!"[30] Attuned to the expectations of the lighter-colored, net-famous mixed child—which he thought would undoubtedly be the case for his born-famous child—Jamie feels relief that the color of his daughter's eyes would save them a space in the algorithm's graces. Except it doesn't. His mother looks at the child and announces that her eyes are assuredly brown, and that all babies are born with blue eyes. Viewers will come to learn that Jamie's mother is correct, but Jamie stands in firm belief of her eye color. Like the "mixed w/ delusion" of his TikTok contemporary, he hallucinates an image of his daughter that is contingent with his expectations of her coloration in tandem with—or in hopes of—their success as net celebrities. The expectation of an artificially white, Cyborg-esque child has found its way into Jamie's psyche. Jamie and Nikki's daughter must be the dream girl. Her eyes must be blue.

Today, her eyes are still brown, her coloration favors her mother, and users leave comments like, "my brain thought [your daughter] was somali,"[31] or ask questions such as, "is [your daughter] just black?",[32] asserting that the girls' blackness bears no resemblance to their white father. Unsurprisingly, their views are not at the level of the fluellenfam's babies nor those of other lighter-colored interracial children, but one of Jamie's March 2022 TikToks does surprisingly well. In it, his daughter is holding a Baby Alive brand doll. With bright blonde hair, hyper-saturated blue eyes, and an uncanny cuteness, this doll's image was apparently captured as the central subject of the video and favored in opposition to the darker-featured human child moving and talking in-frame.[33] With a lighter-colored (fake) girl child at the center of its capture, the TikTok was then pushed through the algorithm

FIGURE 15.8
Jamie and Nikki's daughter, Zoe, with her doll (2022).

with a much greater reach than Jamie and Nikki's human daughters had ever been able to produce.

The "Cyborg cute" aesthetic exposits how, in opposition to the post-racial fantasy championed by interracial internet spaces, predilections towards whiteness are so hardwired into internet practices that the mixed-race child's expected appearance has metamorphosed into a phenotypic monolith. Lighter features are imbued with the innate capacity for a "Goodness" afforded by colorist tendencies, all with the hopes of virality, algorithmic success, and adjunct economic wealth. Anything less is unacceptable, and this monolith's greatest disciples have begun to show signs of succumbing to the "delusions" of this chimera. The effects of this practice on filtered subjects are troubling, to say the very least. Across image and video-based social media sites from Instagram, to Snapchat, to TikTok, the distorted application of AR filters in tandem with their effect on beautification standards have started to pose troubling effects on the human psyche (these negative effects are elevated for young adults, notably late-teenage girls).[34]

This said, studies on AR's negative impact have so far taken place within communities who came of age with social media filters. These subjects, to a degree, are cognizant of the disjuncture between their IRL appearance and the filtered alternate. There was a life before they were held to the standard of their image's filtration. Thus, what does it mean to have been *born* into beauty and appearance standards that are not only difficult to attain, but simply aren't real? For mixed-race youth specifically, what does it mean to have an aesthetic anticipation of non-reality (AR distortion) and to also act as a conceptual figure for societal non-reality (post-raciality)? As society witnesses the childhood and pubescence of generation α, born after the advent of filter technology, questions of new human ontologies must be considered alongside their congenital exposure to anamorphic expectations of the infantile and child self, all of which are mediated through cybernetic feedback loops and digitally manipulated mirages. My argument has taken a small subset of one social media to show how digital manipulation, especially intense for mixed-race children, creates a cyborg-like amalgamation of organic and artificial features. Secondly, it explores how this phenomenon fuels the attention economy, potentially leading to material gains for families, and a re-structuring of the economy in tandem with the child image. Finally, it has uncovered the material consequences of this feedback loop, delving into how digital manipulation leads to unsettling corporeal expectations. The aesthetics of URL have seeped into the IRL. Science, futurology, and speculative fiction anticipate the arrival of the human-cyborg. A quick scroll through TikTok shows that they might already be here.

NOTES

1. Cannon, *School Dance*

2. Henceforth, all linguistic variations and grammatical errors in-quote are verbatim to the source material.

3. williams. "School Dance Prison Visit (PLEASE Subscribe)" (YouTube, 2022).

4. @stormangel611 (TikTok, 2022).

5. @__madssss (TikTok, 2022).

6. Nakamura and Chow-White, eds., *Race after the Internet,* 7.

7. *Loving v. The State of Virginia* (June 12, 1967), was a Supreme Court case that allowed federal interracial marriage. Celebrations of its anniversary lend to increased uploads and posts from #Swirl couples, typically under the metadata hashtags #swirl and #LovingDay. The court case's name has also become synonymous with the #swirl community. A tweet from @trestiffany11 states, "If you watch a Loving v Virginia video for longer than 3 seconds, TikTok will flood your feed with them."

8. Consider Liisa Malkki's assertion that "[children] are central to widely circulating representations of 'humanity' that are foundational to the whole affective and semiotic apparatus of concern and compassion for 'the human,'" rendering them ideal figures for the utopic, post-racial phantasm. Malkki, "3. Figurations of the Human. Children, Humanity, and the Infantilization of Peace."

9. If these media gain users' attention, they can include advertisements and partnership icons on their hosting site.

10. Attention should be given to Donna Haraway's seminal text, "A Cyborg Manifesto" (1985), which curiously does not discuss children. I find that this is because Haraway's sense of the *cyborg* exists outside of a purely corporeal genealogy; all humans are cyborgs born of systems as opposed to created in-utero or via legal procedures such as adoption. This understanding of cyborgs as innate also helps transcend gender-sex binaries, as well as the physical contours (and violences) of categories such as race. For Haraway's work, the parent/child binary collapses since the child as standalone category, defined by the ideologies and standards of global politics, tends towards heteronormativity. With Haraway's text in mind, this piece similarly points to the *cyborgian* as one that questions structural impacts (such as capitalism and racism) on the human. However, it does so by measuring those effects aesthetically in a peculiar contingent of the internet's image economy—children.

11. Pateezy (@Pateezy_). Twitter, 2022.

12. Zhang and Liu, "A Commentary of TikTok Recommendation Algorithms in MIT Technology Review 2021."

13. Google Developers. "Collaborative Filtering | Recommendation Systems."

14. TikTok, "How TikTok Recommends Videos #ForYou."

15. Lin, "Inside TikTok's Algorithm: A WSJ Video Investigation"

16. Although it is a bit outside of the scope of this chapter, I cannot ignore the massive presence and popularity of Asian children on the app, especially Asian children who are seen engaging in "cute" and infantile activities in tandem with Asian cultural practices. These child images are largely housed within #asianbaby or #kawaii metadata, even if the child isn't of Japanese descent.

17. Ngai, "The Cuteness of the Avant-Garde," 813.

18. Kung, "What Internet Outrage Reveals about Race and TikTok's Algorithm."

19. Hodgson, "Childhood of the Race," 39.

20. BD Editors, "Dominant Trait."

21. Critical_Ad5778, "Anyone else get stared at in public?"

22. @Zein77777 under video by Jamie Perkins, "ZOE IS BECOMING SO CHEEKY!"

23. Musolf, "Why We Stare, Even When We Don't Want To."

24. Javornik et al., "'What Lies behind the Filter?'"

25. Ryan-Mosely, "How Digital Beauty Filters Perpetuate Colorism."

26. May-Chahal et al., "Safeguarding Cyborg Childhoods."

27. Lammel, "Les 'Cyborg Child'."

28. @yatziriallison. 2022. TitTok, n.d. Although the original video was deleted, it was "stitched" by @a.leg.sis on July 17, 2022. "Stitching" refers to the act of a TikTok user reacting to a source TikTok by creating a split-screen TikTok (split between the original TikTok on the right and the reaction TikTok on the left). The original video is preserved by the stitcher's account, even if the source account deletes. A.leg.sis's TikTok stitch expresses dismay and disapproval at the original video's sentiments. Comments are pulled from A.leg.sis's stitch.

29. @andrenickatina. 2022 (TikTok, 2022).

30. "AVA SARAH IS HERE! | Birth Vlog—Dailymotion Video" (Dailymotion, 2016). The original video uploaded by Jamie and Nikki Perkins has been deleted from their YouTube channel. The video has since been downloaded and re-uploaded to video-hosting site *Dailymotion* by an anonymous source.

31. @chef_bambit (TikTok, 2022).

32. @itsme_abdijabar (TikTok, 2022).

33. @jamieperkinsfilms (TikTok, 2022).

34. Condé Nast, "Social Media Filters Are Changing How Young People See Themselves."

BIBLIOGRAPHY

@andrenickatina. "Mixed w/ delusion." TikTok, July 17, 2022.

@chef_bambit. "my brain thought Zoey was somali." TikTok, February 25, 2022.

@itsme_abdijabar. "where they from." TikTok, March 24, 2022.

@jamieperkinsfilms. "Zoe insisted I post this and that her version was better than mine #kidsoftiktok #kimkardashian #fyp #canteatthis #babyalivedoll." TikTok, March 27, 2022.

@__madssss. "omg I love her blonde hair and blue eyes she is so beautiful." TikTok, June 3, 2022.

@stormangel611. "[3 laughing emojis] staaaaaap [1 laugh emoji] each one got lighter tho [2 laugh emojis] beautiful family [5 smiley emojis surrounded by hearts]." TikTok, December 15, 2022.

@trestiffany11. "If you watch a Loving v Virginia video for longer than 3 seconds, TikTok will flood your feed with them." TikTok.

@yatziriallison. TitTok, N.d.

"AVA SARAH IS HERE! | Birth Vlog—Dailymotion Video." 2016. *Dailymotion*. March 23, 2016. https://www.dailymotion.com/video/x3zozz5.

Cannon, Nick, dir. *School Dance*. Los Angeles: N'Credible Entertainment, 2014.

Condé Nast. "Social Media Filters Are Changing How Young People See Themselves." *Teen Vogue*. January 25, 2022. https://www.teenvogue.com/story/social-media-filters-how-young-people-see-themselves.

Critical_Ad5778. "Anyone else get stared at in public?" *Reddit*, March 8, 2022. https://www.reddit.com/r/mixedrace/comments/t9gh6d/anyone_else_get_stared_at_in_public/.

BD Editors. "Dominant Trait." *Biology Dictionary*. January 13, 2017. https://biologydictionary.net/dominant-trait/.

"Collaborative Filtering | Recommendation Systems." n.d. *Google Developers*. https://developers.google.com/machine-learning/recommendation/collaborative/basics.

Haraway, Donna. "A Cyborg Manifesto: Science, Technology, and Socialist-Feminism in the Late Twentieth Century." In *Simians, Cyborgs, and Women: The Reinvention of Nature*, 149–181. New York: Routledge, 1991.

Hodgson, Lucia. "Childhood of the Race: A Critical Race Theory Intervention into Childhood Studies." In *The Children's Table*, 38–48. University of Georgia Press, 2013.

Javornik, Ana, Ben Marder, Jennifer Brannon Barhorst, Graeme McLean, Yvonne Rogers, Paul Marshall, and Luk Warlop. "'What Lies behind the Filter?' Uncovering the Motivations for Using Augmented Reality (AR) Face Filters on Social Media and Their Effect on Well-Being." *Computers in Human Behavior* 128 (March 2022): 107–126.

Kung, Jess. "What Internet Outrage Reveals about Race and TikTok's Algorithm." *Georgia Public Broadcasting*, 2014. https://www.gpb.org/news/code-switch/2022/02/14/what-internet-outrage-reveals-about-race-and-tiktoks-algorithm.

Lammel, Annamária. "Les 'cyborg child': Les effets des cyber technologies sur le développement humain." *Champ Psychosomatique* 22, no. 2 (2001): 51–69.

Lin, Liza. "Inside TikTok's Algorithm: A WSJ Video Investigation." *Wall Street Journal*, July 21, 2021, sec. Tech. https://www.wsj.com/articles/tiktok-algorithm-video-investigation-11626877477.

Malkki, Liisa H. *The Need to Help: The Domestic Arts of International Humanitarianism*.. Durham, NC: Duke University Press, 2020.

May-Chahal, C., C. Mason, A. Rashid, J. Walkerdine, P. Rayson, and P. Greenwood. "Safeguarding Cyborg Childhoods: Incorporating the On/Offline Behaviour of Children into Everyday Social Work Practices." *British Journal of Social Work* 44, no. 3 (April 2014): 596–614.

Musolf, Dianne. "Why We Stare, Even When We Don't Want To." *Wired*, May 25, 2009. Accessed September 15, 2022, https://www.wired.com/2009/05/freaks/.

Nakamura, Lisa, and Peter Chow-White, eds. *Race after the Internet*. New York: Routledge, 2012.

Ngai, Sianne. "The Cuteness of the Avant-Garde." *Critical Inquiry* 31, no. 4 (Summer 2005): 811–47.

Pateezy (@Pateezy_). "Your TikTok algorithm most probably knows you better than your closest friend." Twitter, September 29, 2022, 8:17 A.M. https://twitter.com/Pateezy_/status/1575505113644408832.

Perkins, Jamie. "ZOE IS BECOMING SO CHEEKY!" n.d. Www.youtube.com. Accessed January 15, 2023. https://www.youtube.com/watch?v=t3zr5kkXbQ4&ab_channel=JamiePerkins.

Ryan-Mosely, Tate. "How Digital Beauty Filters Perpetuate Colorism." *MIT Technology Review*. August 15, 2021. https://www.technologyreview.com/2021/08/15/1031804/digital-beauty-filters-photoshop-photo-editing-colorism-racism/.

TikTok. "How TikTok Recommends Videos #ForYou." *Newsroom | TikTok*. August 16, 2019. https://newsroom.tiktok.com/en-us/how-tiktok-recommends-videos-for-you/.

Williams, Eman. "School Dance Prison Visit (PLEASE Subscribe) Https://Youtu.be/NAn2Ai6eprw." n.d. YouTube. Accessed September 15, 2022. https://www.youtube.com/watch?v=QwezOnucNVI&t=65s&ab_channel=emanwilliams.

Zhang, Min, and Yiqun Liu. "A Commentary of TikTok Recommendation Algorithms in MIT Technology Review 2021." *Fundamental Research* 1, no. 6 (November 2021): 846–47.

16 YELLOW WOMEN'S WHITE FACES IN CHINESE *LITTLE-FRESH* SELFIES

JIANQING CHEN

On February 3, 2021, Guinness World Records announced on its YouTube Channel that the Chinese vlogger and Internet celebrity Li Ziqi, with 14.2 million subscribers, had set the record for "Most Subscribed Chinese Language Channel on YouTube." One month later, Li stepped into the spotlight at the 2020 Weibo Awards Ceremony, receiving the accolade of Sina Weibo's "Person of The Year in Social Media" for her remarkable fame and notability on the Chinese Internet. These awards catapulted Li Ziqi to the zenith of her career as an Internet influencer, which she had started five years prior. With her videos presenting the pastoral beauty of rural China and the simple joys of gourmet living, Li gained domestic recognition as a media content creator in 2016. She soon rose to international fame after she began to post her videos on YouTube, bypassing China's Great Firewall that blocks many Western video sharing and social media platforms. During the COVID-19 pandemic, the many days spent by the global masses cooking at home in quarantine, extended the appeal of Li's videos and made her lifestyle more desirable. *New York Times* food columnist Tejal Rao lauded Li as the "Quarantine Queen" while millions of avid fans on YouTube adored her as their "Chinese Princess."[1]

But this Chinese princess is not one of the idle rich. In her videos, Li Ziqi wears cotton and linen clothing and is constantly at work: chopping wood, harvesting corn, weaving straw baskets to kill time while braising meat, and hunting wild mushrooms in the woods. Lasting no more than fifteen minutes, Li's videos seem short and simple—unmediated even. However, closer scrutiny of these videos reveals how Li Ziqi employs a set

FIGURE 16.1
Li Ziqi. *Peach Blossom* (2020). 19 min. Li Ziqi's whiteface in *little-fresh* portraiture.

of contradictory visual techniques to convey a rich layering of meaning and aesthetic pleasure. Among these visual elements, the close-ups of Li's face that are interspersed throughout the video hold a peculiar fascination and tempt us to pause and consider. In the video *The Life of Tomato*, Li Ziqi's face remains fair and "clean," providing a sharp contrast to her hands, firm, roughened, and inevitably smeared with mud. Bright sunlight hits her face without giving her a sunburn or a tan. Amid flowers and plants, her face remains serene, almost blank, providing a perfect foil for her busy and deft hands. These close-ups interrupt the flow of the video and lend the video a certain feminine glamour that paradoxically enhances its rustic charms. These close-ups of her face explain why she always looks so effortlessly graceful, even when doing heavy manual labor. This is how Li Ziqi is presented as strong and delicate at the same time and her life as a desirable combination of leisure and labor.

These shots of Li's face, and their visual aesthetic, provide a starting point for my analysis of Chinese women's digitally produced whiteface in *little-fresh* portraiture. In this chapter, I contend that Chinese *little-fresh* selfie makers' production of "fair and flawless" faces through beauty apps performs a controversial gendered and racialized ideal which negates the socialist portrayal of women featuring technologically-enhanced rosy cheeks and their assigned gender roles in socialist agricultural modernization

FIGURE 16.2
Li Ziqi's fair face and muddy hands in her self-made video *The Life of Tomato* (2021). 17 minutes, 43 seconds.

and industrialization. Simultaneously confirming and contesting white supremacist logic, the digital production of white face is also symptomatic of the dilemma confronting East Asians when encountering Western skin tone-based racial hierarchies. The investigation of the Chinese case serves not only to fill a theoretical lacuna in feminist theory, critical race studies, and digital media studies by moving beyond the white/black binary; it also expands the critical inquiry of race from the North Atlantic setting to the transnational and cross-cultural context, which is so much needed in investigating today's intensely globalized media environment.

"Little fresh," my translation of the Chinese term *xiao qingxin*, is not Li Ziqi's original creation but a digital selfie-style popular among East Asian women. A quick Google search shows the sheer volume of *little-fresh* selfies, which, when taken en masse, manifests this portraiture's elusive visual style: these images situate the female face (and upper body) within a natural environment of flowers and plants. The woman's face, in either frontal or side view, appears to be fair and flawless, displaying no strong emotion other than a demure smile. The extensive use of desaturated colors, shallow depth of field, backlighting by the sun, and slight overexposure envelop the woman and her natural surroundings in serenity and ethereality.

Paradoxically, *little-fresh* selfies present the woman in an ostensibly "natural" state through the active intervention and intensive operation of digital technology. *Little-fresh* selfies visualize a woman in the natural state not only through the rural scenery surrounding her, but also by making this figure synonymous with the no-makeup makeup or *au naturel* look. However, rather than showing a bare face devoid of makeup, the *little-fresh* selfie presents the fair-skinned female face free of spots, blemishes, pores, and fine lines. In doing so, the *little-fresh* selfie is complicit with the Internet photographic trend on Chinese social media in idealizing "white" as the desirable, natural skin tone of Chinese women. The fair and flawless face is produced—and arguably only producible—through user-friendly manual editing or automatic facial retouching functions enabled by popular beauty apps; by the same token, the poetic crystallization of nature is a techno-aesthetic construct: the fairytale-like natural surroundings, which smack of serenity and ethereality, result from the desaturation of color in such images through the application of digital filters.

Popular beauty apps have radically altered the process of (post-)producing digital photos, thereby naturalizing the retouching of self-images. Touchscreen-based beauty apps, such as Chinese apps Meitu and BeautyCam and their international versions such as AirBrush, not only integrate photo-shooting and editing into one continuous image-making process using the same equipment and app; they also enable manual image editing in an increasingly simplified, nearly intuitive fashion by directly touching the image—to wit, retouching images via direct touch. To retouch a selfie in a

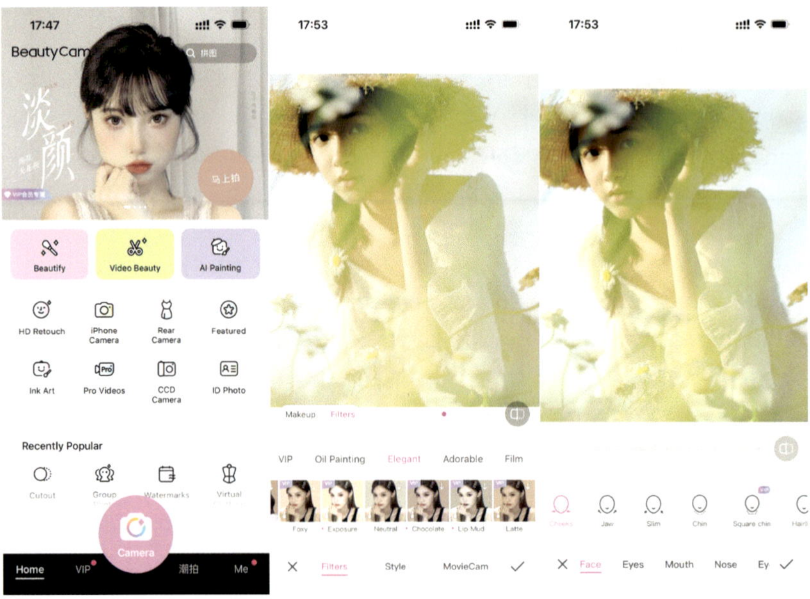

FIGURE 16.3
Screenshots of the interfaces for touchups with BeautyCam, one Chinese touchscreen-based beauty app.

beauty app, selfie makers use pinch-to-zoom gestures to enlarge or shrink the face-image to edit the final product at the granular level. Next, selfie makers use tapping and swiping gestures to remove acne, clean sunspots, and erase dark circles. Finally, the dragging gesture allows the maker to push her jawline upward, raise or lower her cheekbones, or make her face less round. In addition to direct touches, the beauty camera further naturalizes the digitally produced self-image by supplementing, if not supplanting, manual operations with new possibilities of retouching and reshaping facial features automatically. Including multiple beauty filters to carry out predetermined beauty decisions, the automated beauty camera gives the selfie face an instant, standardized makeover—larger eyes, a higher nose, finer and lighter skin—at the swipe of a screen. With these beauty filters, the beauty camera automatizes the alteration of skin tone and facial features *before* the images are recorded, i.e., in the camera's live view. With direct touches and an automated beauty camera, the process of selfie-taking and editing thus enacts consubstantiation of the touch-sensitive interface with the face, (con)fusion between the selfie with the self, and thus naturalization of the digital production of self-image.

Further blurring the distinction between natural and technological, what the beauty app naturalizes is a mediated self-image, a digital mask. In beauty apps, the virtual beauty filter that smooths and lightens skin conflates the photographic filter with the filter used in electronic signal processing, converting the camera accessory—an optical apparatus that selectively transmits light of different wavelengths—to an algorithm that removes noises or undesired components from a signal. In the retouching process, the beauty app treats spots and acne as image noises and uses a smoothing filter (e.g., the Gaussian blur or the bilateral filter) to remove noise data, blur the image, and whiten the face. Beauty filters thus see spots and acne as unsightly nonessential attributes that can be removed just as skin disorders can be cured. By extension, skin colors, as well as phenotypic traits, are no longer regarded as natural signs of race, but the semiotic noise generated by race as a medium, or an intervening substance, that we see through. This "noise" interferes with our perception of human faces and needs to be modified or filtered out. The application of beauty filters thus forms a digital mask that can be put on, played upon, or removed at will. It can be fabricated manually or applied automatically but does not have to be put on the face physically. This digital mask is, then, a palliative, cosmetic mask. Stated differently, it is a technological extension of, or a substitute for, makeup, skincare products, or even plastic surgeries that have already encrusted the Chinese female face with layer after layer of aesthetic expectations.

The digitally produced face in *little-fresh* selfies forms a mode of gendered and racialized self-imaging and self-fashioning that I call "whiteface." This term "whiteface" inevitably resonates with "blackface" (as a performance of "blackness" by white people in exaggerated costumes and make-up in minstrelsy and vaudeville shows) and "yellowface" which refers to the performance of an Asian or Asian-American role by a non-Asian actor, often a Caucasian actor, on American stages and screens.[2] This reverberation is deliberately designed to emphasize the constructed nature and performative aspects of Chinese women's manufacture of idealized whiteface by borrowing skin pigmentation and physical traits not attributed to their ethnic group. However, unlike blackface and yellowface, wherein racial and cultural controversies lie in white people's configuration and embodiment of a black or an Asian countenance, this whiteface's problematic arises from the inverse: as the "yellow race" or non-white Other, Chinese women stylistically construct a white skin or appearance in disregard of their assigned racial category. A simple reversal of the conclusions drawn from previous studies of yellowface, notably the notions of passing as white, would not be insufficient to explain this practice of whiteface. Chinese women's whiteface demands that we rethink the existing discourse while attending to the specific social, historical, and ideological struggles that give rise to this peculiar practice.

At first glance, this digital whiteface reflects Chinese women's notorious yearning for a fair and flawless complexion. The aspiration for bigger eyes, double eyelids, and a straight nose in addition to white skin as manifested in Chinese women's stylization of their faces is the irrefutable evidence of this yearning. Popular media outlets are quick to condemn this obsession with white skin as it reflects Chinese women's internalized white supremacist logic and their refusal to "face up" to their own yellowness. In this reading, whiteface reveals a form of psychological damage that Western imperialism has exerted on East Asian subjects, and in return, reifies the privileges granted to whiteness. Even though this racialized critique is rarely accepted by Chinese women themselves, it is not unwarranted. However, while criticizing Chinese women's creation of digital whiteface as complicit with white supremacy, we can view this whiteface also as a symbolic destabilization, if not an outright refusal, of the skin-color-based racial hierarchy in western racial thinking and its historical racialization of the color "yellow" as a designation for East Asian people. As Michael Keevak's book *Becoming Yellow* shows (an excerpt of which appears in part I), East Asians did not refer to themselves as belonging to a "yellow race" until the late nineteenth century, when Western racial paradigms, along with many other concepts from scientific racism, were imported into China and Japan.[3] In other words, it was nineteenth-century Western race science that colorized East Asians as yellow. Yellow then not only became an "intermediate" shade between white and black, but one that was closely associated with adjectives like ugly, sick, sallow, jaundiced, and dangerous.[4] This racial science, or scientific racism, validated a racial hierarchy that buttressed white supremacy from the outset, and it continues to this day to fuel the Western imagination and characterization of East Asians as physiologically ill, culturally primitive, and racially threatening. Despite the pejorative connotations associated with the color yellow in Western discourse, in Chinese culture, yellow is a color of positive cultural signification (for example, the Yellow Emperor and Yellow River), but is rarely used to describe skin tone.[5] The anti-colonialism movements in early 20th century China transformed yellow from a racial slur in Euro-American traditions to a positive term of Chinese self-identification.

More than a symbolic disruption of the Western skin-color-based racial hierarchy and the assigned yellow race, the digital production of whiteface in the *little-fresh* selfie is also an aesthetic defiance of the racialized figuration of the yellow woman associated with ornamentation. In her theorization of the yellow woman, Anne Anlin Cheng proposes "ornamentalism" as a conceptual framework for understanding Asian femininity's expansive embroilment with the ornamental and Oriental.[6] Collapsing the Oriental logic of ornamentation and the ornamental technologies of Orientalism, ornamentalism describes a process of racialization "whereby personhood is named and conceived through ornamental gestures, which speak through the minute, the sartorial, the prosthetic, and the decorative."[7] Producing incarnations ranging from the iconographic images of Anna May Wong on the Hollywood screen to the resplendent Ming Dynasty vase dress in the Metropolitan Museum of Art's "China: Through the Looking Glass," this Asiatic female ornamentality is ubiquitously deployed in Euro-American visual cultures—so ubiquitous that it goes uncontested, stirring no further critical engagement or public debate than being recognized as a pathology.

It was after the flagrant manifestations of ornamentalism went viral on the Chinese Internet as Dolce & Gabbana ads and fashion photos at the Dior Exhibition in Shanghai that this yellow decorativeness was finally greeted with public outrage. Taken by Chinese fashion photographer Chen Man, the Dior photo features a Chinese woman with sallow, freckled skin and slanted eyes, holding a Dior purse and wearing embellished flower hair accessories and enamel nail guards. It shows the imbrication of a racialized female corporeality and aesthetic thingness, a combination of beauty and savagery. By the same token, wearing a shimmering red sequined dress and elaborate gold jewelry, the Chinese female model in the D&G video displays a sartorial and decorative extravagance. The thick makeup and subdued lighting accentuate her monolid eyes and tint her skin yellow, giving her a stereotypical Chinese look in Western eyes. This ornamental configuration of the Chinese woman becomes blatantly sexist and racist as the woman manages a sheepish smile when she is asked by a ribald male voiceover to pick up and consume Italian food items, e.g., pizza, cannoli, or a bowl of spaghetti, with her chopsticks.

This ornamental schema of the racialized depiction of Chinese women infuriated the Chinese Internet and sparked public outcry. Different from Asian American scholarly efforts, such as Cheng's, to theorize ornamentalism as an alternative ontological condition that produced out the strange embroilment of thingness and personness, the Chinese public vehemently rejected this oriental female objectification as neither ontologically nor indexically related to the "real" lives of Chinese women.[8] The predominance of this ornamental schema not only obscures alternative representations of Chinese women but also allures them into an uncritical reenactment of racial stereotypes imposed by Western racial imagery. *Little-fresh* portraiture thus mounts a stylistic negation of Asian ornamentality that aesthetically configures the yellow woman with natural tropes. These natural tropes, both the digitally produced whiteface as a Chinese women's natural look and a sense of intimacy with her natural surroundings, form an aesthetic strategy of de-ornamentation purporting to reanimate the lived experience of Chinese womanhood.

Beyond the realm of representation, the Chinese woman's digitally produced whiteface is also a response to the technology of photography and film that takes the Caucasian-white face as the norm. Despite their presumed neutrality, photographic media assume, privilege, and construct whiteness.

FIGURE 16.4
Dolce & Gabbana. *Eat with Chopsticks* (2018). 2 minutes, 20 seconds. A Chinese woman, wearing a red sequined dress and elaborate gold jewelry, eats a slice of pizza with chopsticks.

Richard Dyer's study of whiteness and photographic media argues forcefully that film stocks, cameras, and lighting were developed by taking the white face as a standard or a touchstone.[9] According to Dyer, the aesthetic technologies of photography and film are involved in the production of images of whiteness, and whiteness as the norm participated in the orientation of technological innovations in photographic and cinematic history.[10] If this white centricity is so ingrained as to go unnoticed, a glimpse at the so-called "China girl" test image makes visible this white norm.

The China girl, most often a Caucasian woman juxtaposed with a color bar, is a type of quality control measure used to calibrate color standards for film and television. As a test leader, the China Girl is only visible in the theater through a glitch, a momentary breakdown in the cinematic machine. However, these images are subjected to the most careful scrutiny in film

FIGURE 16.5
Dolce & Gabbana advertisement. Photograph by Man Chen (2021). In this advertisement, posted to Dior's Weibo feed and displayed at the Art'N Dior Exhibition at Shanghai, a Chinese woman with sallow, freckled skin and slanted eyes holds a Lady Dior purse.

laboratories since technicians rely on woman's ideal appearance to calibrate the desired exposure and color balance of film reels. Genevieve Yue describes the China girl as "a kind of face kept *outside* the field of vision afforded by the screen" that "speaks to a film's most fundamental conditions of visibility."[11] The "China" modifier in the name refers not to the Chinese woman but the China doll or porcelain-like skin of the model, as well as the colorful embellishment of the test film. The "China girl" test image thus embodies the fundamental ornamentalism, or the uncanny yellowface masquerade, underlying the construction of visual representations of whiteness. As her skin is always white, the "China girl" inscribes the racial and aesthetic bias that presumes whiteness to be the neutral skin tone mediating the technical management of color.

The outcome is that photographic and cinematic apparatuses have worked better for light-skinned people, to the extent that photographing non-white people has presented a problem. If the exposure is correct for a pale white face, that might mean that darker complexions will suffer from the so-called "problem" of underexposure.[12] This problem has beset professional photographers and cinematographers and now troubles Chinese selfie makers. The vibrant online discussions about phone cameras' varying degrees of distortion of Asian flesh tones (e.g., the California-designed iPhone vs the China-made Huawei or Oppo) indicate their savvy about the racial bias embedded in photographic technology. Wrestling with the problem of underexposure from which Asian skin often suffers, practitioners in the Chinese and American film and television industries have developed different technical and aesthetical strategies to compensate for the underexposure of Asian faces, such as using light bounce boards and adding incandescent ring lights.[13]

In a sense, the whitening and smoothing filters used in *little-fresh* selfies could be seen as a continuation and digitalization of lighting strategies developed to illuminate Asian skin. The digitally produced whiteface compensates, or perhaps over-compensates, for the camera eye's racialized gaze which darkens and distorts Chinese faces. Black media scholars' critiques of digital technologies have argued forcefully that today's computer algorithms and facial recognition technologies replicate whiteness normativity and the gendered and racialized codes of looking formulated in scopic regimes to the degree that renders black faces invisible.[14] Sharing a similar concern but espousing a more techno-optimist view, Chinese women embrace digital facial retouching technology as an opportunity to disrupt the color-based racial categorization inscribed into photographic and cinematic apparatuses. Because they use these selfies as training datasets, deep learning-based beauty apps, in close interaction and co-evolution with Chinese selfie makers, seem to promise a reinvention of aesthetic technology capable of recalibrating itself in response to diverse skin tones.

To justify their digital production of whiteface, *little-fresh* portraiture draws inspiration from ancient Chinese literary and artistic traditions to form a culturally specific discourse of whiteface that is not overdetermined by the definitions of whiteness in Western cultures. For *little-fresh* selfie-makers, their search for an interpretation of their desire for white skin from traditional Chinese aesthetics, historically outside the Western framework, is an attempt to achieve a multiculturalist discourse of "white" as a hue and as a skin color. In Western culture, the term "white" carries moral connotations that are explicitly positive, i.e., purity, chastity, spirituality, cleanliness, and so forth. *Little-fresh* interprets whiteface as a longing for an elusive, disappearing quality in Chinese culture that is often deemed undesirable in the West—blandness. This blandness, or *dan* in Chinese, signifies a lack of flavor and an absence of color, evoking an air of inner detachment. Approximate equivalences in English, such as "flavorless" and "colorless," are indelibly imprinted by the negative connotation of something missing. However, the Chinese notion of blandness suggests a subtle admiration for the presence of absence, the realization of greater richness through relative emptiness, and the pursuit of fresh renewal in erasure, as Francois Jullien's monograph elucidates.[15]

This paradoxical understanding is inseparable from the elevation of blandness to a literary and artistic ideal, as well as a moral and political virtue in the Song Dynasty in the eleventh century. Since then, blandness has permeated and enriched the Chinese aesthetic tradition as a desirable quality showing a refusal of excessive formal refinement and rhetorical amplification. In Chinese ink wash painting, blandness not only refers to the painting techniques—less vivid color, diluted ink, intended white space, and light brushstroke, but also characterizes the overall tone of a painting or its atmosphere. In philosophical culture, blandness was also celebrated as a way of life pursuing austerity and plainness: for example, to savor the most unseasoned dish of the least pronounced flavor, to refrain from appreciating rhetorical flourish and sensual avidity.[16] This lifestyle recovers the capacity for inner detachment, with which one is able to set oneself free from the pressures of the external world and the excitement of the sensations. In this sense, *little-fresh* portraiture resurrects blandness in Chinese art and philosophy. As the name of the selfie style, *qingxin,* the Chinese equivalent for "fresh," describes both the cool air and the aesthetic quality of the limpid and unaffected. It thus addresses both the natural motifs of these photographs, and their visual aesthetic featuring a preference for desaturated or pale colors. The pale complexion, dubbed as *danyan* in Chinese (literally "bland face"), not only emulates the ink-wash visual style but also suggests a life of simplicity and plainness.

More than an alternative to Western racial thinking, the *little-fresh* visual style's turn to the aesthetic traditions of blandness in the distant past also constitutes an effort to reinvent the female figures and nature imagery prevalent in the more recent Maoist past. *Little-fresh* female portraiture stages a radical negation of the model female figures constructed by socialist visual culture, as well as their assigned gender roles in socialist modernization. By model female figures, I mean the rosy-cheeked women appearing in color films, propaganda posters, and magazine covers, ranging from iconic rural female farmers and workers in the Great Leap Forward (1958–1961) to Iron Girls, the symbolic heroines of the Cultural Revolution (1966–1976). These female figures represented the glorified participation of women in collectivized agriculture or nascent heavy industry, taking on physically demanding work that were locally defined as male. Initially a name given to a brigade of adolescent girls in the model village of collective agriculture, Dazhai, Iron Girls consolidated the dual theme of breaking gender boundaries and transforming rural/urban distinctions.[17]

Women's liberation from gender and class constraints through high levels of participation in collective farming dovetailed with the state's need for female labor in self-reliant rural development. Enhanced by a saturated color palette, these women figures' pink cheeks, along with their sturdy features and shining eyes, suggest their bodily force and vigor, ready for hard physical labor and revolutionary commitment. Portrayed together with their sister workers, the heroines featured in photos and posters mobilize women to work out of the house and participate in socialist production, placing collective efforts in agricultural work ahead of caring for individual families. They swing a hammer or drive a farm tractor, proudly boasting the specialized skills they have acquired in the course of playing a central role in agricultural mechanization. These female figures in socialist visual

FIGURE 16.6
Dawei Wang. Chinese Socialist Poster. “Women Can Hold Up Half of the Sky” (1975). The rosy-cheeked model female worker is swinging a hammer to chisel stone block. Other women in background are working on the blocks for an aqueduct, under construction in background, lit up by bright lights. A banner on the scaffolding says “In agriculture, learn from Dazhai.”

culture and symbolic system embody a new feminist gender subjectivity in the pursuit of socialist modernity.

As many scholars have claimed, chromatic aesthetics and technologies should be read both as fundamental aspects of an artwork's formal preferences and as important signifiers of its ideological message and political context. The chromatic Cold War, or the mobilization of distinction between Technicolor and Agfacolor as a well-known trope of ideological difference between capitalism and socialism, is emblematic of color's intimate association with political values. Superficial yet pervasive, the Cold War playbook imposed a "colored" binary that often aligned Technicolor's brilliantly saturated hues with Classical Hollywood's glamour, and the sober palette and muted hues imparted by Agfacolor with socialist aesthetics prevalent in the Eastern Bloc. Under this colored binary, the brightness and saturation of Technicolor became a symbol of American gaudiness and dazzling capitalist consumerism, whereas the greyness of Agfacolor gained socialist connotations, inciting a western view of oppression, poverty, and depression behind the Iron Curtain.[18]

Deviating from this simplistic chromatic binary between the drab colors of socialism and the brilliant hues of capitalism, socialist China manifests an obsession with saturated color across an array of media. Seemingly incompatible with socialism's ideological critique of capitalism, socialist China ardently appropriated the capitalist chromatic schema and apparatus and repurposed them in service of socialist ideals. The vivid kaleidoscope of Technicolor was deployed to depict rural landscapes, offering a spectacularized representation of the socialist utopia closely hewing to the idea of building a new, prosperous countryside. The bright saturation of Technicolor also played a pivotal role in the visual construction of the model female figures: by portraying a shining, ruddy face bathed in light, the vibrant colors, notably the shades of red, focus emphasis on the bright heroism of characters, following the aesthetic rule of *hong guang liang* (in English: red, shiny, and bright).[19] In a scene in the peasant melodrama *Young People in Our Village* (Su Li, 1959), Shuzhen, the female protagonist, cuts off her braids and hides her feminine curves in unisex clothing in order to demand gender equality in labor division as well as to show devotion to hydraulic infrastructure construction. The saturated color palette enhances the healthy, youthful rosiness in her cheeks. Empowered with a hammer and other tools, Shuzhen, like other female villagers in socialist cinema and prints, boasts the strength and skill to participate in rapid agricultural industrialization. The socialist redeployment of Technicolor aesthetics thus contributes to the visual construction of these rosy-cheeked model female figures, the embodiment of socialist ideals for the transformation of gender and labor.

The *little-fresh* visual aesthetic, as Li Ziqi's videos exemplify, thus marks a departure from socialist depictions of the idealized form of female beauty. With their subtle use of desaturated colors, Li's videos build a utopia where people can lead a poetic existence in isolation. Inserted into the video sequence, the wide-angle establishing shots, from above or from afar, display the picturesque scenery of her village, such as the quiet pond rippled by a gentle breeze, trees with peach blossoms, and remote mountains shrouded in mist, redolent of the reclusive utopia of the "Peach Blossom Land" by Chinese poet Tao Yuanming. To conjure up a self-reliant way of life, Li Ziqi makes food from scratch, all on her own, using traditional methods. *The Life of Tomato* starts with her chopping wood to build a greenhouse for tomatoes grown in cold weather. However, to romanticize a life of self-sufficiency, the video skillfully uses formal devices to downplay the lengthy process of producing and preparing food, and the enormous physical labor entailed. With time-lapse photography and vertical subtitles that

FIGURE 16.7
Li Su. *Young People in Our Village* (1959, 103 minutes). The rosy-cheeked female protagonist is cutting her hair to show her determination to participate in the construction of hydraulic infrastructures.

mark solar terms in the lunar calendar, the video condenses the vegetative development of the tomatoes over two to three months into less than twenty minutes of footage. The closeups of her clean, alabaster face lend an air of effortlessness. Her face is smooth, quiet, and free from shine—the mirror effect of sweat. The signifier of Li's video thus not only presents the unpresentable extended intervals between sprouting, flowering, and fruiting, but also alleviates the laborious efforts of raising crops.

By virtue of the *little-fresh* aesthetic, Li Ziqi's food vlogs link together viewing videos, making and eating food, and working in the field in a shared logic of pleasure. They thereby promote an ideal, self-reliant life in rural China. Unlike the self-sufficient national economy once advocated in the socialist era, *little-fresh* visual aesthetics espouses a completely different idea of self and self-sufficiency—self-sufficiency on the individual level. From anonymous selfies to Li Ziqi's popular vlogs, these images show a female subject in solitude, suggesting her independence from industrial commodities, social interaction, or even amorous relationships. This romanticized self-sufficiency excites a re-enchantment with country life, ushering a return or relocation to the countryside by middle-class urban dwellers after the mass exodus to the cities over the past three decades.

In conclusion, Chinese women's seemingly simple yet highly ambivalent digital whiteface in *little-fresh* portraiture mounts a double resistance to the Western imaginary of the yellow woman and the socialist portrayal of the rosy-cheeked female ideal that needs to be examined at the intersection of gender, race, visual aesthetics, and chromatic technology on a historical and global scale. This double negation not only reflects the dilemma that whiteness as a racial category poses to East Asians but also epitomizes the struggles of contemporary Chinese to differentiate themselves from both their socialist past and the Western imaginary to shape their new image in the global mediascape retouched by digital technologies.

NOTES

1. Rao, "The Reclusive Food Celebrity Li Ziqi is My Quarantine Queen."
2. Fuller, *Hollywood Goes Oriental*, 7–8.
3. Keevak, *Becoming Yellow*, 1–9.
4. Keevak, *Becoming Yellow*, 36.
5. Keevak, *Becoming Yellow*, 7.

6. Cheng, *Ornamentalism,* 3. An excerpt of Cheng's article "Ornamentalism: A Feminist Theory for the Yellow Woman" can be found in part I of this volume.

7. Cheng, *Ornamentalism,*18.

8. Cheng, *Ornamentalism,* 3.

9. Dyer, *White,* 93. An excerpt of *White* can be found in part II of this volume.

10. Dyer, *White,* 89.

11. Yue, "The China Girl on the Margins of Film." This essay is also reproduced in part II.

12. Yue, "The China Girl on the Margins of Film," 89.

13. See Sung, "Reading the Light Right."

14. See Benjamin, *Race after Technology,* an excerpt of which can be found in part III of this volume; Browne, "Digital Epidermalization: Race, Identity, and Biometrics."

15. Jullien. *In Praise of Blandness,* 103–116.

16. Jullien. *In Praise of Blandness,* 67.

17. Wang, *Finding Women in the State,* 221–228.

18. Fehérváry, *Politics in Color and Concrete,* 1–26.

19. Dootson and Zhaoyu, "Did Madame Mao Dream in Technicolor?"

BIBLIOGRAPHY

Benjamin, Ruha. *Race after Technology: Abolitionist Tools of the New Jim Code.* Cambridge, UK: Polity, 2019.

Browne, Simone. "Digital Epidermalization: Race, Identity, and Biometrics." *Critical Sociology* 36, no. 1 (January 2010): 131–150.

Cheng, Anne Anlin. *Ornamentalism.* Oxford: Oxford University Press, 2019.

Dootson, Kristy Sinclair, and Zhu Zhaoyu. "Did Madame Mao Dream in Technicolor? Rethinking Cold War Color Cinema through Technicolor's 'Chinese Copy.'" *Screen* 61, no. 3 (Autumn 2020): 343–367.

Dyer, Richard. *White: Essays on Race and Culture.* New York: Routledge, 2013.

Fehérváry, Krisztina. *Politics in Color and Concrete: Socialist Materialities and the Middle Class in Hungary.* Bloomington: Indiana University Press, 2013.

Fuller, Karla Rae. *Hollywood Goes Oriental: CaucAsian Performance in American Film.* Detroit: Wayne State University Press, 2010.

Jullien, François. *In Praise of Blandness: Proceeding from Chinese Thought and Aesthetics.* New York: Zone Books, 2004.

Keevak, Michael. *Becoming Yellow: A Short History of Racial Thinking.* Princeton, NJ: Princeton University Press, 2011.

Rao, Tejal. "The Reclusive Food Celebrity Li Ziqi is My Quarantine Queen." *The New York Times,* April 22, 2020. https://www.nytimes.com/2020/04/22/dining/li-ziqi-chinese-food.html.

Sung, Yu-Lun. "Reading the Light Right: The Exposure of Asian Skin Tones in Cinematography." *Frames Cinema Journal,* no. 17 (June 27, 2020). https://framescinemajournal.com/article/reading-the-light-right-the-exposure-of-asian-skin-tones-in-cinematography/.

Wang, Zheng. *Finding Women in the State: A Socialist Feminist Revolution in the People's Republic of China, 1949–1964.* Berkeley, CA: University of California Press, 2017.

Yue, Genevieve. "The China Girl on the Margins of Film." *October* 153 (July 2015): 96–116.

17 ALGORITHMIC IMAGES AND RECURSIVE EPIDERMALIZATION

QURAN M. KARRIEM

In the context of digital culture, the film-photograph has been dead for decades, having suffered the same fate its own advent inflicted upon "auratic" images just over a century prior. The film-photograph's successor, the photorealistic *algorithmic* image, trades on its social capital without honoring the barest premises under which its capital has been accumulated. Walter Benjamin conceived reproducibility in a dialectical relationship with aura, originality, and "placedness." Positioning painting and sculpture as paradigmatically auratic forms, he asserted that "the masses'" relationship to art emerged only in the age of photography. From there, he theorized that, in contrast to the ritual modes of reception associated with singular art objects in museums, reproducible artworks resemble architecture in that we live "in" them, and in that they influence our habits.[1]

In this chapter, I apply this concept of the photographic image's dissolution into social architecture to the advent of photorealistic images produced using generative artificial intelligence and machine learning techniques. Methodologically, I connect notions of imagistic social architecture derived from Benjamin and Vilém Flusser to Frantz Fanon's discourses on sociogeny and epidermalization. The former refers to the ways linguistic and cultural norms come to be experienced as "instinct" to such a degree that they can rewire the mappings between sense impressions and biochemical reward/punishment pathways in the human body,[2] the latter to how subaltern, dark-skinned, postcolonial subjects can internalize normative cultural assertions of their inferiority. This is an urgent juxtaposition because the memetic, appropriative qualities of algorithmic images bind them more closely to hegemonic ideologies and normative skin color stereotypes than film photography, which maintained a relationship to real events due to its semi-indexical relationship to light rays. In other words, I argue that algorithmic images produced using generative techniques are necessarily more "biased" than the photographs they imitate. Further, they reify race and color stereotypes as a direct result of their grounding in counterfeiting processes, where "believability" replaces "indexicality" as social authentication mechanism.

BACKGROUND: PHOTOGRAPHIC INDEXICALITY AND SOCIAL TRUST

By "photographic indexicality," I refer to the notion that photographs necessarily represent real scenes, situations, or events. Belief in photography's relationship to reality persists, even after the practice of film photography, the source of such beliefs, was largely relegated to nostalgia. In 2008, Ariella Azoulay claimed: "Ever since photography's appearance on the stage of history, any possibility of repudiating what has turned into the self-evidence of photography, or photography as self-evident, has been drastically curtailed."[3] Photography's self-evidence is grounded in the understanding that it represents a "neutral" or "objective" transcription of light rays, even after granting that photographers compose scenes. For Azoulay, this "self-evidence" has posed challenges to official state narratives to such a degree that it has created something akin to a global citizenry. These challenges are symbolized in the photograph's ability to evoke affect and response in addressees—as during various exposures of state atrocities, which contributed to the gradual development of global human rights discourse. For example, photographic evidence of Belgium's systemic acts of mutilation against black Congolese subjects under Leopold II, brutal even by the standards of nineteenth-century European colonial enterprises in Africa, were an early example of the medium's capacity to cultivate international pressure against the racializing prerogatives of state power. Of course, this claim to indexicality has been used not only to check the state, but to legitimate it. As Susan Schuppli observes, paraphrasing Lorraine Daston and Peter Galison, photography initiated "a radical transformation in evidence law . . . contributing to what has been referred to as an episteme of mechanical objectivity."[4] That is, "self-evidence" by way of mechanical capture had profound effects on the adjudication of truth claims in cultural and juridical domains alike.

Whether used to challenge or support constituted power, however, photographic indexicality has never been absolute. As "experts began to face each other across the courtroom,"[5] Schuppli notes, it became clear that even film-photography's "neutral" transcription of light rays would always remain subject to selection, interpretation, and debate. The Czech-Brazilian philosopher Vilém Flusser has rejected the idea of "self-evidence" entirely, warning that the "lack of criticism of technical images is potentially dangerous," precisely because "they appear to be on the same level of reality as their significance."[6] He illustrates this viewpoint by comparing black-and-white to early color photography, drawing attention to the selection of certain color and chemical theories over others in the standardization of photographic protocol: "Black-and-white photographs are more concrete and in this sense more true: They reveal their theoretical origin more clearly, and vice versa: The "more genuine" the colours of the photograph become, the more untruthful they are, the more they conceal their theoretical origin."[7]

In this account, Flusser offers a convincing portrayal of artifice that conceals its status by cleaving to what *seems* perceptually natural—where the black-and-white photograph is an obvious distortion that reveals a mediating process, the color photograph employs a more complex process that implies naturalism. This formulation paradoxically suggests that, as the "surface" image more closely resembles reality, the underlying logics of simulacra become further removed from it.

Work by Adam Broomberg and Oliver Chanarin reinforced the notion of the artifice that underlies photography and highlights its historical relationship to state-directed processes of racialization. For their 2012 exhibit "To Photograph the Details of a Dark Horse in Low Light," the artists used decades-old film stock and cameras to examine "the radical notion that prejudice might be inherent in the medium of photography itself."[8] The work drew attention to color photography's historical conditioning on the representation of lighter skin tones to such a degree that darker tones were often "rendered invisible." Reporting on the exhibit, David Smith wrote:

> The light range was so narrow, Broomberg said, that "if you exposed film for a white kid, the black kid sitting next to him would be rendered invisible except for the whites of his eyes and teeth." It was only when Kodak's two biggest clients—the confectionary and furniture industries—complained that dark chocolate and dark furniture were losing out that it came up with a solution.[9]

FIGURE 17.1
Adam Broomberg and Oliver Chanarin, *Everything Was Beautiful and Nothing Hurt, Installation View*, FoMu, Antwerp (2014). Image courtesy of the late estate of Broomberg and Chanarin.

FIGURE 17.2
Adam Broomberg and Oliver Chanarin, *To Photograph the Details of a Dark Horse in Low Light*. Installation View, Paradise Row (2012). Image courtesy of the late estate of Broomberg and Chanarin.

FIGURE 17.3
Polaroid ID-2 Camera with "Boost Button." Image courtesy of Cold Electric Wares.

Broomberg and Chanarin evoked the 1970s efforts of African American Polaroid employees-turned-human-rights-activists Caroline Hunter and Ken Williams. The pair formed the Polaroid Revolutionary Workers Movement and led a boycott of the corporation, after Hunter uncovered its complicity in the South African apartheid government's efforts to monitor and control the movement of its black population via its "passbook" system.[10] Noting how "Polaroid built an efficient tool for South Africa's apartheid regime to photograph and police black people" with its ID-2 camera, Broomberg and Chanarin discussed the camera's "boost button," a feature that intensified the brightness of the on-board flash, ostensibly to help the regime produce identifying materials for dark-skinned subjects that were previously "rendered invisible."[11]

Figures 17.1–17.3 show that photographs have never been *quite* as "self-evident" as their cultural pedigree and juridical esteem would suggest. Of course, film photography does have some legitimate truth claims. Photographic indexicality, to the extent that it existed, was a result of light-reactant properties of silver halide that preceded camera applications. Polaroid and Kodak could experiment with exposure and color reproduction according to their purposes, but photography's grounding in light sensitivity undergirded its range of possibility and afforded it a high degree of social trust. To summarize, photography has had an ambivalent relation to the humanization and inclusion of dark-skinned subjects, contributing to both the articulation of a universal notion of human rights, and to racialized systems of surveillance and control. These opposing instrumental uses are each grounded in the medium's claim to indexicality.

ALGORITHMIC IMAGES AND THE GENERAL REFERENT

In contrast to photographic imagery, I use the term "algorithmic" to describe photorealistic images produced via algorithmic digital processing rather than direct sampling. Photography, whether captured on film or Secure Digital (SD) card, is fundamentally an act of sampling. To the extent that digital cameras share the indexical qualities of their filmic predecessors, they replace the halide reaction with photovoltaic image sensors: the relationship between light and electron flows is instrumentalized to create electric charges in proportion to light intensity at a given pixel. An analog-to-digital converter samples these voltages and writes them as bits on a hard drive. Once this writing occurs, we are already dealing with a computational simulation. In the shift from digital photograph to algorithmic image, it is only the goal of this simulation that has changed. Early digital cameras and photovoltaics simulated the kind of indexicality that arose from film cameras as closely as possible. With the advent of photorealistic algorithmic images, however, it is as though the practice of digital imaging has embraced the latent mutability associated with its material underpinnings. Algorithmic images display none of the medial qualities upon which the photograph's cultural-juridical pedigree rests. The properties of electrons, as instrumentalized on digital storage devices, are neither fixed nor light-dependent. Lenses, shutters, and photovoltaic sensors have become optional, rather than constitutive, elements in the production of "photorealistic" digital images.

In the context of these images, I examine what I term the *general referent*—the "average" of image data resulting from the labor, attention, and aesthetic values of a mass of photographers, editors and photographic subjects, which is harnessed using machine learning techniques to produce new "originals" that complicate already tenuous notions of "placedness" and indexicality. Where photographs differ from older, "auratic" works (such as painting) in that the position of the observer is placeless, algorithmic images take an additional step in that they detach *depiction* from place. The efforts and visages of these masses are de-individuated to portray general aesthetic tendencies as specific occurrences—non-subjects that appear to be subjects.

The "new" images produced thereafter are conditioned by the general referent. Such depictions lose the last vestiges of their predecessors' moorings to the physical real, becoming utopic. As I will demonstrate, the ontogenetic difference between algorithmic images and photographs lies in the fact of the former's conception *first and foremost* in falsification. That is, while film and digital photographs have always been manipulable, their fundamental utility lies in the fact that they are rooted in light-reactant processes and are therefore accepted as sampling or measurement techniques with *some* claim to the historical real. The algorithmic image I describe below has the exact opposite function: it is fundamentally conceived as a counterfeit, designed from the first to *imitate* the properties we associate with photographs. The most likely and efficient use of algorithmic images, then, is to exploit the lingering cultural belief in photographic indexicality.

The political utility and paranoia surrounding deepfakes is well-documented.[12] However, their relation to racialization and skin color is under-examined, especially insofar as media philosophical perspectives and approaches are concerned. The era of racist cartooning has returned, only this time it is attended by three powerful forms of "objectivity" that make it comparatively difficult to challenge: the ghost of photographic indexicality, the quantitative veneer of data science, and the perceived lack of identifiable image producers to hold accountable. In its coupling of imitation and imagination, computational photograph falsification must necessarily be preoccupied with the *should*. My argument, then, is that algorithmic images are predisposed toward the propagation and reification of racial stereotypes. Regardless of the good or ill intent of the organizations involved in steering these technologies, I estimate that this will be one of their primary cultural effects in the absence of "de-biasing" approaches that, at minimum, will require regular "systematic auditing"[13] of models and human curation of the datasets used to train them.[14]

GANS AS COUNTERFEITING MACHINES

> . . . just as economies are destabilized by counterfeit coinage, so the practices of textual production and exchange through which subjects construct their understandings of the world are subverted by the surreptitious interjection of signifying tokens of uncertain value.[15]
>
> —W. J. T. Mitchell, *The Reconfigured Eye: Visual Truth in the Post-Photographic Era*

The basic idea behind machine learning is to find a function that adequately describes a given dataset and can thereafter be used to classify, make predictions based on, or generate similar kinds of data outside that set. Since they are used to classify unknowns, predict futures, or generate believable counterfeits, it is possible to *overfit* a model such that there is minimal loss regarding the training data, but stagnant or increasing loss in the validation data, which is used to measure a model's performance on data outside the training set. Overfitting means that the function a model discovers describes the training data perfectly (or close to it) but won't perform well on new data. In the case of a generative image model, this might mean that output images simply mimic photographs from the training set—near-direct reproductions rather than believable "originals."

A "Generative Adversarial Network" (GAN) is a deep learning method for training machine learning models to produce believable facsimiles of examples in the training set. The technique was famously used to create photorealistic "portraits" of people who never existed, trained from a set of photographs of real people. The GAN technique pits two separate neural networks against one another: a "generator" and a "discriminator." During training, the generator attempts to produce algorithmic images that the discriminator fails to distinguish from the digital photographs contained in the training set. This process increases the sophistication of the generator's output to such a degree that human beings may also have difficulty distinguishing photographs of real people from deepfakes.

The engineers most directly involved in the innovation behind the GAN technique liken the generative model to "a team of counterfeiters" and the discriminative model to "the police, trying to detect the counterfeit currency."[16] If we extend W. J. T. Mitchell's metaphors on the image as currency, we might liken such saturation to currency inflation. The sheer volume and speed of image production means that we are unable to fully process either the significance or validity of any single image, and are left with an affective, rather than analytical image culture. Saturation and counterfeiting processes are thus close associates that allow us to filter our attentions to images and interpretations that already conform to our habits. In 2016, Ian Goodfellow, the researcher most often associated with the technique, displayed sensitivity to this concern. In a tutorial written to address common questions about the method and facilitate adoption, he admits that "when applied to images, such models seem to merely provide more images, and the world has no shortage of images."[17] Yet he subsequently answers the question "Why study generative modeling?" by highlighting several concurrent developments in image synthesis, such as video frame prediction and image superresolution.

From its introduction, we can see the that entire objective of the algorithmic image is to fool us "into thinking [it] is a photograph."[18] This represents a clear difference from the inscription techniques that form the basis of photography, which Louis Daguerre viewed as "an instrument which . . . gives [Nature] the power to reproduce herself."[19] Though falsification and saturation have long fueled contests over the interpretation of photographs, my reading of the algorithmic finds a distinct image category, fundamentally grounded in believability, imagination, and ideology rather than in the capture of light rays. Despite its discursive association with mechanical objectivity, the algorithmic image constitutes a more durable and renewable source of skin color and race-based stereotyping than the photograph.

RECURSIVE EPIDERMALIZATION AND WHITE UTOPIANISM

> In the white world, the man of color encounters difficulties in elaborating his body schema. The image of one's body is solely negating. It's an image in the third person.[20]
>
> —Frantz Fanon, *Black Skin, White Masks*

From Fanon's work, we understand "epidermalization" to refer to the psychosomatic internalization of inferiority based on skin pigmentation by a member of a subaltern/colonized group, and "epidermal racial schema" as describing the way epidermalization affects the range of bodily action, agency, and freedom experienced by the subject: ". . . the body schema, attacked in several places, collapsed, giving way to an epidermal racial schema."[21] According to Fanon, this internalization of inferiority restricts the physical-material scope of action for the epidermalized subject. Epidermalization is the means through which historical social patterns of exclusion and subalterity premised on skin color—on my read, a historically specific manifestation of the broader notion of sociogeny—come to affect conscious and subconscious agency in the individual. In the attempt to dissociate from the confines of pervasive harmful stereotypes, the subaltern subject often poses conscious or subconscious restrictions on the self. Pervasive prejudices against dark skin create a situation where the black subject is "overdetermined from the outside", living "in fear that their acts will correspond to" stereotypes.[22] For example, Afro-British food writer Melissa Thompson connected racist depictions in D. W. Griffith's 1915 film *The Birth of a Nation* with the reason "some African American friends . . . refused to be seen eating fried chicken on camera" for a Netflix cooking show.[23] Acts once taken for granted by the "black" subaltern are subject to the scrutiny of an internalized "white" colonial inner voice. Insofar as this phenomenon concerns algorithmic images, we have a situation where it potentially becomes recursive. The interpellative aspect of the refrain "Look! A Negro!" from *Black Skin, White Masks* takes on new dimensions when fictive, photorealistic people of color primarily exist within the confines of a range of gestures, poses, settings, and occupations that are believable from, palatable to, and *generated by* the "average" perspective of "average" (mostly white, mostly male) internet users in the Trump era.[24]

Flusser defined *utopia* as a complete abstraction or removal from physical reality, adding that it can be either "positive" or "negative": "*Utopia* means groundlessness, the absence of a point of reference."[25] To avoid the negative, a "fascistic society, centrally controlled by senders,"[26] he envisioned the positive utopia as a mode of "dialogic" image production that immerses humanity into "an artificial life in art, life as play with pictures and sounds."[27] The symbiotic rise of centralized social media and user-generated content in the Web 2.0 era demonstrated that, far from existing in a state of diametric opposition, Flusser's notions of positive and negative image utopia were fully compatible with one another. Even so, the broader concept of an illusory utopia powerfully extends the Benjaminian notion of imagistic social architecture. Many researchers and journalists have found that the skin colors depicted by generative text-to-image systems are normatively associated with certain occupations and settings.[28] My argument is that this is largely because algorithmic images must be optimized for "believability" as cognized from hegemonic social perspectives which potentially intensify the imposition of a stratified, majoritarian utopianism on marginal body schemas. Technical utopia is not purely "groundless," as Flusser claimed, but grounded in hegemonic sociogeny.

The unmooring of so-called photorealism from chemical or photovoltaic processes results in a situation whereby harmful skin color stereotypes are not only reified but exacerbated. The algorithmic image must maintain a correlative association with the real by appealing to the believable. Such a shift is massively consequential for those outside or at the margins of the utopia, whose horizons of believability are largely determined by skin color. Changes in the receptivity to photorealism by global publics, habituated to regard it as a self-evident arbiter of truth, history, and memory since the nineteenth century, seem all but certain to lag behind the ability of GANs and GPTs (generative pre-trained transformers) to flood media channels with skin color stereotypes. The leveraging of apparent self-evidence insofar as it relates to the depiction of dark-skinned nonsubjects is likely to further concretize color bias in the collective unconscious. Ultimately, the algorithmic image does not merely *appeal* to the believable through a counterfeit form of self-evidence. It also, left unchecked, potentially changes the very *scope* of the believable, cognizable, and imaginable for large swaths of society.

As a case in point, a 2023 Bloomberg article quantified relationships between skin color, gender, occupation, and criminality in AI-generated

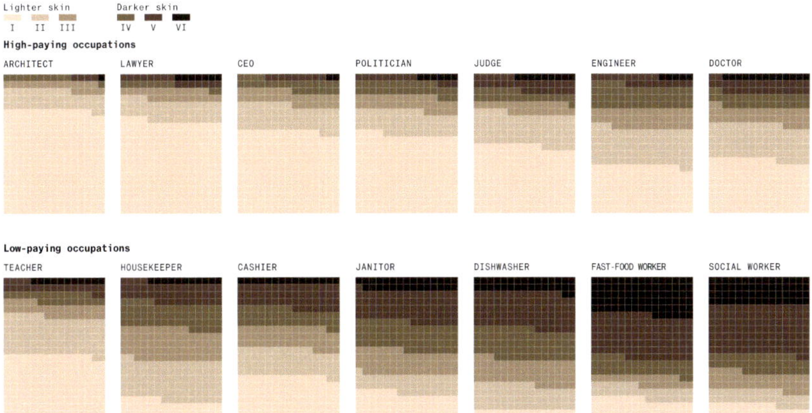

FIGURE 17.4
Skin colors in generated imagery by occupation; Bloomberg analysis of Stable Diffusion. Used with permission of Bloomberg L.P. © 2024. All rights reserved.

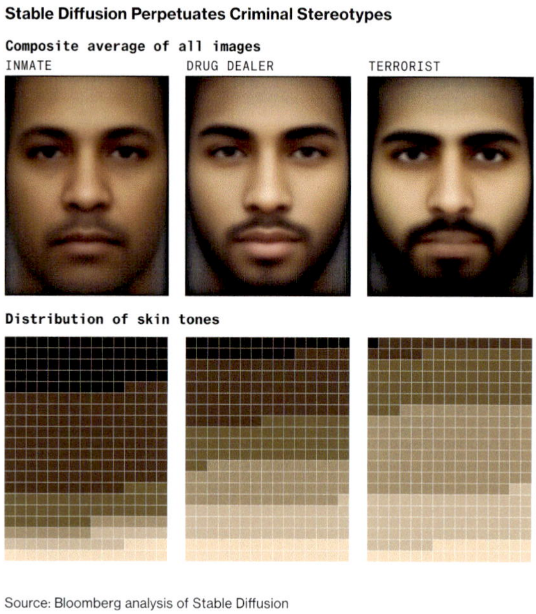

FIGURE 17.5

"Average face" of an "INMATE," a "DRUG DEALER," and a "TERRORIST" generated during a Bloomberg analysis of Stable Diffusion. Used with permission of Bloomberg L.P. © 2024. All rights reserved.

imagery.[29] The study's materials were created using Stable Diffusion, a widely available "text-to-image" model, like those of competitors Dall-E and midjourney.[30] Journalists Leonardo Nicoletti and Dina Bass produced 5,100 images, split into seventeen categories containing three hundred algorithmic images each. Seven categories represented "high-paying" jobs, seven represented "low-paying" jobs, and three represented criminal classifications. Afterward, they analyzed the skin color of the generated faces using the Fitzpatrick phototyping scale, a dermatological classification system that separates the range of human skin pigmentation into six types. The study's results, perhaps unsurprisingly, showed that "image sets generated for every high-paying job were dominated by subjects with lighter skin tones." Additionally, Nicoletti and Bass concluded that, according to Stable Diffusion, "men with dark skin commit crimes, while women with dark skin flip burgers." The authors go on to uncover several specific distortions in the ways Stable Diffusion reproduces color bias, a pair of which I've selected below:

> More than 80% of the images generated for the keyword "inmate" were of people with darker skin, even though people of color make up less than half of the US prison population, according to the Federal Bureau of Prisons.
>
> When prompted to generate images of a "terrorist," the model consistently rendered men with dark facial hair, often wearing head coverings—clearly leaning on stereotypes of Muslim men . . . far-right extremists, including White supremacists, committed nearly three times as many [deadly terrorist attacks on US soil since September 11, 2001].[31]

The reporters consistently find that generative AI produces racially stereotypical representations with statistical distributions that are even "worse than reality"—supporting the notion that they are a product of a social imaginary. Christoph Schuhmann, co-founder of LAION, an organization that manages the dataset used to train Stable Diffusion, concurs: "This will give you the average stereotype of what an average person from North America or Europe thinks. You don't need a data science degree to infer this."[32] Schuhmann, however, never comments on the social implications of flooding the internet with stereotyped counterfeits, potentially at an order of magnitude greater than the sum of all images produced using all other methods. Researchers at Europol estimate that generative AI will account for as much 90 percent of all content found on the web by 2026.[33]

Essentially, now that the "floodgates" are open, there is no clear path to providing enough "unbiased data" to retrain models. Because they rely on such a large mass of pre-existing material for their training, model "de-biasing" efforts have proved challenging, with AI researchers attempting to discover systemic de-biasing approaches more optimal than the current "whac-a-mole dilemma"—where the mitigation of one bias creates others that may or may not be detected.[34] Post-hoc de-biasing efforts are likely to merely conceal the most visible manifestations of the algorithmic propagation of stereotypes. Billions of questionably labeled (if not overtly racist, sexist, ableist, homophobic, transphobic, or pornographic) text-image pairs are the foundational template for trillions or quadrillions of potential algorithmic images. Even assuming substantial progress is made regarding broad cultural understanding of the damage caused by harmful skin color stereotypes at mass scale, internet culture potentially becomes mired in a non-terminating loop.

Due to the scale and speed with which they produce images, generative techniques operate closer to the realm of collective imagination—or, at least a snapshot of it as manifested in the early 2020s internet—than was possible with any previous imaging technology. And, as Achille Mbembe notes, imagination stands at the core of the social process of racialization:

> For it to operate as affect, impulse, and speculum, race must become image, form, surface, figure, and—especially—a structure of the imagination. And it is as a structure of the imagination that it escapes the limitations of the concrete, of what is sensed, of the finite, even as it participates within and manifests itself most immediately through the senses.[35]

Because wholly fictive constructions of Otherness and inferiority can now be given photorealistic form, it is as though Mbembe's distinction between imaginative projection and sensory immediacy in the process of racialization has collapsed. Hegemonic racial imaginaries are no longer restricted to the utterances of avowed racists, the habitual lapses in the performativity of white liberal guilt, or secret aspirations to post-racial political transparency held by dark-skinned subjects wearing light-skinned masks. They are instead made material, energetic, and visual on servers and silicon distributed across the globe.

Much like the "general referent" or "average face," the epidermal racial schema can be cognized as a restriction of freedom issuing from the imposition of fictive categories of race imposed upon the psyche of the individual. Describing the phenomenological experience of having his body schema replaced by an epidermal racial schema, Fanon wrote, "I was responsible not only for my body but also for my race and my ancestors."[36] With the proliferation of algorithmic images into domains of judgement and adjudication typically reserved for photographs, subaltern subjects become additionally responsible: not only for their ancestors, race, or possible descendants, but for synthetic manifestations of the white gaze that happen to resemble them:

> One potential way police might use the technology is to create photo-realistic composite images of suspects. "Showing someone a machine-generated image can reinforce in their mind that that's the person even when it might not be—even when it's a completely faked image," said Nicole Napolitano, director of research strategy at the Center for Policing Equity.[37]

Here, immediate sense perceptions replace signifying dog whistles, Flusser's negative utopia made manifest in the purest sense. In addition to an image culture "centrally controlled by senders" in the technical-economic sense of centralized platform governance, we also now have visual-material-energetic manifestations of "average" or "general" ideological perspectives—a distributed, apparently subjectless gaze that lends corporate actors a veil of objectivity by mere virtue of its scale and replicability. "Big" data somehow becomes "better" data, even though each labeled text-image pair contributing to algorithmic color associations comes from a specific subject position and social context. The association of dark skin and subalterity is authenticated by modes of faux-objectivity that function as confirmation bias at scale—a giant filter bubble that has gained new means to project its assumptions outward.

CONCLUSIONS: BEYOND THE MIRROR

In 2016, Joy Buolamwini used facial recognition techniques to construct aspirational reflections of the viewer/self for her media artwork "Aspire Mirror." A dark-skinned woman, she found that the algorithm wouldn't detect her face and chose to don a white mask to make herself visible to it for the project.[38] In a 2018 paper, Buolamwini and Timnit Gebru analyzed the source of this disparity, undertaking an intersectional approach to show how three image datasets crucial to the development of widely used facial recognition and classification technologies underrepresented female and dark-skinned subjects. In their analysis of models that used these datasets, they found "that all classifiers performed best for lighter individuals and males overall. The classifiers performed worst for darker females,"[39] concluding that "[t]he most improvement is needed on darker females specifically."[40]

However, Ramon Amaro critiqued Buolamwini's seeming lack of attention to the history of the teleology of inclusion within Western classification schema, claiming that "Coders like Buolamwini . . . fold seamlessly into the desire for representation": "Although the immediacy of computation's lack of diversity—in terms of institutional value and algorithmic function—cannot be understated, a call to make black technical objects compatible to computer vision algorithms risks the further reduction of the lived potentiality of black individuals."[41]

Amaro's critique reflects the notion that representational accuracy within the broader project of computer vision won't necessarily offer a net benefit to darker-skinned subjects—and indeed, it is primarily the proprietors of such systems that stand to gain. He challenges the notion of "cohesion" in analytics systems, given the ways they position a "normal" referent and stratify individuals according to their conformity to it:

> By prioritizing cohesion, algorithmic processes erode the potential for human difference and self-actualization . . . [computer vision] is assigned the role of interpellator, assigning value (in terms of visibility) to the individual only in as much as he/she/they can be measured against a universalizing concept of being.[42]

Instead, Amaro advocates for systems that "value the dissonance that emerges in relation with analytics," following Fred Moten and Stefano Harney's articulation of a vision of blackness "that does not 'want to be correct' or 'corrected'" because "[t]here is nothing wrong with blackness."[43] Here, Buolamwini and Amaro are representing two parallel yet necessary motions: Amaro's call for technical systems that value difference is vital yet beyond the proximate horizon of a mass society that *is* still premised upon totalizing hierarchies of normalization—the cultural sediment of centuries. It is difficult to imagine, for example, the utility of a database or machine learning model that treats race non-hierarchically and values skin color difference for its own sake to a mass society so heavily and historically premised on color stratification—the technical and the social must exhibit compatibility for adoption to occur, and at some level, the purpose of the models is *precisely* to reify and propagate pre-existing values. As Amaro himself argues, the problem resides in the Western racial imaginary, which has been premised primarily upon skin color at least since the colonial Atlantic distinction between black slaves and white indentured servants. This being the case, change requires the shifting of human attitudes at an enormous scale—not only a wholesale reconsideration of Western epistemology since the advent of modernity, but an activist evangelism that brings the relevant critiques into the realm of practice, alongside a host of technical experiments that test the assumptions of alternative approaches and their effects on human eusociality along the way—a "new science of the human," as Aimé Césaire and Sylvia Wynter put it.[44] The sheer scale of development and investment mean that the necessary changes are as much about culture, education, finance, sociogeny, and raw political power as technical implementation or philosophical investigation. Relatedly, Philip Agre wrote on the processual positivism that substitutes for critique within the culture of AI research: "I have often encountered an emphatic, explicitly stated injunction against 'criticizing other people's work,' the idea being that the only legitimate form of critical argument is that 'my system performs better than your system on problem X.'"[45]

In the 1960s, cyberneticists and systems theorists realized that it was impossible to observe feedback systems without participating in them—observers are necessarily participants in the systems they observe, are changed by those systems, and change them in turn.[46] If we consider that the networks which epidermalizing color associations move through are not purely technical, but also human, sensory, and sociogenic, we can begin to conceptualize them within the realms of imagination and aesthetics. The color associations and biases that we witness in this technical utopia produce affirming and/or negating biochemical effects in our living bodies, influenced to a profound degree by our experiences, environments, and social contexts. This means that technical solutions are important, but only one part of the puzzle, and are conditioned upon the prevailing sociality. It may well be the case that algorithmic images accurately represent average stereotypes and skin color associations held by average North American and European internet users—at least the subset of that group inclined to upload, caption, and comment on images on sites targeted by web crawlers such as LAION's. But the effects of propagating color-based stereotypes using photorealistic aesthetic forms long associated with "self-evidence" on human subjectivity and race bias—especially in the context of a near-totalizing imagistic social architecture—remain to be seen.

The totalizing tendency of the algorithmic image utopia, a result of massive scale and the aspiration to create "unbiased" or "de-biased" systems, is a likely dead-end insofar as the reduction of color stereotypes reified using generative techniques is concerned. Intentional image curation, continuous maintenance, auditing, and smaller, purpose-built models that take the needs and perspectives of specific user groups into account, as Bender et al. (2021) advocated regarding large language models, seem much less likely to reproduce harmful color stereotypes.[47] Realizing that post-hoc de-biasing will likely always be a game of whack-a-mole—not least because cultural sensibilities regarding harmful biases will change over time—it seems best to affirmatively state the biases we desire in our systems (such as fairness, equity, consent, repair), and to design toward those rather than continuing to pretend that 'objectivity' will somehow spontaneously manifest through the mere collection of image data at unreviewable scales. As such, the path(s) through this morass must include robust, liberatory, and speculative imaginaries of the end goal alongside concrete, small-scale iterative demonstrations of systems that "perform better" on issues of fairness, repair, equity, and consent. The aim of these experiments and speculations is that, broadly speaking, these values are no longer considered problems to be addressed after a system is already built and released, but, rather, the universal solutions and first principles for every technique we introduce that impacts the life outcomes of others.

NOTES

1. Benjamin, "The Work of Art in the Age of Its Technological Reproducibility," 40.

2. See Wynter, "Toward the Sociogenic Principle," for an elaboration and extension of Fanon's notion of sociogeny I adopt here.

3. Azoulay, *The Civil Contract of Photography,* 154.

4. Schuppli, "Can the Sun Lie?", 57. See also Lorraine Daston and Peter Galison, *Objectivity.*

5. Schuppli, "Can the Sun Lie?", 58.

6. Flusser, *Towards a Philosophy of Photography,* 15.

7. Flusser, *Towards a Philosophy of Photography,* 44.

8. Broomberg and Chanarin, "The Polaroid Revolutionary Worker's Movement." This artwork is also discussed in Benjamin's *Race after Technology,* the relevant excerpt of which is republished in part III of this volume.

9. Smith, "'Racism' of Early Colour Photography Explored in Art Exhibition."

10. African Activist Archive, "Polaroid Revolutionary Workers Movement."

11. The artists echo a claim that "black skin absorbs 42 percent more light" to support their conclusion that the boost button's 42 percent flash intensification was predicated on capturing black subjects during apartheid. While some researchers claim that the melanin in "dark-skinned" subjects absorbs between 30 to 40 percent more ultraviolet rays than in "light-skinned" subjects, it is uncertain whether this observation has any relation to the *visible* light spectrum. A 2024 study by Setchfield et al. finds a much greater range in variability, from 6 to 74 percent absorption. It is possible that the deciding parties at Polaroid operated under the same assumptions as Broomberg and Chanarin.

12. Mirsky and Lee, "The Creation and Detection of Deepfakes"; Graham Meikle, *Deepfakes.*

13. Fatma-Elzahraa Eid et al., "Systematic Auditing Is Essential to Debiasing Machine Learning in Biology."

14. Human curation, of course, would reduce the size of training sets and the speed with which they are deployed. Gebru et al. suggested this very approach in 2021 in

"On the Dangers of Stochastic Parrots: Can Language Models Be Too Big?" The paper's conclusions seemingly led to her controversial departure from Google.

15. Mitchell, *The Reconfigured Eye*, 191.

16. Goodfellow et al., "Generative Adversarial Nets," 1.

17. Goodfellow, "NIPS Tutorial: Generative Adversarial Networks," 3.

18. Mitchell, *Reconfigured Eye*, 194.

19. L. J. M. Daguerre, quoted in Silverman, *Miracle of Analogy*, 26.

20. Fanon, *Black Skin, White Masks*, 91.

21. Fanon, *Black Skin, White Masks*, 92.

22. Fanon, *Black Skin, White Masks*, 95.

23. Melissa Thompson: "I've always loved fried chicken. But the racism surrounding it shamed me."

24. Fanon, *Black Skin, White Masks*, 91.

25. Flusser, *Into the Universe of Technical Images*, 3.

26. Flusser, *Into the Universe of Technical Images*, 171.

27. Flusser, *Into the Universe of Technical Images*, 172.

28. See Tiku et al., "These Fake Images Reveal How AI Amplifies Our Worst Stereotypes"; Turk, "How AI Reduces the World to Stereotypes"; and Small, "Black Artists Say A.I. Shows Bias."

29. Nicoletti and Bass, "Humans Are Biased."

30. Stable Diffusion's open data and models are likely why it is overrepresented in journalistic and scholarly critique compared to its contemporaries in generative AI.

31. Nicoletti and Bass, "Humans Are Biased"

32. Tiku et al., "AI Amplifies Our Worst Stereotypes"

33. Europol, "Facing Reality? Law Enforcement and the Challenge of Deepfakes."

34. Zhiheng Li et al., "A Whac-A-Mole Dilemma"

35. Mbembe, *Critique of Black Reason*, 32.

36. Fanon, *Black Skin, White Masks*, 92.

37. Nicoletti and Bass, "Humans Are Biased."

38. Buolamwini, "Aspire Mirror."

39. Buolamwini and Gebru, "Gender Shades."

40. Buolamwini and Gebru, "Gender Shades," 11.

41. Amaro, "As If."

42. Amaro, "As If."

43. Moten and Harney, *The Undercommons*, 47–48, cited in Amaro, "As If."

44. Césaire, "Poetry and Knowledge", xxix; Wynter, "Toward the Sociogenic Principle," 60.

45. Agre, "Toward a Critical Technical Practice," 150.

46. See Mead, "Cybernetics of Cybernetics" and Maturana and Varela, *Autopoesis and Cognition*.

47. Bender et al., "On the Dangers of Stochastic Parrots," 619.

BIBLIOGRAPHY

African Activist Archive. "Polaroid Revolutionary Workers Movement." Accessed December 12, 2018. http://africanactivist.msu.edu/organization.php?name=Polaroid+Revolutionary+Workers+Movement.

Agre, Philip. "Toward a Critical Technical Practice: Lessons Learned in Trying to Reform AI." In *Social Science, Technical Systems, and Cooperative Work: Beyond the Great Divide*, edited by Geoffrey C. Bowker, Susan Leigh Star, William Turner, and Les Gasser, 131–158. Mahwah, NJ: Lawrence Erlbaum Associates, 1997.

Amaro, Ramon. "As If." In *e-flux Architecture: Becoming Digital*. February 2019. Accessed February 6, 2022. https://www.e-flux.com/architecture/becoming-digital/248073/as-if/.

Azoulay, Ariella. *The Civil Contract of Photography*. Translated by Ruvik Danieli and Rela Mazali. New York: Zone Books, 2008.

Bender, Emily, Timnit Gebru, Angelina McMillan-Major, and Shmargaret Shmitchell. "On the Dangers of Stochastic Parrots: Can Language Models Be Too Big?" FAccT, March 2021, Canada (virtual event), 610–623. https://doi.org/10.1145/3442188.3445922.

Benjamin, Ruha. *Race after Technology: Abolitionist Tools for the New Jim Code*. Cambridge, UK: Polity, 2019.

Benjamin, Walter. *The Work of Art in the Age of Its Technological Reproducibility, and Other Writings on Media*. Edited by Michael W. Jennings, Brigid Doherty, and Thomas Y. Levin. Translated by Edmund Jephcott, Rodnet Livingstone, Howard Eiland, et al. Cambridge, MA and London: Belknap Press of Harvard University Press, 2008.

Broomberg, Adam, and Oliver Chanarin. *The Polaroid Revolutionary Worker's Movement*. 2013. Accessed December 12, 2018. http://www.broombergchanarin.com/polaroid-revolutionary-workers#.

Broomberg, Adam, and Oliver Chanarin. *To Photograph the Details of a Dark Horse in Low Light*. 2012. Accessed December 12, 2018. http://www.broombergchanarin.com/polaroid-revolutionary-workers#.

Buolamwini, Joy. "Aspire Mirror." https://www.aspiremirror.com. Accessed May 8, 2019.

Buolamwini, Joy, and Timnit Gebru. "Gender Shades: Intersectional Accuracy Disparities in Commercial Gender Classification." In *Proceedings of Machine Learning Research,* 81 (2018), 77–91, https://proceedings.mlr.press/v81/buolamwini18a/buolamwini18a.pdf.

Césaire, Aimé. "Poetry and Knowledge" (1945). In *Aimé Césaire: The Collected Poetry*. Translated by Clayton Eshleman and Annette Smith. Los Angeles: University of California Press, 1983.

Daston, Lorraine, and Peter Galison. *Objectivity*. Edited by Peter Galison. Cambridge, MA: Zone Books, 2007.

Eid, Fatma-Elzahraa, Haitham A. Elmarakeby, Yujia Alina Chan, et al.. "Systematic Auditing Is Essential to Debiasing Machine Learning in Biology." *Communications Biology* 4, no. 1 (2021): 183. https://doi.org/10.1038/s42003-021-01674-5.

Europol. "Facing Reality? Law Enforcement and the Challenge of Deepfakes: An Observatory Report from the Europol Innovation Lab." Luxembourg: Publications Office of the European Union, 2022.

Fanon, Frantz. *Black Skin, White Masks* (1952). Translated by Richard Philcox. New York: Grove Press, 1967.

Flusser, Vilém. *Into the Universe of Technical Images* (1985). Translated by Nancy Ann Roth. Minneapolis: University of Minnesota Press, 2011.

Flusser, Vilém. *Towards a Philosophy of Photography*. Translated by Anthony Mathews. London: Reaktion Books, 2000.

Originally published in Czech in 1983.

Goodfellow, Ian, Jean Pouget-Abadie, Mehdi Mirza, Bing Xu, David Warde-Farley, Sherji Ozair, Aaron Courville, and Yoshua Bengio. "Generative Adversarial Nets." *arXiv,* June 10, 2014. https://arxiv.org/abs/1406.2661.

Goodfellow, Ian. "NIPS 2016 Tutorial: Generative Adversarial Networks." *arXiv*, December 31, 2016. arXiv:1701.00160v4.

Harney, Stefano, and Fred Moten, *The Undercommons: Fugitive Planning & Black Study*. New York: Minor Compositions, 2013.

Li, Zhiheng, Ivan Evtimov, Albert Gordo, Caner Hazirbas, Tal Hassner, Cristian Canton Ferrer, Chenliang Xu, and Mark Ibrahim. "A Whac-A-Mole Dilemma: Shortcuts Come in Multiples Where Mitigating One Amplifies Others." *arXiv,* March 21, 2023. arXiv:2212.04825.

Maturana, Humberto R., and Francisco J. Varela. *Autopoesis and Cognition: The Realization of Living.* Boston: D. Reidel Publishing Company, 1928.

Mbembe, Achille. *Critique of Black Reason* (2013). Translated by Laurent Dubois. Durham, NC: Duke University Press, 2017.

Mead, Margaret. "Cybernetics of Cybernetics." In Heinz von Foester, ed., *Purposive Systems: Proceedings of the First Annual Symposium of the American Society for Cybernetics,* 1–11. New York: Spartan Books, 1968.

Meikle, Graham. *Deepfakes*. Cambridge, UK: Polity Press, 2022.

Mirsky, Yisroel, and Wenke Lee. "The Creation and Detection of Deepfakes: A Survey." *ACM Computing Surveys* 54, no. 1 (January 2021): 1–41. https://doi.org/10.1145/3425780.

Mitchell, W. J. T. *The Reconfigured Eye: Visual Truth in the Post-Photographic Era*. Cambridge, MA: MIT Press, 1992.

Nicoletti, Leonardo, and Dina Bass. "Humans Are Biased. Generative AI Is Even Worse." *Bloomberg*. June 9, 2023. Accessed March 13, 2024. https://www.bloomberg.com/graphics/2023-generative-ai-bias/.

Schuppli, Susan. "Can the Sun Lie?" In *Forensis: The Architecture of Public Truth*, 56–64. London and Berlin: Forensic Architecture and Sternberg Press, 2014.

Silverman, Kaja. *The Miracle of Analogy, or The History of Photography, Part I*. Palo Alto, CA: Stanford University Press, 2015.

Small, Zachary. "Black Artists Say A.I. Shows Bias, with Algorithms Erasing Their History." *The New York Times*, July 4, 2023. Accessed February 22, 2024. https://www.nytimes.com/2023/07/04/arts/design/black-artists-bias-ai.html.

Smith, David. "'Racism' of Early Colour Photography Explored in Art Exhibition." *The Guardian*, January 25, 2013. Accessed December 12, 2018. https://www.theguardian.com/artanddesign/2013/jan/25/racism-colour-photography-exhibition.

Thompson, Melissa. "I've Always Loved Fried Chicken. But the Racism Surrounding It Shamed Me." *The Guardian*, October 13, 2020. https://www.theguardian.com/food/2020/oct/13/ive-always-loved-fried-chicken-but-the-racism-surrounding-it-shamed-me.

Tiku, Nitasha, Kevin Schaul, and Szu Yu Chen. "These Fake Images Reveal How AI Amplifies Our Worst Stereotypes." *The Washington Post*, November 1, 2023. Accessed February 22, 2024. https://www.washingtonpost.com/technology/interactive/2023/ai-generated-images-bias-racism-sexism-stereotypes/.

Turk, Victoria. "How AI Reduces the World to Stereotypes." *Rest of World*, October 10, 2023. Accessed February 22, 2024. https://restofworld.org/2023/ai-image-stereotypes/.

Wynter, Sylvia. "Toward the Sociogenic Principle: Fanon, Identity, the Puzzle of Conscious Experience, and What It Is Like to Be 'Black.'" In *National Identities and Socio-Political Changes in Latin America*, edited by Antonio Gomez-Moriana and Mercedes Duran-Cogan. London and New York: Routledge, 2001.

18 *CYBERTYPES* TWENTY YEARS LATER: LISA NAKAMURA TALKS WITH LIDA ZEITLIN-WU

LIDA ZEITLIN-WU: Thanks for agreeing to have this conversation on *Cybertypes*, twenty-one years after its publication in 2002. The book is now a cornerstone in both digital media studies and critical race studies, and one of the first explorations of color graphics and interfaces in a world of new media framed at that point in time as "colorblind." Can you talk about how you conceived this project?

LISA NAKAMURA: At the time, Lev Manovich's book *The Language of New Media* [2001] was the hottest thing, and it's a wonderful book that turned the field towards formalism.[1] I still teach it. But I saw really no discussion of technology in ethnic studies and no discussion of race, ethnicity, or even gender in digital media studies. I was drawn to this topic because the Internet was a news story that wouldn't end. But it seemed that no newspaper or magazine could stay away from the topic, and they always came out with the same conclusion: this is a radical opportunity to reimagine democracy and eliminate racism and sexism. Everybody is going to get a chance to be equal because nobody can see who you are. I started *Cybertypes* with an image of the *New Yorker* cartoon, "On the Internet, Nobody Knows You're a Dog" [1993], because it made the Internet seem like a double-edged sword: because you can hide who you are, you are now required or expected to hide who you are.

I wanted to understand why it was such a good thing for there to be no race in the future, because that was the idea of what the Internet would do. People would leave their bodies behind and then they would compose a one life, but the act of composition in Internet space is always an ideological and computational act. Early on, creating a digital self meant writing out what you looked like, which always had some implied identity robbery—by not mentioning race, everyone assumed you were white. By mentioning race, you were assumed to be whatever phantasmic ethnicity you wanted to create. Then we had virtual worlds with graphical interfaces like The Palace and Second Life—or basically any videogame now, where you take a slider and select all of your physical attributes. Race, or color, was just another slider and to this day it still doesn't make sense to people why there's no connection between the color of your avatar and the way your avatar looks in other ways. Digital color becomes a property or gradient independent of other things.

The idea that race is separable from other parts of visual identity, or composable, seemed both an opportunity, but also the beginning of the neoliberal separation of the visual from the economic and material. I was writing *Cybertypes* in 1996–2000, before the Internet was very visual—but it was starting to be. The people who were making choices about digital color and virtual bodies in that way were kids, engineers, and geeky people. They weren't the majority of Internet users, who were still using ISP's like AOL or Prodigy. It was a marginal practice I happened to know a lot about because I was a grad student with too much time on my hands, writing about this so as not to not write my dissertation.

LZW: Just to confirm, this was *not* what you wrote your dissertation about?

LN: Oh, no. My dissertation was on Henry James, Edith Wharton, Thomas Hardy, Rudyard Kipling, and racial and gender passing.

LZW: So there are still some interesting overlaps.

LN: Definitely. I was writing about Edward Said's *Orientalism*, and I also cite Kipling in *Cybertypes* a little bit. How Kim is this kind of touristic character [1901] but always in the service of the Empire, and he can pass as any race, but Kipling stresses that his blood is Irish, which is one of the many martial races that he thought would defend the Empire. So, this mutability that biracial or multiracial people were thought to have is also in my reading of *The Matrix* [1999], about Keanu Reeves. Circa 2002, the Internet was not only going to replace race by qualities you could choose yourself using a slider or menu, but it would also incorporate multiracial and multicultural people as totems of the future. Race would disappear because it would be hybrid and everywhere. It was about color as something that you could choose in a kind of aesthetic way that was just about what you wanted to look like, and had no political connotations. It was about the idea of race as always disappearing. Which actually the way that Jodi Melamed puts it—racism as always disappearing.[2] So I was looking at the way race was always disappearing in all the cyberpunk and science-fiction representations of the Internet, where the Internet would make us all both deracinated but also every race at the same time.

LZW: That resonates with me as a multiracial person. I also think about the ability to change the skin color of our emojis now. In 2023, if you're not Black, but you choose an emoji that's coded as Black, that would be seen as "digital blackface." How do you think the promise of passing and identity tourism has shifted since you wrote the book?

LN: The idea that you need to have an emoji that corresponds with your actual identity speaks to the relatively new acceptance that race *does* exist online, that it does represent an actual person. To choose the wrong color emoji online is to misrepresent yourself. By contrast, in the 1990s there was no notion of misrepresentation; it was only representation. Everything everyone chose was equally valid and interesting because it was like a beta test for the technology itself. So I critiqued it a lot, not for the same reason that people critique what you might call "minstrel emojis," where people think using a Black emoji is racial appropriation or making light of or mocking Black identity. But in the 1990s, people thought these choices were inconsequential because "nothing that happened on the Internet was real." It was obvious that it was real to me, as someone who was using it and investing lots of time in it. But also, while people are supposedly wasting time and choosing their emoji colors and trying to make dreadlocks out of various pixels, *Time* magazine, *Newsweek,* and the *New York Times* were publishing a new article every week about how "this is going to change things, it's gonna be the new globalization, it's gonna be the new economic engine." And they weren't far off. What they didn't understand was that individuals were going to use it to create their own identities.

At the same time, ideas about the Oriental and the primitive were being used to sell networking products by IBM and WorldCom, because people didn't know what networking was. So, to explain something very abstract that's ultimately about financialization, the visual was needed to make people understand it. You know—beautiful children of color all using the Internet, or the MCI commercial [1997],[3] were a huge provocation for me. It was this Benetton idea of "the Internet will fix all of our bodily misfortunes and disabilities." I would write that very differently now. It *was* about the disabled and improperly raced body and how we can erase it and hide it, and again, not seen as problematic. Now I think it's very problematic. To go back to the emoji question, everybody gets the same Homer Simpson yellow if they don't do anything. There's always a choice that's being made, even if it's not one that always strikes people as political at the time.

LZW: Once you introduce race as a set of "real" skin tones into that option, it's very hard not to see the yellow default as racialized. Like, are we all "yellow Asians" now? This is why color as a way of talking about race is so slippery. Yellow here becomes the default whiteness in a strange way.

LN: That's really smart.

LZW: You've brought up the question of the visual. Rereading *Cybertypes*, I was struck by how visual it is. Even your definition of cybertypes operates around how images construct race on the Internet and the idea that there's no misrepresentation, only representation. How do you think this

FIGURE 18.1
MCI Internet commercial (1997). Video still compilation by editors. On the Internet, "there is no race. There is no gender . . ."

has changed in terms of the way that the general public thinks about the role of images? Your more recent work, like your 2014 article "Indigenous Circuits"[4] and your forthcoming book on digital racial capitalism and the invisible labor of women of color, looks more at questions of infrastructure and points to a larger turn in media studies.

LN: I was trying to get there in that book's conclusion, which is always where you put the things you wish you had actually written about but about didn't have time. It was 1995–2000, and the Internet was so different. The last chapter is called "Keeping It (Virtually) Real: The Discourse of Cyberspace as an Object of Knowledge," and this use of parentheses I'm now very embarrassed by, but it was what people were doing. I was talking about how there's representation, but there's also material inequalities between who even gets to use the Internet, and also who gets to write about it. Who gets to be the authority, to speak in a way that's listened to? Partly because I felt I wasn't being listened to, as a woman and a scholar of color, but also as someone who didn't have a technical background. So I would give these talks and often they would be completely hated by people, I think because they were more dystopian than they should have been. But also, I was trying to get to a more material critique. That's why I'm trying to talk about economics in that book a little bit without a lot of tools for talking about it. There's a lot of citations of scholars like Judith Butler and people like that, but there's also a lot of references to journalists and what they were trying to say, because they were making in some way the truth of what the Internet was for people. Most people hadn't even used it yet. I wanted to talk about who's making things beneath the visible surface, because what we do has everything to do with what we see and aren't allowed to see.

That is a great turn in media studies,. It's a way to ground critique, which tends to be very abstract. When you write about images, you can get very divorced from people and what they're actually doing. But people make images. That's what the Internet is about in a lot of ways. So, when you were talking about the yellow emoji, I was thinking about being in Zooms with white people and how they always pick the white thumb—like, there is a white thumb, and that's an option for them. It's weird that they're the ones who have to make a choice, whereas Asian people don't have to choose. I look at them and I wonder, "is this the right color?" And I'm like, "yes, that is the right color. I wonder how that happened." These things are so often about decisions that have nothing to do with representation, but have to do with a poor grad student engineer stuck with the job of making that choice. What kinds of resources are there to support color and emoji choice in Zoom? Zoom is stuck trying (or not trying that hard) to address zoombombing and safety and other things, while at the same time, there's a long and fascinating history of individuals creating new emojis like the beaver emoji that Internet researcher Joan Donovan helped to add to the lexicon.[5]

LZW: That's interesting that you say most white people feel the need to customize because, as an Asian person, I wouldn't default to the yellow thumb. I would choose a tan one. We could probably talk about emojis forever . . . But the question of labor—who's making the images that we get to see—is also obviously a really important one. Going back to *Cybertypes,* you cite the line from Du Bois's *The Souls of Black Folk* [1903]: "The problem of the Twentieth Century is the problem of the color-line." You then observe, "At the end of the Internet epoch and the advent of the twenty-first century, [the color line] is *still* the problem that haunts cyberspace" (xviiii). Scholars of race and ethnicity like Eduardo Bonilla-Silva[6] and Jodi Melamed, who you mentioned, have also critiqued the "color-blind" ideology of the 1990s and argued that the color line of Reconstruction morphed from a black-and-white binary into something messier, in an attempt to deal with the plurality of ethnicity beyond black and white. Do you remember Crayola's "Multicultural Crayons," released in the 1990s?

LN: I remember that.

LZW: There used to be a crayon just called Flesh, and Crayola realized they needed to address this. They released their "Multicultural" pack, with illustrations of "diverse" children on the box. But there were still only eight colors, and they didn't call them multiracial—they called them multicultural. So, I was interested in why you thought in the mid-'90s that the color line still remained a useful metaphor for thinking about race online, over hundred years after Du Bois's initial quote.

LN: Talking about race was seen as incredibly rude in the tech community while I was writing this book.,. It was not something that you were supposed to do, and it was seen as being itself the problem. This is something a lot of other scholars have written about, that just mentioning race was to be viewed as a racist yourself, because you're seeing race. Just like if you see color, that's also seen to be racist, right? But seeing color is a sensory capacity that people have. I think what they mean is attaching values to certain

FIGURE 18.2
Crayola Multicultural Crayons, 8-count (2023).

kinds of colors. So when you mentioned the multicultural pack, I thought about the controversy in game studies that happened after this, where avatars became more diverse—there was a Black avatar, an Asian avatar, etc. But they were kind of like Mortal Kombat avatars—they kind of looked better, but they were basically just recreating racial stereotypes. They were missing the mark even more than before, showing their ass about their own racial typecasting. The whole '90s and early 2000s were the programmers showing their ass, because they were the ones making these decisions. They would often say, "I'm just an programmer." They still say this! "I'm just an programmer, I'm just trying to make it work, so none of the decisions I make have any implication because they're not intentional—they have no aim and no purpose."

The big difference now is that you can critique this racism. When people critique woke-ism, I think what they're saying is, "Ten years ago you would've gotten into trouble for talking about race and now I'm getting in trouble for refusing to talk about it."

LZW: Color is something that often doesn't get discussed because, of course, reducing race to color is problematic, so there's a desire to get away from color, as a way of not essentializing race.

In your avatar chapter on identity tourism, you discuss how color functions as a superficial thing you just pick. For example, when you describe your own experience creating your avatar, somebody laughs and says, "Your face color doesn't match your body color," and so you changed it. What's interesting here is that differences in skin, hair, and eye color have historically been used by nineteenth-century "race science" manuals to classify and oppress people. Here, it's like your online identity can have some other phenotype that isn't actually yours. It complicates the way that race is typically constructed in relation to skin color. Could you speak a bit more about the relationship between digital color, as in the graphics that are rendered, and the complex relationship between race and phenotype in these spaces?

LN: That's such a good question. I wasn't studying this at the time, but a lot of the people who were making the Internet were very involved in anime and Japanese visual culture. The Japanese anime tradition of having characters with blue eyes and blonde hair who were Japanese and wearing a Japanese schoolgirl or schoolboy uniform was kind of confounding to a lot of people. But fans just accepted it, like "this is the way this visual style works." It may or may not be a sign that Asians think the white beauty standard is better, but I think that those kinds of unreal racializations were endemic to this culture. So, when people would go into online spaces and see somebody whose body didn't match their face, or somebody who was Black when they shouldn't be Black . . . that was another story I told, where I was given a Black avatar by default, which was very unusual then. I don't think it's happened much since. There was a video game called *Rust,* which randomly generated starting avatars for people, and some of them were Black women, and players hated it. They had a big protest saying, "This is so unfair and racist to me. I paid $39 for this game, and I don't want to be a Black woman. I get to choose who I am." What I think they're really saying is: "I would pay $39 to be a white man because that's what I've always been, but if I have to be a Black woman, I'm not into it." But the game is the same—it doesn't matter what your avatar is.

I think there's always been some detachment between color and race in the community of people who built the Internet, in the kind of media they liked to watch and the kind of places they liked to immerse themselves. It's often been this Asian fantasy idea. It's this perennial question: is it racist or not to have blue-eyed anime characters? I don't know, but I think that tradition of seeing color as just part of a different aesthetic tradition, that isn't about race, doesn't quite come from there.

LZW: That's a great connection. *Sailor Moon* [1992–1997] was my all-time favorite show growing up, and I watched it in both Chinese and in English. I understood the characters were meant to be Japanese, even though Sailor Moon has blonde hair and blue eyes and other characters had turquoise hair. The North American dub was infamous for a lot of reasons, including the fact that *Sailor Moon* is a very queer show and they made the same-sex couples heterosexual. But they also went out of their way to make the show not seem Japanese, like flipping images of cars so the steering wheel was on the other side. The result was that American and North American viewers were like, "Wait, these characters are supposed to be Japanese? I thought they were white." But of course, these are animated characters. There were a couple of live action *Sailor Moon* things, and it creates this big dissonance where you realize that it actually looks really strange to have these kinds of hairstyles . . .

LN: I don't think we talk about this enough. Emojis are one example, but there are also bitmojis—those little self-portraits people make for social

FIGURE 18.3
Box cover for RPG video game *Bishoujo Senshi Sailor Moon: Another Story* (1995).

FIGURE 18.4
Pretty Guardian Sailor Moon cast (live-action adaptation, 2003).

media, Instagram, and so on. Those aren't that popular anymore, but it is so much about animation and artificial color and not photographed representation. It's understudied.

LZW: You mentioned Lev Manovich. He wrote in 2001 that when you do [computer-generated imagery] and color grading, it's no longer cinema because it's not indexical—it's a special form of animation. This was in line with the discourse surrounding the crisis of the digital image in the late '90s and early 2000s that animation and live action were getting closer and closer.

LN: Yes, that's right. Whenever you make an image digital, you can adjust its color, just like you can autotune digital sound. It's not hard to do; it's already numbers, and all you have to do is change some of the numbers. That's a huge difference from film. I know people could paint on film and change color, like a lot of D. W. Griffith's films, but the fact that this is a permanent feature of digital images is really important. I agree with that fundamental idea: it is all animation once it gets turned into numbers. That's just how it works, and some look more like photographs and some don't, but they're all manipulable that way, they're all modular. Some are ostentatiously cartoony—like emojis could look more realistic than they do, but we just don't like them to.

LZW: Absolutely. That's another key contribution *Cybertypes* makes: What does it mean when we start thinking of identity as numerical and boxed-up in what you call "menu-driven identities"? Of course, the identity boxes today are maybe more flexible than they were, but I've seen ones where you check "Asian," and then it's like, what kind of Asian? There's Chinese, South Asian, Japanese, and eventually there's a box that just says "Other Asian."

LN: Yeah [laughing].

LZW: This is key to what cybertyping is. Race online is much more reductive than it is IRL, so to speak. How you think that plays out in an era of online hate speech?

LN: I think it's still bullshit, and it's still happening for sure. The reason that census and demographic boxes want us to say whether we're East Asian or Southeast Asian or South Asian has to do with the reality that a lot of people from Southeast Asia came here as refugees from wars sustained by US empire and are often disproportionately super poor, so they should not always be lumped in. But, yeah, menu-driven identities are still the way the world works and still the way the Internet works.

LZW: One last question about OpenAI's ChatGPT, which we can't seem to get away from these days. I doubt this, but are you part of the cultural panic around it? Or are you excited about the promises that it could bring? Do you see or anticipate racial bias manifesting in these kinds of chatbots?

LN: I think like other chatbots before it, if it's being trained on racist text, it's going to produce racist text. I read something quite appalling. Sama, a global company formerly known as Samasource, outsources tech work to women in the Global South, had hired a bunch of Nigerian women to do a contract for Microsoft, cleaning the ChatGPT data specifically to get rid of racism and sexism so as to make it more usable. A bunch of them had to quit due to [posttraumatic stress disorder] from the horrible stuff they were exposed to, just looking at text. So that's the way I see the issue with ChatGPT: it's an issue that it produces racist results, but who has to look at those racist results to get rid of them so we don't have to look at them? I think there's a whole layer of women of color doing that kind of work, which I'm super interested in, and it often shows up as a news story but doesn't get to be part of the conversation about ChatGPT. In order to make something less racist, you have to do extremely racist neocolonial things. Which is kind of the history of the Internet as well, right? Who makes devices, who gets rid of them, so that we can have these postracial anonymous conversations? It's always somebody who's already on the other side of the color line that global capitalism produces.

NOTES

1. Manovich, *The Language of New Media.*
2. Melamed, *Represent and Destroy: Rationalizing Violence in the New Racial Capitalism.*
3. MCI Internet commercial, https://www.youtube.com/watch?v=Ed0V3glPNZA.
4. Nakamura, "Indigenous Circuits."
5. The proposal for the beaver emoji was submitted in March 2019 to the Unicode Consortium, the organization that standardizes emojis and text. The full document is available at https://www.unicode.org/L2/L2019/19110-beaver-emoji.pdf.
6. Bonilla-Silva, *Racism without Racists.*

BIBLIOGRAPHY

Bonilla-Silva, Eduardo. *Racism without Racists: Color-Blind Racism and the Persistence of Inequality in America*. Lanham, MD: Rowman & Littlefield, 2003.

Manovich, Lev. *The Language of New Media*. Cambridge, MA: MIT Press, 2001.

Melamed, Jodi. *Represent and Destroy: Rationalizing Violence in the New Racial Capitalism*. Minneapolis: University of Minnesota Press, 2011.

Nakamura, Lisa. "Indigenous Circuits: Navajo Women and the Racialization of Early Electronic Manufacture." *American Quarterly* 66, no. 4 (December 2014): 919–941.

CONTRIBUTORS

BOOK EDITORS

Carolyn L. Kane is the author of *Chromatic Algorithms: Synthetic Color, Computer Art, and Aesthetics After Code* (University of Chicago Press, 2014); *High-Tech Trash: Glitch, Noise, and Aesthetic Failure* (University of California Press, 2019); and *Electrographic Architecture: New York Color, Las Vegas Light, and America's White Imaginary* (University of California Press, 2023). She is a professor of visual communication at Toronto Metropolitan University.

Lida Zeitlin-Wu is an assistant professor in the Department of Film, Television, and Theatre at the University of Notre Dame. She is currently writing a book titled *How Color Became a Technology: The Making of Chromatic Capitalism.* With the DISCO (Digital Inquiry, Speculation, Collaboration, and Optimism) Network, a collective of scholars and artists challenging digital social and racial inequalities, she recently collaborated on the multivoiced monograph *Technoskepticism: Between Possibility and Refusal* (Stanford University Press, 2025).

CONTRIBUTING AUTHORS

Ruha Benjamin is an associate professor of African American Studies at Princeton University. Her books include *People's Science: Bodies and Rights on the Stem Cell Frontier* (Stanford University Press, 2013); *Race after Technology: Abolitionist Tools for the New Jim Code* (Polity Press, 2019); and *Viral Justice: How We Grow the World We Want* (Princeton University Press, 2022).

Jianqing Chen is an assistant professor in Film & Media Studies and East Asian Languages & Cultures at Washington University in St. Louis. She is currently completing a book manuscript titled *Touchscreen Media: The Touch and Its User-Spectators in Twenty-First Century China*.

Anne Anlin Cheng is a professor of English and affiliated faculty in the Program in American Studies, the Program in Gender and Sexuality Studies, and the Committee on Film Studies at Princeton University. She is the author of three books (all published by Oxford University Press): *The Melancholy of Race: Psychoanalysis, Assimilation, and Hidden Grief* (2000); *Second Skin: Josephine Baker and the Modern Surface* (2011); and *Ornamentalism* (2019).

Frederick Douglass (1817–1895) was a social reformer, abolitionist, intellectual, statesman, orator, and writer. After escaping slavery in the US South, he moved northwards to assume a prominent role as leader of the abolitionist movement. He is the author of three autobiographies including: *Narrative of the Life of Frederick Douglass, an American Slave* (1845), *My Bondage and My Freedom* (1855), and *Life and Times of Frederick Douglass* (1881, 1892).

W. E. B. Du Bois (1868–1963) was an American sociologist, civil rights activist, and author of *The Souls of Black Folk: Essays and Sketches* (1903) and *Dusk of Dawn: An Essay Toward an Autobiography of a Race Concept* (1940), in addition to being the first African American to earn a doctorate from Harvard University and one of the founding members of the National Association for the Advancement of Colored People (NAACP).

Richard Dyer is professor emeritus of Film Studies at King's College London and has held visiting professorships at institutions across Europe and North America. His books (all from Routledge) include *Heavenly Bodies: Film Stars and Society* (1986); *White: Essays on Race and Culture* (1997); *The Culture of Queers* (2001); and *In the Space of a Song: The Uses of Song in Film* (2011).

Ali Feser is a cultural anthropologist and interdisciplinary artist. She was trained at Bard College and the University of Chicago and is currently a visiting assistant professor of health care at Clarkson University. She is writing a book about the material history of Kodak film, its ecological aftereffects in Rochester, New York, and the erosion of utopian fantasies of the capitalist "good life."

Nicholas Gaskill is an associate professor of American literature at the University of Oxford. He works at the intersection of philosophy and literary studies, specifically the relationship between American pragmatism and modernist literature, and is the author of *Chromographia: American Literature and the Modernization of Color* (University of Minnesota Press, 2018).

Quran M. Karriem is an experimental musician, artist, and scholar working primarily with electronic and algorithmic media. He holds a PhD in computational media, arts, and cultures from Duke University and is currently a postdoctoral fellow at the Samuel DuBois Cook Center on Social Equity at Duke University. His work is motivated by a concern about the circulation of emergent forms of capital in networked and automated systems, particularly insofar as they reproduce hegemonic power and racial stratification.

Michael Keevak is a professor of foreign languages at National Taiwan University. He is the author of *Becoming Yellow: A Short History of Racial Thinking* (Princeton University Press, 2011), which traces how East Asians became "yellow" in the Western imagination.

Lisa Nakamura is Gwendolyn Calvert Baker Collegiate Professor in the Department of American Cultures at the University of Michigan. She is a founding member of the DISCO (Digital Inquiry, Speculation, Collaboration, and Optimism) Network, and her books include *Cybertypes: Race, Ethnicity, and Identity on the Internet* (Routledge, 2002); *Digitizing Race: Visual Cultures of the Internet* (University of Minnesota Press, 2007); *Technoprecarious* (MIT Press, 2020); *Racist Zoombombing* (Routledge, 2021), and *Technoskepticism: Between Possibility and Refusal* (Stanford University Pres, 2025).

Tina Post is an assistant professor of English at the University of Chicago. Arguing that inexpression is a gesture that acquires distinctive meanings in concert with Blackness, her recent book *Deadpan: The Aesthetics of Black Expression* (NYU Press, 2023) tracks instances and meanings of deadpan—a vaudeville term meaning "dead face"—across literature, theater, visual and performance arts, and the performance of self in everyday life.

Aileen Robinson is an assistant professor of theater and performance studies at Stanford University and affiliated with the Program in Science, Technology, and Society. Her current monograph project, "Technological Wonder: The Theatrical Fashioning of Scientific Practice, 1780–1905," explores the contribution of theater and magic performance to emerging practices of science communication in the nineteenth century.

Michael Rossi is an associate professor of the history of medicine and the conceptual and historical studies of science at the University of Chicago and the author of *The Republic of Color: Science, Perception, and the Making of Modern America* (University of Chicago Press, 2019), an account of color science's transformative impact on American social and political ideals in the late nineteenth century.

Lorna Roth is a Canadian media scholar and professor emerita in communication studies at Concordia University. Her research focuses on screen technologies and representations of race and indigeneity. She is the creator of the Colour Balance project, an online database of how cultural and racial bias are embedded within technologies and everyday objects ranging from crayons to prosthetics.

Amber Sweat is a PhD candidate in French at the University of California, Berkeley, with emphases in media studies and global urban humanities. Her dissertation analyzes the lives of girls in the Parisian banlieue—notably descendants of the Caribbean and Black African diaspora—as they engage in processes of self-actualization, self-fabrication, and network creation.

Genevieve Yue is an associate professor of culture and media at the New School, where she directs the Screen Studies Program. Her book *Girl Head: Feminism and Film Materiality* (Fordham University Press, 2020) shows how gender has played a persistent role in film production processes well before the image ever appears onscreen.

INDEX

Note: Photos, figures, and tables are indicated by italicized page numbers

Publisher contact:
The MIT Press
Massachusetts Institute of Technology
77 Massachusetts Avenue, Cambridge, MA 02139
mitpress.mit.edu

EU Authorised Representative:
Easy Access System Europe, Mustamäe tee 50,
10621 Tallinn, Estonia
gpsr.requests@easproject.com

Printed by Integrated Books International,
United States of America